Student Study Guide and Solutions Manual

to accompany

Guinn/Brewer's Essentials of General, Organic, and Biochemistry

Rachel C. Lum, Ph.D.

W. H. Freeman and Company
New York

ISBN-13: 978-1-4292-2432-1
ISBN-10: 1-4292-2432-0

Printed in the United States of America

First Printing

W.H. Freeman and Company
41 Madison Ave
New York, NY 10010

www.whfreeman.com/gob

For Mateja and Alexander

About the Author

Rachel Lum received her S.B. in Chemistry from the Massachusetts Institute of Technology and her Ph.D. in Organic Chemistry from Harvard University. As a graduate student, she received the American Chemical Society Division of Organic Chemistry Fellowship sponsored by Procter & Gamble. She was postdoctoral fellow at the Georgia Institute of Technology. She was an adjunct professor at Regis University where she taught courses in analytical chemistry, general chemistry, and environmental chemistry. She was an instructor at the University of Colorado at Boulder where she taught courses in organic chemistry. She currently resides in Boulder with her husband, Al, and her two children, Alexander and Mateja, and works as a consultant.

Table of Contents

Preface

A helpful way to learn chemistry is to practice many problems. This study guide has been designed toward that goal. Each chapter starts with a brief summary of the material covered in the chapter. For each section, there is a summary of the key points, followed by a Worked Example. You can either read the answers in the Worked Example or attempt to answer the question yourself and compare your answer to the Worked Example. The Worked Example is complemented by a Try It Yourself question, which is a similar question to the Worked Example broken down into steps for your answer. At the end of each section there are practice problems to perfect your skills. At the end of each chapter there is a quiz to test your knowledge. So pick up your pencils and start practicing!

July 2009

Acknowledgments

I would like to thank Denise Guinn and Becky Brewer for giving me the opportunity to work with them on this project. I would like to thank Kathryn Treadway, Dave Quinn, and Susan Moran at W. H. Freeman for their editorial help. Finally, I would like to thank my husband, Al, for his unwavering support.

Chapter 1

Measurement, Atoms, and Elements

Chapter Summary

In this chapter, you have learned about how scientists measure matter. When scientists measure matter, they use the metric system. They also use significant figures and conversion factors to manipulate units. Understanding how to use conversion factors and dimensional analysis will help with other calculations used within this textbook and calculations that you may encounter in the medical field (i.e., dosages). You learned about elements and the properties of the atom and how to read the periodic table. You also learned that the properties of the atom are determined by their valence electrons. Understanding valence electrons will help you understand properties of atoms, which in turn will help you understand the reactions of atoms and molecules.

Section 1.1 Matter and Measurement

In a Nutshell: Units and the Metric System

Every measurement consists of both a number and a unit. The metric system is most often used in scientific measurements and consists of several base units and a number of prefixes that represent a multiple of ten of the base unit: giga, kilo, deci, centi, milli, micro, nano, pico. The meter is the base unit of length and the gram is the base unit of mass in the metric system. The most common unit of volume in the metric system is the liter. The volume of a solid is often measured by determining how much water the solid displaces.

$V_{object} = V_{final} - V_{initial}$

Worked Example #1

Which length is bigger: 5 pm or 5 fm?

Using Table 1-2, you see there are 1×10^{-12} m in 1 pm, while there are 1×10^{-15} m in 1 fm. Therefore, a picometer is bigger than a femtometer, and 5 pm is bigger than 5 fm.

Try It Yourself #1

Which mass is bigger: 10 kg or 10 mg?

Tools: Use Table 1-2

There are ___1×10^3___ g in 1 kg, while there are ___1×10^{-3}___ g in 1 mg.

A ___kg___ is bigger than a ___mg___.

Worked Example #2

When a rock is placed in graduated cylinder containing 50.0 mL of water, the water level rises to 54.5 mL. What is the volume of the rock in cm^3?

$V_{initial}$ = 50.0 mL V_{final} = 54.5 mL

The volume of the rock is equal to the difference in the volume before and after the rock is submerged. $V_{rock} = V_{final} - V_{initial}$

Substituting the known variables into the equation:

V_{rcok} = 54.5 mL – 50.0 mL

V_{rock} = 4.5 mL, *which corresponds to 4.5 cm^3.*

Try It Yourself #2

When a gold nugget is placed in a graduated cylinder containing 80.0 mL of water, the water level rises to 97.2 mL. What is the volume of the rock in cm^3?

V_{inital} = ___80.0 mL___ V_{final} = ___97.2 mL___

Equation for the volume of the gold nugget:

$V_{gold\ nugget}$ = $V_{final} - V_{initial}$

Substuting the known variables into the equation:

$V_{gold\ nugget}$ =

___17.2___ mL = ___17.2___ cm^3

In a Nutshell: Precision and Accuracy

Precision is a measure of how close together repeated measurements are to each other; accuracy is a measure of how close the measurement is to the "true" value.

Worked Example #3

A person weighed himself on two different scales three times. The man weighs 77.3 kilograms. Which scale is precise and accurate?

Scale A	Scale B
77.6 kg	80.5 kg
77.0 kg	80.6 kg
77.3 kg	80.7 kg

Scale A is precise and accurate. The average weight on scale A is 77.3 kg and all the measurements on the scale are within ± 0.3 kg. The measurements on scale B are precise, but not accurate.

Try It Yourself #3

A premature baby was weighed three times on two different scales. Which set of measurements has more precision?

Scale A	Scale B
1619 g	1619 g
1622 g	1659 g
1616 g	1579 g

The average weight for scale A _____✓_____

The average weight for scale B_____

In a Nutshell: Significant Figures

The numerical value obtained from any measurement always contains some degree of uncertainty. The number of digits reported conveys information about the error in a measurement. Significant figures are all the certain digits as well as one uncertain digit in a value. The following digits are significant: nonzero digits, zeros between nonzero digits, zeros at the end of a number containing a decimal point, and all digits expressed in scientific notation are significant. If there is no decimal point, zeros following a nonzero digit are not significant, they are merely place holders. Zeros that appear before nonzero digits, whether or not there is a decimal point are not significant figures. The zeros merely serve as place holders. Exact numbers contain an infinite number of significant figures because they carry no uncertainty.

Worked Example #4

How many significant figures are there in the following measured values?

    a. 0.002030 L

    b. 3.20×10^4 nm

    c. 2,400. g

Solutions

    a. *There are four significant figures in 0.002030. The zeros before the non-zero digits are place holders and are not significant. The zero after the three is significant.*

    b. *There are three significant figures in 3.20×10^4 nm. The zero after the two is significant.*

    c. *There are four significant figures in 2,400. The decimal point after the zeros indicates that the two zeros are significant.*

Try It Yourself #4

How many significant figures are there in the following measured values?

    a. 0.0004108 m

        4

    b. 7.10×10^{-3} g

        3

    c. 500. Seconds

        3

In a Nutshell: Significant Figures in Calculations

When multiplying or dividing measured values, the final calculated answer cannot have more significant figures than the measurement with the fewest number of significant figures. When adding or subtracting measured values, the final answer can have no more decimal places than the measured value with the fewest decimal places.

Worked Example #5

Perform the following calculations and round the final answer to the correct number of significant figures.

a. $21.34 \div 0.105 =$

Type of calculation: division

Number of significant figures in 21.34: four

Number of significant figures in 0.105: three

The calculation involves division, so the final answer cannot have more significant figures than the measured value with the fewest significant figures.

Measured value with the fewest significant figure: 0.0105 with three significant figures

Calculator result: $21.34 \div 0.105 = 203.2380952$

*Rounded calculator result to three significant figures: **203**.*

b. $9.23 + 3.314 - 21 =$

Type of calculation: addition and subtraction

Last decimal place in 9.23: hundredths place

Last decimal place in 3.314: thousandths place

Last decimal place in 21: ones place

The calculation is an addition and subtraction, so the final answer should have no more decimal places than the measured value with the fewest decimal places.

Measured value with the fewest decimal places: 21 with ones place

Calculator result: $9.23 + 3.314 - 21 = -8.456$

*Rounded calculator result to the ones place: **-8**.*

Try It Yourself # 5

Perform the following calculations and round the final answer to the correct number of significant figures.

a. 34.56 × 0.0042 =

   Type of calculation: mult.

   Number of significant figures in 34.56: __4__

   Number of significant figures in 0.0042: __2__

   The calculation involves __mult.__, *so the final answer cannot have more significant figures than the measured value with the fewest significant figures.*

   Measured value with the fewest significant figures: __2__

   Calculator result: 34.56 × 0.0042 = 0.145152

   Rounded calculator result: __0.15__

b.   31.5678 + 25.34589 =

   Type of calculation: __add.__

   Last decimal place in 31.5678: _____

   Last decimal place in 25.34589: _____

   *The calculation is*_____, *so the final answer should have no more decimal places than the measured value with the fewest decimal places.*

   Measured value with the fewest decimal places: _____

   Calculator result 31.5678 + 25.34589 = _____

   Rounded calculator result : __56.9137__

In a Nutshell: Conversions between Metric Units

To convert one metric unit to another metric unit, we use a process called dimensional analysis. Step 1: Identify the conversion(s). Step 2: Express each conversion as two possible conversion factors. Step 3: Set up the calculation so that the supplied units cancel. As a final step in any calculation, you should double check that you have reported the correct number of significant figures in the final numerical answer.

Worked Example #6

The human ovum (egg cell) is about 150 μm in diameter. What is this diameter in mm?

Identify the conversion(s):

$1 \ \mu m = 1 \times 10^{-6} \ m$

and

$1 \ mm = 1 \times 10^{-3} \ m$

Express each conversion as two possible conversion factors:

1st *conversion factor:* $\dfrac{1 \ \mu m}{1 \times 10^{-6} \ m}$ *or* $\dfrac{1 \times 10^{-6} \ m}{1 \ \mu m}$

2nd *conversion factor:* $\dfrac{1 \ mm}{1 \times 10^{-3} \ m}$ *or* $\dfrac{1 \times 10^{-3} \ m}{1 \ mm}$

Set up the calculation so that the supplied units cancel:

$$150 \ \cancel{\mu m} \times \dfrac{1 \times 10^{-6} \ \cancel{m}}{1 \ \cancel{\mu m}} \times \dfrac{1 \ mm}{1 \times 10^{-3} \ \cancel{m}} = 0.15 \ mm$$

Check that the answer has the correct number of significant figures.

Try It Yourself #6

How many kilograms are there in 2.34 pg?

Identify the conversion(s).

$$1 \ pg = 1 \times 10^{-12} \ g$$

Express each conversion as two possible conversion factors.

$$\dfrac{1 \ pg}{1 \times 10^{-12} \ g}$$

Set up the calculation so that the units cancel.

$$? \ kg = 2.34 \ pg \ \left| \ \dfrac{10^{-12} \ g}{1 \ pg} \ \right| \ \dfrac{10^{3}}{1 \ g}$$

2.34 pg ×   =

In a Nutshell: Conversion between Metric and English Units

Dimensional analysis is also used to convert between metric and English units. Use the same steps as described above.

Worked Example #7

The Eiffel tower in Paris, France is 300.65 meters tall. What is the height of the Eiffel Tower in feet?

Identify the conversion(s):

$1\ m = 39.37\ inches$ and $1\ ft = 12\ inches$

Express each conversion as two possible conversion factors:

1st *conversion factor:* $\dfrac{1\ m}{39.37\ inches}$ or $\dfrac{39.37\ inches}{1\ m}$

and

2nd *conversion factor:* $\dfrac{1\ ft}{12\ inches}$ or $\dfrac{12\ inches}{1\ ft}$

Set up the calculation so that the supplied units cancel:

$$300.65\ \cancel{m} \times \frac{39.37\ \cancel{inches}}{1\ \cancel{m}} \times \frac{1\ ft}{12\ \cancel{inches}} = 986\ ft,\ \text{rounded to 990 ft}$$

Try It Yourself #7

In the United States, a can of soda has 12 oz. Is this can of soda larger or smaller than the 333-mL cans of soda sold in Europe?

Identify the conversion(s).

Express each conversion as two possible conversion factors.

Set up the calculation so that the supplied units cancel.

Worked Example #8

Amoxil, an antibiotic, is prescribed for an infant who weighs 7.2 lbs. The dosage is 30 mg per kilogram of body weight per day. How many mg should one dose for this child contain?

Identify the conversion(s).
The conversion for the patient's weight from pounds to metric units is 1 kg = 2.205 lbs.

Express each conversion as two possible conversion factors.
English to metric units for the patient's weight are:

$$\frac{1\,kg}{2.205\,lbs} \quad or \quad \frac{2.205\,lbs}{1\,kg}$$

The dosage is the conversion factor between the mass of medicine in mg and the weight of the patient in kg per dose:

$$\frac{30\,mg}{kg \cdot dose} \quad or \quad \frac{kg \cdot dose}{30\,mg}$$

Set up the calculation so that the supplied units cancel:

$$7.2\ \cancel{lbs} \times \frac{1\ \cancel{kg}}{2.205\ \cancel{lbs}} \times \frac{30\,mg}{\cancel{kg} \cdot dose} = 98\ mg/dose\text{, rounded to 100 mg/dose}$$

Try It Yourself #8

A child who weighs 47 lbs needs to take Tylenol. The dosage is 10 mg per kilogram of body weight every 4 hours.

     a.   How many mg of Tylenol are in one dose for this child?

     b.   How much Tylenol would the child take in one day (24 hrs.)?

Identify the conversion(s).

Express each conversion as two possible conversion factors.

Set up the calculation so that the supplied units cancel.

In a Nutshell: Density

Density represents a physical property. The density of a material is defined as its mass (m) divided by its volume (V). It can also be used as a conversion factor to calculate the mass of a substance whose volume and density are known, or the volume of a substance whose mass and density are known.

Worked example #9

What is the volume of 69.3 mL of alcohol? The density of alcohol is 0.79 g/mL.

Identify the conversion(s).
The conversion factor is the given in the physical quantity of density, so skip to the next step.

Express each conversion as two possible conversion factors:

$$\frac{0.79 \text{ g}}{1 \text{ mL}} \quad or \quad \frac{1 \text{ mL}}{0.79 \text{ g}}$$

Set up the calculation so that the supplied units cancel:

$$69.3 \text{ mL} \times \frac{0.79 \text{ g}}{1 \text{ mL}} = 54.7 \text{ g}, \text{ rounded to 55 g.}$$

Try It Yourself #9

What is the volume in mL of 10 lbs of ice? The density of ice is 0.9167 g/cm³.

Identify the conversion(s).

Express each conversion as two possible conversion factors.

Set up the calculation so that the supplied units cancel.

Practice Problems for Matter and Measurement

1. How many mg are there in 0.607 kg?

2. The Gateway Arch in St. Louis, Missouri is 630 ft tall. What is its height in meters?

3. A 4.4 kg infant received a fentanyl infusion for pain relief at dosage of 5.7 μg/kg per hour. What was the total dose in a 24 hour period?

Section 1.2: Elements and the Structure of the Atom

An atom is the smallest intact component of all matter. An element is a substance composed of only one type of atom, which cannot be broken down into any simpler form of matter. A compound is a substance composed of two or more different types of atoms held together in a unique proportion, which can be broken down into its elements through a chemical reaction.

In a Nutshell: The Parts of an Atom

An atom has three subatomic particles: protons, electrons, and neutrons. A proton has a positive charge +1; an electron has a negative charge −1; and a neutron has no charge. Protons and neutrons are concentrated in a small volume at the center of the atom, known as the nucleus. Electrons occupy a larger volume of space, known as the electron orbital. The mass of a subatomic particle is measured in units called atomic mass units, amu. The atomic number is the number of protons in an atom. The identity of an element is determined by the number of protons in its atoms—its atomic number.

In a Nutshell: Physical and Chemical Properties of the Elements

Every element has two properties that we can observe on the macroscopic scale: 1) physical properties and 2) chemical properties. The physical properties of an element are those properties we can observe or measure without considering how the element interacts with other substances. The chemical properties of an element refer to its behavior in the presence of other chemical substances.

In a Nutshell: Isotopes of an Element

Isotopes are atoms with the same number of protons, but a different number of neutrons. Most elements have more than one naturally occurring isotope, and additional isotopes can sometimes be prepared artificially. The mass number is number of protons plus the number of neutrons.

Worked Example #10

How many protons and neutrons are there in potassium-41 with an atomic number of 19?

The number of protons is the same as the atomic number. The number of protons is 19.
The neutrons can be calculated by subtracting the number of protons from its mass number
(41).

> *# neutrons = mass number − atomic number*
> *# neutrons = 41 − 19 = 22 neutrons*

Try It Yourself #10

What are the mass number, number of protons, and number of neutrons in iron-57? The
atomic number is 26.

> *The atomic number = _____*

> *The mass number = _____*

> *# neutrons = _____ − _____*

In a Nutshell: Average Atomic Mass

The mass of an atom is determined by the number of protons and neutrons that it contains.
The average atomic mass of an element is a weighted average of the mass of its isotopes
based on their natural abundance. The term "average atomic mass" is often simplified to
"atomic mass."

Worked Example #11

Germanium has five naturally occurring isotopes: germanium–70, germanium–72,
germanium–73, germaniume–74, and germanium–76. The table below shows the natural
occurrence of each of these isotopes as a percentage of all germanium isotopes.

Germanium Isotope	Natural Abundance
Germanium–70	21.234%
Germanium–72	27.663%
Germanium–73	7.731%
Germanium–74	35.942%
Germanium–76	7.442%

a. Which of these isotopes has the greatest mass and why?

b. Which of these isotopes is the most abundant isotope of germanium on Earth?

c. Which isotope has the least number of neutrons?

Solutions

a. *The isotope that has the greatest mass is germanium–76. It has the highest mass number of the isotopes listed.*

b. *The most abundant isotope is germanium–74. It has the highest percentage of natural abundance.*

c. *The isotope with the least number of neutrons is germanium–70. It has the smallest mass number. The mass number is the number of protons and the number of neutrons. Since all the isotopes have the same number of protons, the isotope with the smallest mass number will have the smallest number of neutrons.*

Try It Yourself #11

Tungsten has five naturally occurring isotopes: tungsten–180, tungsten–182, tungsten–183, tungsten—184, and tungsten–186.

The table below shows the natural occurrence of each of these isotopes as a percentage of all tungsten isotopes.

Tungsten Isotope	Natural Abundance
Tungsten–180	0.134%
Tungsten–182	26.32%
Tungsten–183	14.31%
Tungsten–184	30.6715%
Tungsten–186	28.62%

a. Which of these isotopes has the smallest mass and why?

b. Which of these isotopes is the least abundant isotope of germanium on Earth?

c. Which isotope has the greatest number of neutrons?

Practice Problems for Elements and the Structure of the Atom

Chromium has four naturally occurring isotopes: chromium–50, chromium–52, chromium–53, and chromium–54. Complete the table below:

Isotope	Mass number	Atomic number	Number of protons	Number of neutrons
Chromium—50		24		
Chromium–52		24		
Chromium–53		24		
Chromium–54		24		

a. What do all chromium isotopes have in common?

b. How are the isotopes of chromium different?

c. Which isotope of chromium has the greatest mass? Explain.

Section 1. 3: The Periodic Table of Elements

In a Nutshell: How to Read the Periodic Table

Each box on the periodic table of elements displays the element symbol, atomic number, and the average atomic mass of a single element. Information about isotopes and their mass number does not appear in the periodic table. The elements are listed in the periodic table in order of their atomic number: from lowest atomic number to highest atomic number from left to right and from top to bottom across the table.

In a Nutshell: Groups and Periods

Within the periodic table a column of elements is known as a group or family of elements. Elements in the first two (columns numbered 1A and 2A) and last six columns (columns numbered 3A to 8A) are known as main group elements. The middle section (columns numbered 1B to 8B) contains the transition metal elements. Four of the groups have common names: 1) group 1A are the alkali metals; 2) group 2A are the alkaline earth metals; 3) group 7A are the halogens; 4) group 8A are the inert gases or noble gases. Elements within the same group exhibit similar chemical and physical properties.

Rows of elements are known as periods. There are seven periods in the table. Sections of periods 6 and 7 are always separated from the rest of the periodic table, and appear below the main table in two rows. These two rows actually belong in the gap after the elements La and Ac, and they are known as the lanthanides and actinides, respectively.

In a Nutshell: Metals, Nonmetals, and Metalloids

The elements can also be classified as being a metal or nonmetal. Metals appear to the left of the bold zigzag that runs diagonally from boron to polonium. Nonmetals appear to the right of the bold line. The elements located along the zigzag line are called metalloids. Metalloids display characteristics of both metals and nonmetals.

In a Nutshell: Important Elements in Biochemistry and Medicine

Some of the elements, classified as building block elements, macronutrients, micronutrients, and radioisotopes in nuclear medicine, play an important role in medicine and biochemistry. The building block elements include carbon, hydrogen, nitrogen, oxygen, phosphorus, and sulfur. These are the elements that make up the chemical structure of the majority of molecular compounds found in living organisms. Essential nutrients (macronutrients and micronutrients) must be supplied through the diet. Macronutrients are required in large quantities (more than 100 mg a day) and micronutrients are required in less than 100 mg a day.

Worked Example #12

Answer the following questions about the element selenium.

    a. Determine the atomic number and the element symbol.

b. State the group number or family name, if one exists.

c. Is selenium a metal or nonmetal?

Solutions

a. *Locate the element symbol on the periodic table. The atomic number will be located in the box as well. It will always be a whole number. Selenium, Se, atomic number = 34.*

b. *Selenium is in group 6A—it does not have a family name.*

c. *Selenium is a nonmetal because it falls to the right of the blue line.*

Try It Yourself #12

Answer the following questions about the element krypton.

a. Determine the atomic number and the element symbol.

Locate the element symbol on the periodic table.

The element symbol is _____. The atomic number is _____.

b. State the group number or family name, if one exists.

Krypton is in _____. The family name is _____.

c. Is krypton a metal or nonmetal?

Krypton lies to the (left or right) of the blue line.

Krypton is a _____.

Practice Problems for The Periodic Table of Elements

For the following elements, determine the atomic number, the element symbol, the group number, family name (if it exists). Also state if it is a nonmetal or metal

a. silver

b. lithium

c. xenon

   d. iodine

Section 1.4: Electrons

In a Nutshell: Electron Shells

Electrons can be characterized by the shape of their orbitals and their energy. . Electrons within an atom that belong to the same energy level are said to be in the same shell. Electrons occupy different energy levels, or shells, n, where the maximum number of electrons in a given shell is $2n^2$. The lower the value of n, the closer the electrons are to the nucleus.

In a Nutshell: Valence Electrons and Electron Shells

Valence electrons are the outermost electrons found in the highest most n value. Every element within a group has the same number of valence electrons which correspond to the group number. In general, atoms with valence electrons in higher energy levels will have a larger diameter than atoms with valence electrons in lower energy levels, because their valence electrons are in larger orbitals, which occupy a volume farther from the nucleus.

Worked Example #13

Fill in the table below.

Element	Symbol	Atomic number	Group number	Number of $n = 1$ electrons	Number of $n = 2$ electrons	Number of $n = 3$ electrons	Number of valence electrons
Phosphorus	P						
				2	8	3	

Solutions

The periodic table shows the atomic number for each element. Table 1-12 provides electron arrangements for some of the elements. The number of valence electrons equals the number of electrons in the outermost shell.

Element	Symbol	Atomic number	Group number	Number of $n = 1$ electrons	Number of $n = 2$ electrons	Number of $n = 3$ electrons	Number of valence electrons
Phosphorus	P	15	5	2	8	5	5
Aluminum	Al	13	3	2	8	3	3

Try It Yourself #13

Fill in the table below.

Element	Symbol	Atomic Number	Group Number	Number of $n = 1$ electrons	Number of $n = 2$ electrons	Number of $n = 3$ electrons	Number of valence electrons
	Li						
				2	8	0	

Practice Problems for Electrons

Fill in the table below.

Element	Symbol	Atomic number	Group number	Number of $n = 1$ electrons	Number of $n = 2$ electrons	Number of $n = 3$ electrons	Number of valence electrons
	S						
				2	8	1	
		17					

Chapter 1 Quiz

1. When a gold nugget is placed in a graduated cylinder containing 50.0 mL of water, the water level rises to 68.2 mL.

    a. What is the volume of the gold nugget in liters?

    b. What is the mass of the gold nugget in kilograms? Use the density supplied in Table 1-5.

2. How many nanometers are there in 728 cm?

3. How many mL are there in a 12 oz. cup of coffee? There are 32 oz. in a quart.

4. A 37 lb child is prescribed Advil. The dosage is 10 mg/kg.
   How many mg should be given per dose?

5. Answer the following questions about the element potassium. Potassium has three isotopes: potassium–39 (with an abundance of 93.258144%), potassium–40 (with an abundance of 0.01171%), and potassium–41 (with an abundance of 6.730244%).

    a. What is the symbol for the element? _____

    b. What is its atomic number? _____

    c. What is its mass number? _____

    d. Which isotope is more abundant on Earth? _____

    e. How many neutrons are present in each isotope of potassium? _____

    f. Is potassium a metal or nonmetal? _____

    g. Identify the group number and family for potassium. _____

    h. How many valence electrons does potassium have? _____

6. Fill in the following table

Element	Symbol	Atomic number	Mass number	Group number	Number of protons	Number of neutrons	Number of valence electrons
	O						
		35					
					5	6	

7. Classify the following as metals or nonmetals.

    a. mercury: _____

    b. sulfur: _____

    c. barium: _____

    d. radon: _____

8. How many valence electrons does Group 2A have? _____

9. How many electron shells do period 2 elements (i.e., Li, Be, B, C, N, O, F, and Ne) have? _____

10. Answer the following questions about iron.

    a. What is its atomic number? _____

b. Is it a metal or nonmetal?   _____

c. Is it a building block element, micronutrient, macronutrient, or a radioisotope used in
    nuclear medicine?   _____

Chapter 1
Answers to Additional Exercises

1.57   a. A skyscraper is on a macroscopic scale (you can see it). b. A skin cell is on the microscopic scale (you need a microscope to see it). c. DNA is on the nanoscale (it is too small to be clearly visible in a microscope). d. A red blood cell is on the microscopic scale (you can see it with a microscope).

1.59   pico (10^{-12}), nano (10^{-9}), micro (10^{-6}), kilo (10^3)

1.61   a.   10 mm is shorter than 1 m.

b.   10 mm is the same length as 1 cm.

c.   1 cm is shorter than one 1 dm.

d.   1 dm is shorter than 15 cm.

1.63   The volume of the lead ball is equal to the difference in volume before and after the lead is submerged: V_{lead} = 16.5 mL − 15.0 mL = 1.5 mL

1.65   a. 10^4 is smaller than 10^8.

b. 10^{-6} is smaller than 10^{-3}.

c. 3.7×10^{-4} is smaller than 3.7×10^4.

1.67   a.   1 ng is smaller than 1 mg.

b.   100 mg is smaller than 1 g.

c.   1000 mg is the same as 1g.

1.69   a. 1×10^{18}

b. 2.305×10^9

c. 1.5×10^{-12}

d. 2.08×10^{-2}

1.71   a.   100,000

b.   0.0024

c.   165

1.73   a.   False

b.   True

c.   True

d.   False

1.75   a.   three

b.   one

   c. three

   d. three

1.77   a. 33,000 + 910 = 34,000

   b. 0.333 g x 0.22 = 0.073 g

   c. (37.55 mL + 22.2 mL) x 56.66 = 59.8 mL x 56.66 = 3390 mL

1.79   a. exact number   b. not exact   c. not exact

1.81   a. $6.000 \; \cancel{L} \times \dfrac{1000 \text{ mL}}{1 \; \cancel{L}} = 6000. \text{ mL}$

   b. Density = m/v

$$\text{Density} = \frac{3.8 \text{ g}}{2.0 \text{ cm}^3} = 1.9 \text{ g/cm}^3$$

1.83 $\dfrac{1 \times 10^3 \text{ m}}{1 \text{ km}}$ $\dfrac{1 \text{ km}}{1 \times 10^3 \text{ m}}$

1.85 $200. \; \cancel{\text{mg}} \times \dfrac{1 \times 10^{-3} \text{ g}}{1 \; \cancel{\text{mg}}} = 0.200 \text{ g}$

1.87 $2.000 \; \cancel{\text{minutes}} \times \dfrac{60 \text{ seconds}}{1 \; \cancel{\text{minutes}}} = 120.0 \text{ seconds}$

1.89 $150. \; \cancel{\text{kg}} \times \dfrac{2.205 \text{ lbs}}{1 \; \cancel{\text{kg}}} = 330. \text{ lbs}$

1.91 $86 \; \cancel{\text{gallons}} \times \dfrac{4 \; \cancel{\text{qt}}}{1 \; \cancel{\text{gallon}}} \times \dfrac{1 \text{ L}}{1.057 \; \cancel{\text{qt}}} = 330 \text{ L}$

1.93 $12 \; \cancel{\text{lb}} \times \dfrac{1 \; \cancel{\text{kg}}}{2.205 \; \cancel{\text{lbs}}} \times \dfrac{20 \text{ mg}}{\cancel{\text{kg}} \cdot \cancel{\text{day}}} \times \dfrac{1 \; \cancel{\text{day}}}{3 \text{ dose}} = 40 \text{ mg/dose}$

1.95   a. A brick has the greater density. It has more mass than a loaf of bread for the same volume.

   b. A bowling ball has the greater density. It has more mass than a soccer ball and they both have the same volume.

   c. The bucket of concrete is denser. It has more mass than the bucket of water.

1.97   volume 2.3 mL

$$2.3 \; \cancel{\text{mL}} \times \frac{19.32 \text{ g}}{1 \; \cancel{\text{mL}}} = 44 \text{ g}$$

1.99   Volume = (length of side)^3

   Volume = (2.20 cm)^3

   Volume = 10.648 cm^3

$$10.648 \; \cancel{cm^3} \times \frac{19.32 \; g}{\cancel{cm^3}} = 205 \; g$$

1.101   Convert all measurements to a common unit.

$$5{,}000 \; \cancel{\mu L} \times \frac{1 \times 10^{-6} \; \cancel{L}}{1 \; \cancel{\mu L}} \times \frac{1 \; mL}{1 \times 10^{-3} \; \cancel{L}} = 5 \; mL$$

$$0.5000 \; \cancel{L} \times \frac{1 \; mL}{1 \times 10^{-3} \; \cancel{L}} = 500.0 \; mL$$

$$8.000 \; cm^3 = 8.000 \; mL$$

Largest to smallest 0.5000 L, 50.00 mL, 8.000 cm^3, 5,000 μL

1.103   The gold sphere with a mass of 15 g would have a greater volume than the 6 g sphere. Therefore, the diameter of the sphere would be larger.

1.105   The electron is the lightest of the three subatomic particles.

1.107   Helium has the smaller diameter. It has fewer electrons. As the number of electrons increases, the outermost electrons spend more of their time in larger orbitals farther from the nucleus causing the atom to be bigger in size.

1.109   a. carbon, C. b. aluminum, Al. c. americium, Am. d. platinum, Pt. e. cobalt, Co.

1.111   The atomic number equals the number of protons in an atom. The number of protons in an atom equals the number of electrons in an atom. a. atomic number = 58; 58 protons, 58 electrons. b. atomic number = 75; 75 protons, 75 electrons. c. atomic number = 25; 25 protons, 25 electrons.

1.113   arsenic

1.115   radium

1.117   Selenium is bigger. It has more electrons, which take up more space around the nucleus.

1.119   The mass number is the sum of the number of protons plus the number of neutrons.

1.121 The atomic number equals the number of protons. The mass number = # protons + # neutrons.

Isotope	Mass number	Atomic number	Number of protons	Number of neutrons
Sulfur-32	32	16	16	16
Sulfur-33	33	16	16	17
Sulfur-34	34	16	16	18
Sulfur-36	36	16	16	20

1.123 Sulfur-32 is the lightest isotope, because it has the fewest number of neutrons.

1.125 $^{16}_{8}O$, $^{17}_{8}O$, $^{18}_{8}O$

1.127 a. Chlorine-35 has mass number 35. Chlorine-37 has mass number 37. b. The atomic number for both isotopes is 17. c. Chlorine-35 has 18 neutrons, while chlorine-37 has 20 neutrons.

1.129 The physical properties of an element are the physical characteristics of the element, such as color and consistency, that it has on its own. The chemical properties of the element are defined by the manner in which it interacts with other elements or substances.

1.131 Elements in the first two and last six columns of the periodic table are the main group elements. The transition metals are located in columns 1B-8B.

1.133 The alkali metals are located in group 1A of the periodic table, while alkaline earth metals are located in group 2A of the periodic table.

1.135 a. physical properties b. physical properties

1.137 a. nonmetal b. metalloid c. nonmetal

1.139 a. Element in column 1, row 4 of the periodic table—potassium. b. Element in column 8A, row 6 of the periodic table—radon.

1.141 The electron has the greatest effect on the chemical and physical properties of an element.

1.143 a. shell n = 1, 2 electrons, shell n = 2, 3 electrons. Three valance electrons. b. shell n = 1, 2 electrons, shell n = 2, 8 electrons, shell n = 3, 5 electrons. Five valance electrons.

1.145 The s orbital is spherical in shape and the p orbital has a two lobed shape, similar to a dumbbell.

1.147 group 4A

1.149 The macronutrients are Na, K, Mg, Ca, P, S, and Cl. The micronutrients are V, Cr, Mn, Fe, Co, Cu, Zn, Mo, Si, Se, F, and I. Macronutrients are required in our diet in large quantities—more than 100 mg a day, while micronutrients are required in quantities of less than 100 mg a day.

1.151 Your body needs micronutrients in trace quantities, less than 100 mg a day.

1.153 Iron is part of hemoglobin that transports oxygen in red blood cells. Iron is also found in many other oxygen transport molecules and enzymes that are involved in extracting energy from the foods you eat.

Chapter 2

Compounds

Chapter Summary

In this chapter, you have learned about how atoms combine with other atoms to form a compound. Atoms can interact in two different ways to form compounds: They can be electrostatically attracted to form ionic compounds or they can be held together by covalent bonds. You also learned the important skill of writing Lewis dot structures. Lewis dot structures are a type of electron bookkeeping that tells you how atoms are connected to one another in a molecule. Lewis dot structures will also enable you to build three-dimensional representations of molecules consider how molecules interact with each other. You have also learned to go from the atomic scale to the macroscopic scales using moles, molar mass, and Avogadro's number. Understanding the bonding between atoms will help you understand the properties of compounds and molecules, which in turn will help you understand how these molecules interact with other molecules. Understanding moles and molar mass will help you understand blood test results, drug dosages, and IV solutions.

Section 2.1 Ionic Compounds

A compound is substance composed of two or more different atoms in a defined whole-number proportion. A substance composed of only one compound or element is classified as a pure substance. There are two basic types of compounds: ionic and covalent. An ionic compound is formed when a metal atom transfers some or all of its valence electrons to a nonmetal atom, creating charged species known as ions. A positively charged ion and a negatively charged ion exhibit a strong force of attraction known as an electrostatic attraction. The electrostatic between oppositely charged ions is the glue that holds ions together to form an ionic compound.

In a Nutshell: Ions

An ion is an atom that has lost or gained one or more valence electrons. Therefore, an ion has an unequal number of protons and electrons, which give it a positive (+) or a negative (−) charge. The magnitude of the charge on an ion is equal to the difference between the number of protons and electrons in the ion. When the number of electrons is greater than the number of protons, the ion is negatively charged and is known as an anion. When the number of electrons is less than the number of protons, the ion is positively charged and known as a cation. As a general rule, metals tend to lose electrons to become cations, while nonmetals tend to gain electrons to become anions.

When a main group metal loses all of its valence electrons, the ion produced has a full outermost shell—an arrangement of electrons similar to the noble gas in the period above it. Transition metals can lose a variable number of electrons. Nonmetal elements in groups 5A through 7A have a tendency to gain electrons and form anions These elements gain the number of electrons needed to achieve a full valence shell—typically eight electrons (although hydrogen achieves a full valence shell with two electrons).

Worked Example #1

Write the ions that can be formed from the following elements, and indicate whether the elements are metals or nonmetals. Indicate the number of protons and electrons in each ion.

    a. magnesium

    b. sulfur

    c. mercury

Solutions

    a. *Looking at the periodic table, you see that magnesium is in group 2A. Group 2A elements lose two electrons (all of the valence electrons) to become an ion with a full outermost shell of electrons. Magnesium forms Mg^{2+} as an ion. It is a metal. Mg^{2+} has 12 protons and 10 electrons.*

    b. *Sulfur is in group 6A. Group 6A elements gain 2 electrons to become an ion with a full outermost shell of electrons. Sulfur forms S^{2-} as an ion. It is a nonmetal. S^{2-} has 16 protons and 18 electrons.*

    c. *Mercury is a transition metal. Figure 2-5 shows that mercury loses two electrons or one electron. Mercury can form Hg^{2+} or Hg^+ as ions. Hg^{2+} has 82 protons and 80 electrons. Hg^+ has 81 protons and 80 electrons.*

Try It Yourself #1

Write the ions that can be formed from the following elements, and indicate whether the elements are metals or nonmetals.   Indicate the number of protons and electrons in each ion.

    a. rubidium

    b. arsenic

    c. aluminum

    a. *Rubidium is in group: _____*

       Rubidium will (lose or gain) _____ electrons.

       How many electrons will be lost or gained? _____

       Ion formed: _____

       Metal or nonmetal: _____

       Number of protons: _____

       Number of electrons: _____

    b. *Arsenic is in group: _____*

       Arsenic will (lose or gain) _____ electrons.

       How many electrons will be lost or gained? _____

       Ion formed: _____

       Metal or nonmetal: _____

       Number of protons: _____

       Number of electrons: _____

    c. *Aluminum is in group: _____*

       Aluminum will (lose or gain) _____ electrons.

       How many electrons will be lost or gained? _____

       Ion formed: _____

       Metal or nonmetal: _____

       Number of protons: _____

       Number of electrons: _____

In a Nutshell: The Ionic Lattice

An ionic compound is formed as the result of mutual attraction between cations and anions. The electrostatic attraction that holds ions together is also called an ionic bond. Most ionic compounds are brittle solids at room temperature. On the atomic scale, ionic compounds exist as a crystalline lattice. In the lattice structure, every cation is surrounded by anions and every anion is surrounded by cations.

In a Nutshell: Electrolytes

When most ionic compounds are placed in water, the lattice structure falls apart and each cation and anion become surrounded by water molecules. The ions become separated from one another and are instead surrounded by water molecules; the salt is said to be dissolved in water. Ions dissolved in water are often referred to as electrolytes.

In a Nutshell: The Formula Unit

A pure ionic compound has a definite and unique composition that is defined by its formula unit. The formula unit of an ionic compound gives the ratio of cation to anion in the lattice. The formula unit of an ionic compound is written according to the following set of rules:

- The chemical symbol of each ion in the compound is written, with the cation listed first, followed by the anion. The charges of the ions are not included in the formula unit.

- A numerical subscript following each chemical symbol is used to indicate the ratio of cation to anion in the compound. The subscript is understood to be "1" when none is shown. The lowest whole number ratio is used for the subscripts.

- The sum of the positive and negative charges of the individual ions of an ionic compound must always add up to zero, so that the ionic compound is neutral overall.

- When the magnitude of the charges on the cation and the anion are different, subscripts must be added to indicate the ration that results in an electrically neutral compound.

The following guidelines can determine the formula unit: 1) Determine the charge on the cation; 2) determine the charge on the anion; 3) insert subscripts; 4) if the subscripts can be divided by a common divisor, do so; and 5) double check that the formula unit indicates a neutral compound.

In a Nutshell: Naming Ionic Compounds

To name an ionic compound, write the name of the cation first followed by the name of the anion. A cation always has the same name as the element from which it is derived. For transition metals that have variable charged forms, the charge on the cation is indicated by placing a Roman numeral, corresponding to the magnitude of the charge, within parentheses following the name of the cation. The anion is named by changing the ending on the element name to "ide."

Worked Example #2

Write the formula unit for gallium chloride.

Solution

Determine the charge on the cation.

Gallium is in group 3; the charge on gallium is +3. The ion formed is Ga^{3+}

Determine the charge on the anion.

Chlorine is in group 7A; the charge on chlorine is −1.

Insert the subscripts.

$Ga^{3+} \quad Cl^- \quad \rightarrow \quad GaCl_3$

If the subscripts can be divided by a common divisor, do so.

No common divisor exists for the subscripts 1 and 3.

Double check that the formula gives a neutral compound.

(cation charge × cation subscript) + (anion charge × anion subscript) = 0

$(+3 \times 1) + (-1 \times 3) =$

$(+3)..+..(-3) = 0$

Try It Yourself #2

Write the formula unit for titanium(II) selenide.

Determine the charge of the cation.

Determine the charge of the anion.

Insert subscripts.

If the subscripts can be divided by a common divisor, do so.

Double check that the formula gives a neutral compound.

Worked Example #3

What is the name of the compound with the formula unit PbF_4?

Solution

First, we must determine the charge on the cation. You can determine the charge of the cation from the formula unit, knowing that the compound must be neutral overall:

$(? \times 1) + (-1 \times 4) = 0$

$(? \times 1) + (-4) = 0$

$? \times 1 = 4$

$? = +4$

Second, name the cation first followed by the anion: lead(IV) fluoride.

Try It Yourself #3

What is the name of the compound with the formula unit Cu_3P?

Determine the charge on the cation.

Name the compound.

Practice Problems for Ionic Compounds

1. Write the ions that can be formed from the following elements, and indicate whether the elements are metals or nonmetals. Indicate the number of protons and electrons in each ion.

   a. strontium

   b. gallium

   c. selenium

2. Write the formula unit for the following compounds.

   a. potassium sulfide

   b. aluminum oxide

   c. lead(IV) oxide

   d. copper(II) chloride

   e. sodium phosphide

3. Write the names of the following compounds.

   a. FeF_2

   b. $CaCl_2$

   c. $CuCl$

   d. BCl_3

e. Mg_3P_2

Section 2.2 Covalent Compounds

Covalent compounds consist of molecules. A molecule is a discrete entity composed of two or more nonmetal atoms held together by covalent bonds. A covalent bond is formed when two nonmetal atoms come together and share some of their valence electrons. By sharing electrons, nonmetal elements achieve a full outermost shell of eight electrons. One exception is hydrogen, which achieves a full outermost shell when it shares two electrons.

In a Nutshell: The Molecular Formula

The molecular formula describes the definite and unique composition of a covalent compound. In a molecular formula each atom type is listed, usually in alphabetical order, followed by a subscript indicating how many atoms of that type are in the molecule. A molecular formula cannot be altered without also changing the identity of the compound.

In a Nutshell: The Covalent Bond

When two atoms form a covalent bond, each atom achieves a full outermost shell of electrons (like the noble gases). Hydrogen will share only one valence electron, while the elements in period 2 will share one to four valence electrons in order to achieve a full outermost shell of eight electrons. Many atoms follow the octet rule; they have eight electrons in the outermost shell. When two atoms form a covalent bond, each atom shares an electron with the other atom to form a covalent bond. A shared pair of electrons is known as a covalent bond. When only one pair of electrons is shared between two atoms, the bond is known as a single bond. The distance between two nuclei joined by a bond is known as the bond length. Two atoms may also share four electrons to form a double bond or six electrons to form a triple bond. Atoms do not necessarily share all of their valence electrons. The electrons that are not shared are known as nonbonding electrons.

Lewis dot structures serve as an electron bookkeeping tool that provides a two–dimensional representation of a molecule. In Lewis dot structure a pair of shared electrons, a single bond, is usually represented as a line. A double bond will be represented by two lines and a triple bond will be represented by three lines. Nonbonding electrons, also known as lone pairs, are always represented as pairs of dots, written on the atom they belong to, but with no second atom attached.

In a Nutshell: Writing Lewis Dot Structures

You can determine the Lewis dot structure from a simple molecular formula using the following guidelines:

1) Determine which atom in the molecule is the central atom and which atoms are the surrounding atoms. Place a single bond between each pair of atoms.

2) Add up the total number of valence electrons and determine the number of electrons remaining to be distributed as nonbonding electrons.

3) Perform a preliminary distribution of the remaining electrons as nonbonding pairs. Do this by placing pairs of dots around each atom so that each atom has an octet (8), except hydrogen, which should have a duet (2).

4) Turn nonbonding electrons into multiple bonds if any atoms are short of an octet. If after distributing all the valence electrons, there are still some atoms short of an octet, turn a nonbonding pair of electrons on an adjacent atom into a multiple bond to the atom that is short, creating a double bond. If the atom is still short of an octet, a triple bond may be required.

5) Double check each atom in the molecule against Table 2-2 to check that each atom is surrounded by the expected number of bonding and nonbonding electrons.

Worked Example #4

Write the Lewis dot structure for NBr_3.

Determine which atom in the molecule is the central atom and which atoms are the surrounding atoms. Place a single bond between each pair of atoms.

$$\text{Br}-\text{N}-\text{Br}$$
$$|$$
$$\text{Br}$$

Nitrogen is the central atom because it is the atom closest to the center of the Periodic Table.

Add up the total number of valence electrons and determine the number of electrons remaining to be distributed as nonbonding electrons.
The total number of valence electrons is nitrogen—5, bromine—7: 5 + (3 × 7) = 26 total valence electrons. Subtract the six electrons used in the N-Br single bonds: 26 − 6 = 20 electrons remain to be distributed.

Perform a preliminary distribution of the remaining electrons as nonbonding pairs. Do this by placing pairs of dots around each atom so that each atom has an octet (8), except hydrogen, which should have a duet (2).

$$:\!\overset{..}{\underset{..}{Br}}\!-\!\overset{..}{\underset{|}{N}}\!-\!\overset{..}{\underset{..}{Br}}\!:$$
$$:\!\overset{..}{\underset{..}{Br}}\!:$$

Turn nonbonding electrons into multiple bonds if any atoms are short of an octet.

All of the atoms have an octet; therefore, multiple bonds are not needed.

Double check each atom in the molecule against Table 2-2 to check that each atom is surrounded by the expected number of bonding and nonbonding electrons.

The nitrogen atom has three bonds and one nonbonding pair and the bromine atoms have one bond and three nonbonding pairs.

Try It Yourself #4
Write the Lewis dot structure for $CHCl_3$.

Determine which atom in the molecule is the central atom and which atoms are the surrounding atoms. Place a single bond between each pair of atoms.

Add up the total number of valence electrons and determine the number of electrons remaining to be distributed as nonbonding electrons.

Perform a preliminary distribution of the remaining electrons as nonbonding pairs. Do this by placing pairs of dots around each atom so that each atom has an octet (8), except hydrogen, which should have a duet (2).

Turn nonbonding electrons into multiple bonds if any atoms are short of an octet.

Double check each atom in the molecule against Table 2-2 to check that each atom is surrounded by the expected number of bonding and nonbonding electrons.

In a Nutshell: Expanded Octets

The key building block elements, carbon, nitrogen, and oxygen will always have an octet in a molecule. Atoms in period 3, phosphorus and sulfur, are sometimes found with an expanded octet; these atoms may be surrounded by more than eight electrons in the outermost shell.

Worked Example #5

Evaluate each of the following molecules to determine if it contains an atom with an expanded octet. Indicate the number of electrons surrounding each expanded octet.

a. The sulfur atom has an expanded octet; there are 12 electrons around the sulfur atom.

b. The phosphorus atom has an expanded octet; there are 10 electrons around the phosphorus atom.

Try It Yourself #5

Evaluate each of the following molecules to determine if it contains an atom with an expanded octet. Indicate the number of electrons surrounding each expanded octet.

a.

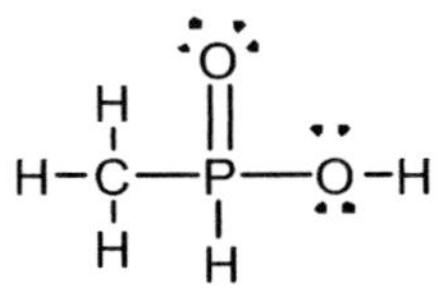

b.

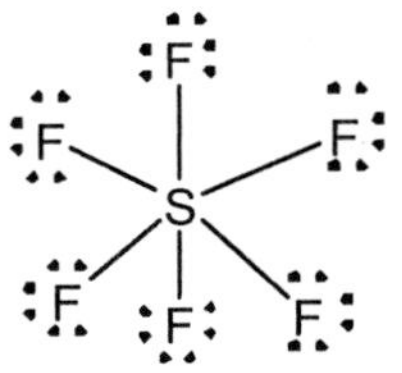

a. Atom with expanded octet: _____

Number of electrons surrounding the expanded octet: _____

b. Atom with expanded octet: _____

Number of electrons surrounding the expanded octet: _____

In a Nutshell: Naming Simple Binary Covalent Compounds

Naming covalent compounds containing only two different types of nonmetal elements—binary compounds—involves a process that is similar to the naming of ionic compounds. Begin by naming the first element in the formula according to its element name, followed by the second element in the formula, but change the ending to "ide." If more than one atom of an element is present as indicated by subscripts, a prefix is inserted before the element name to indicate the number of atoms of that type present in the molecule.

Worked Example #6

Name the compound PBr_3.

Phosphorus tribromide. There is one phosphorus atom, named after the element: phosphorus. There are three bromine atoms, so the prefix "tri" is inserted before the element name and the ending is changed to "ide."

Try It Yourself #6

Name the compound H_2S.

Number of hydrogen atoms:_____

Prefix for hydrogen:_____

Number of sulfur atoms:_____

Prefix for sulfur:_____

Name of compound:_____

Worked Example #7

Write the molecular formula for diphosphorus pentasulfide.

P_2S_5. The prefix "di" indicates that there are two phosphorus atoms; the prefix "penta" indicates that there are five sulfur atoms.

Try It Yourself #7

Write the molecular formula for carbon tetrabromide.

The two elements are:_____

Number of atoms of each element:_____ and _____

Molecular formula:_____

Practice Problems for Covalent Compounds

1. Write the Lewis dot structure for the following molecules:

    a. HBr

    b. ClF

    c. BrCN

2. Write the Lewis dot structure for nitrogen triiodide.

3. Write the Lewis dot structure for phosphorus pentabromide.

4. Write the Lewis dot structure for CS_2. What is the name of this compound?

5. Write the Lewis dot structure for COS.

Section 2.3 Compounds Containing Polyatomic Ions

In a Nutshell: Polyatomic Ions

A third category of compounds exists that has characteristics of both ionic and covalent compounds. These are compounds containing one or more polyatomic ions. A polyatomic ion is an ion formed when a molecule rather than a single atom, gains or loses one, two or three electrons. Covalent bonds join the atoms, but the imbalance of protons and electrons overall creates a net charge on the molecule. The charge can be localized on one atom or it can be spread over several atoms in the polyatomic ion.

In a Nutshell: The Formula Unit and Naming

Writing the formula unit of a compound containing a polyatomic ion is similar to the process described for monoatomic ions, except that If there is a subscript following the polyatomic ion in the formula unit, the entire polyatomic ion must be enclosed in parentheses. When naming an ionic compound containing a polyatomic ion, name the cation followed by the anion. The names for polyatomic cations and anions are given in Table 2-4.

Worked Example #8

Answer the questions below about the carbonate ion:

$$\left[\begin{array}{c} \ddot{\ddot{O}} \\ \| \\ :\ddot{O}-C-\ddot{O}: \end{array} \right]^{2-}$$

Carbonate ion

a. The charge on the carbonate ion is spread over all three oxygen atoms. What is the total charge on the carbonate ion?

b. How many covalent bonds does this polyatomic ion contain?

c. Why is the carbonate ion classified as a polyatomic ion and not a monatomic ion?

d. How is the carbonate ion different from a molecule?

e. What is the formula unit for sodium carbonate?

f. What is the formula unit for calcium carbonate?

Solutions

   a.   The total charge is −2 as indicated outside the brackets.

   b.   There are four covalent bonds: two single bonds and one double bond.

   c.   Carbonate is a polyatomic ion because there are several atoms joined by covalent bond, and collectively they carry a net −2 charge.

   d.   Carbonate is an ion because there is a charge on the collection of ions.

   e.   Determine the charge on the cation.

   The cation is the sodium ion. The sodium ion has a +1 charge.

   Determine the charge on the anion.

   The anion is the carbonate ion. The carbonate ion has a −2 charge.

   Insert the subscripts.

   $Na^+ \quad CO_3^{2-} \quad \rightarrow \quad Na_2CO_{3.}$

   If the subscripts can be divided by a common divisor, do so.

   No common divisor exists for the subscripts 1 and 2.

   Double check that the formula gives a neutral compound.

   (cation charge × cation subscript) + (anion charge × anion subscript) = 0

$(+1 \times 2) + (-2 \times 1) = 0$

$+2 + (-2) = 0$

f. *Determine the charge on the cation.*
    The cation is the calcium ion. The calcium ion has a +2 charge.

Determine the charge on the anion.
The anion is the carbonate ion. The carbonate ion has a −2 charge.

Insert the subscripts.
Ca^{2+} CO_3^{2-} → $CaCO_3.$

If the subscripts can be divided by a common divisor, do so.

Double check that the formula gives a neutral compound.
(cation charge × cation subscript) + (anion charge × anion subscript) = 0
$(+2 \times 1) + (-2 \times 1) = 0$
$+2 + (-2) = 0$

Try It Yourself #8

Answer the questions below about the cyanide ion:

$$\left[: C \equiv N : \right]^{-}$$

Cyanide ion

a. What is the total charge on the cyanide ion?
b. How many covalent bonds does this polyatomic ion contain?
c. Why is the cyanide ion classified as a polyatomic ion and not a monatomic ion?
d. How is the cyanide ion different from a molecule?
e. What is the formula unit for sodium cyanide?
f. What is the formula unit for calcium cyanide?

Solutions

a. The total charge:_____

b. Number of covalent bonds:_____

c. Number of different atoms in cyanide:_____

d. The cyanide ion is different from a molecule_____

e. Determine the charge on the cation.

Determine the charge on the anion.

Insert the subscripts.

If the subscripts can be divided by a common divisor, do so.

Double check that the formula gives a neutral compound.

(cation charge × cation subscript) + (anion charge × anion subscript) = 0

f. Determine the charge on the cation.

Determine the charge on the anion.

Insert the subscripts.

If the subscripts can be divided by a common divisor do so.

Double check that the formula gives a neutral compound.

(cation charge × cation subscript) + (anion charge × anion subscript) = 0

Practice Problems for Compounds Containing Polyatomic Ions

1. Write the formula unit for the following ionic compounds:

   a. sodium hydrogen carbonate (also known as baking soda)

   b. ammonium hydroxide

   c. sodium sulfate

   d. calcium phosphate

2. Name the following ionic compounds:

    a. Na_2HPO_4

    b. $Fe(NO_3)_3$

    c. NaOCl

    d. $Al(OH)_3$

Section 2.4 Formula Mass, Molecular Mass, and Molar Mass

In a Nutshell: Formula Mass and Molecular Mass

If you know the mass of an atom, you can determine the number of atoms present from the mass of a macroscopic sample of matter. For a covalent compound, the mass of one molecule, the molecular mass, is the sum of the individual average atomic masses of its component atoms. For an ionic compound, the mass of one molecule, the formula mass, is the sum of the atomic masses of the elements in the formula unit. The mass of an ion is the same as the element from which it was derived.

Worked Example #9

Calculate the molecular mass of acetaminophen (Tylenol), $C_8H_9NO_2$.

The substance is a covalent compound, so the terms "molecular mass" and "molecular formula" are used.

Atom type	# of atoms		Atomic mass (from Periodic Table)		Total
C	8	×	12.01	=	96.08 amu
H	9	×	1.008	=	9.072 amu
N	1	×	14.01	=	14.01 amu
O	2	×	16.00	=	32.00 amu

The molecular mass = 96.08 + 9.072 + 14.01 + 32.00 = 151.16 amu/molecule. Remember to avoid rounding errors; do not round the atomic mass values derived from the Periodic Table until the end of the calculation.

Try It Yourself #9

Naproxen is a nonsteroidal anti-inflammatory drug. Calculate the molecular mass of naproxen, $C_{14}H_{14}O_3$.

Tools: The Periodic Table

Fill in the table below.

Atom type	# of atoms		Atomic mass (from Periodic Table)		Total
		×		=	
		×		=	
		×		=	

*The molecular mass:*_____

Worked Example #10

Calculate the formula mass of calcium phosphate, $Ca_3(PO_4)_2$.

The substance is an ionic compound, so the terms "formula mass" and "formula unit" are used. Remember, the atomic mass of atoms within parentheses must be multiplied by the subscript outside of the parentheses, as well as the subscript within the parentheses. In this example, there are eight oxygen atoms, as indicated by the subscript "4" following the symbol for oxygen and the subscript "2" after the parentheses.

Atom type	# of atoms		Atomic mass (from Periodic Table)		Total
Ca	3	×	40.08	=	120.24 amu
P	2	×	30.97	=	61.94 amu
O	8	×	16.00	=	128.00 amu

The formula mass = 120.24 + 61.94 + 128.00 = 310.18 amu/formula unit.

Try It Yourself #10

Calculate the formula mass of calcium hydroxyapatite, $Ca_5(PO_4)_3OH$.

Tools: The Periodic Table

Fill in the table below

Atom type	# of atoms		Atomic mass (from Periodic Table)		Total
		×		=	
		×		=	
		×		=	

The formula mass: _____

In a Nutshell: The Mole

To go from the atomic scale to the macroscale, scientists increase from one atom, ion, or molecule to 6.02×10^{23} atoms, ions, or molecules, known as Avogadro's number. Avogadro's number represents a quantity known as a mole. By using Avogadro's number, the numerical value for the mass, in grams, of one mole of any element is the same numerical value as the average atomic mass, in amu, of that element. Likewise, the numerical value for the mass of one mole of any compound is the same numerical value as the molecular mass or formula mass of that compound. The mass of one mole of an element or compound is known as its molar mass.

Worked Example #11

Which has a greater mass: 1 mol of helium or 1 mol of oxygen? Explain why.

Using the Periodic Table, we see that a mole of helium has a mass of 4.00 g/mol and a mole of oxygen has a mass of 16.00 g/mol. Therefore, a mole of oxygen has a greater mass than a mole of helium. Oxygen has a greater average atomic mass.

Try It Yourself #11

Which has a greater mass: 1 mol of titanium or 1 mol of platinum? Explain why.
Tools: The Periodic Table

Average atomic mass titanium: _____

Average atomic mass platinum: _____

Molar mass titanium: _____

Molar mass platinum: _____

In a Nutshell: Converting between Units of Mass and Moles of Any Substance: Grams → Moles or Moles → Grams

Scientists routinely convert between mass and moles. Using conversion factors provides a reliable and straightforward process for converting between mass and moles. The following steps are used to convert between mass and moles or moles and mass. 1) If the substance is an element, look up the molar mass of the element on the Periodic Table. If the substance is a compound, calculate the molar mass from the atomic masses of its constituents. 2) Express the molar mass as two possible conversion factors. 3) Set up the calculation so that the units cancel. Double check that the final answer has been rounded to the correct number of significant figures.

Worked Example #12

How many moles of sucrose are in 4.2 grams (1 teaspoon)? The molecular formula for sucrose is $C_{12}H_{22}O_{11}$.

For a compound, calculate the molar mass from its constituent masses, using the Periodic Table.

Calculate the molar mass of sucrose, $C_{12}H_{22}O_{11}$, from the atomic masses of its constituent atoms:

12 C + 22 H + 11 O = (12 × 12.01) + (22 × 1.008) + (11 × 16.00) = 342.30 g/mol

Express the molar mass as two possible conversion factors:

$$\frac{342.30 \text{ g}}{1 \text{ mol}} \quad \text{or} \quad \frac{1 \text{ mol}}{342.30 \text{ g}}$$

Set up the calculation so that the supplied units cancel:

$$4.2 \text{ g} \times \frac{1 \text{ mol}}{342.30 \text{ g}} = 0.012 \text{ mol}$$

Try It Yourself #12

How many moles of ibuprofen, $C_{13}H_{18}O_2$, are in 400 mg?

Tools: The Periodic Table

Calculate the molar mass of ibuprofen from its constituent mass.

Express the molar mass as two possible conversion factors. You will also need to express the conversion between mg and g as two possible conversion factors.

Set up the calculation so that the supplied units cancel.

Worked Example #13

How many milligrams of gold are there in 0.15 mol of gold?

Look up the molar mass of the element on the Periodic Table.
The molar mass is 197.0 g/mol.

Express the molar mass as two possible conversion factors:

$$\frac{197.0\ g}{1\ mol} \text{ or } \frac{1\ mol}{197.0\ g}$$

We will also need to convert grams to milligrams using the metric conversion. Express the metric conversion as two possible conversion factors:

$$\frac{1\times10^{-3}g}{1\ mg} \text{ or } \frac{1\ mg}{1\times10^{-3}g}$$

Set up the calculation so that the supplied units cancel:

$$0.15\ \text{mol Au} \times \frac{197.0\ g}{1\ \text{mol Au}} \times \frac{1\ mg}{1\times10^{-3}\ g} = 3.0\times10^4\ mg$$

Try It Yourself #13

How many grams of copper are in 2.5 mol of copper?

Tools: The Periodic Table

Look up the molar mass of the element in the Periodic Table.

Express the molar mass as two possible conversion factors.

Set up the calculation so that the units cancel.

Practice Problems for Formula Mass, Molecular Mass, and Molar Mass

1. A patient's blood test shows that she has 130 mg/dL of cholesterol ($C_{27}H_{46}O$) in her blood. How many moles of cholesterol are there in every deciliter of her blood?

2. How many grams of $Al(OH)_3$ are there in 1.7 mol $Al(OH)_3$?

3. Which has more atoms: half a mole of zinc or half a mole of chromium?

Extension Topic 2-1: Converting between Mass and Number of Particles

Sometimes you may want to consider how many atoms, ions, or molecules are present in a given sample. This type of conversion requires Avogadro's number. Avogadro's number serves as a conversion factor between moles and the number of particles. In order to determine how many atoms, ions, are in or molecules a given sample, you must first convert the mass of the sample into moles of the sample, and then convert moles of the sample into atoms, ions, or molecules of the sample.

Worked Example #14

How many silver atoms are there in 0.45 mol of silver?

Express the conversion as two possible conversion factors:

$$\frac{6.02 \times 10^{23} \text{ atoms}}{1 \text{ mol}} \quad \text{or} \quad \frac{1 \text{ mol}}{6.02 \times 10^{23} \text{ atoms}}$$

Set up the calculation so that the units cancel:

$$0.45 \ \cancel{mol} \times \frac{6.02 \times 10^{23} \ \text{atoms}}{1 \ \cancel{mol}} = 2.7 \times 10^{23} \ \text{atoms}$$

Try It Yourself #14

How many carbon dioxide molecules are there in 2.4 mol of CO_2?

Express the conversion as two possible conversion factors.

Set up the calculation so that the units cancel.

Worked Example #15

How many sucrose ($C_{12}H_{22}O_{11}$) molecules are there in 25 g of sucrose?

You first need to convert g to moles and then moles to molecules.

Calculate the molar mass of sucrose, $C_{12}H_{22}O_{11}$, from the atomic masses of its constituent atoms:

12 C + 22 H + 11 O = (12 × 12.01) + (22 × 1.008) + (11 × 16.00) = 342.30 g/mol

Express the conversions as two possible conversion factors:
There are two conversions here: g to moles (use the molar mass) and moles to molecules (use Avogadro's number).

$$1^{st} \ conversion: \quad \frac{342.30 \ \text{g}}{1 \ \text{mol}} \quad or \quad \frac{1 \ \text{mol}}{342.30 \ \text{g}}$$

2nd *conversion:* $\dfrac{6.02 \times 10^{23}\ \text{molecules}}{1\ \text{mol}}$ or $\dfrac{1\ \text{mol}}{6.02 \times 10^{23}\ \text{molecules}}$

Set up the calculation so that the supplied units cancel:

$$25\ \cancel{g} \times \dfrac{1\ \cancel{\text{mol}}}{342.30\ \cancel{g}} \times \dfrac{6.02 \times 10^{23}\ \text{molecules}}{1\ \cancel{\text{mol}}} = 4.4 \times 10^{22}\ \text{molecules}$$

Try It Yourself #15

How many $CaCO_3$ molecules are there in 32.3 g of calcium carbonate?

Calculate the molar mass of $CaCO_3$ from its constituent atoms.

Express the conversions as two possible conversion factors.

Set up the calculation so that the supplied units cancel.

Practice Problems for Converting Between Mass and Number of Particles

1. How many platinum atoms are in 2.34 mmol of platinum?

2. How many molecules of Na_3PO_4 are there in 8.23 g of Na_3PO_4?

3. What is the mass of 3.7×10^{20} carbon dioxide (CO_2) molecules?

Chapter 2 Quiz

1. Identify the compounds shown below as either an ionic compound or a covalent compound:

    a. NF_3

    b. $FeCl_2$

    c. CH_3CH_2OH

    d. NaBr

    e. COS

    f. SO_2

2. Name the following compounds:

    a. V_2O_5

    b. HBr

    c. PbI_2

    d. SO_3

    e. RbCl

3. Identify the ions are present in the following formula units. Provide the name and the symbol of the ions.

    a. $Mg(OH)_2$

    b. $AgNO_3$

      c. Na_2SO_3

      d. NH_4OH

      e. $Ca_5(PO_4)_3F$

4. Draw the Lewis dot structure for the following compounds:
      a. HI

      b. OCS

      c. acetylene, C_2H_2

      d. CF_2Cl_2

5. Draw the Lewis dot structure for the following compounds:
      a. CH_3NH_2

      b. ClF

c. CH_3Cl

d. propane, C_3H_8

6. How many grams of ammonia, NH_3, are there in 0.34 mol of NH_3?

7. A mole of tin has more mass than a mole of magnesium. Explain.

8. How many moles are in 250 mg of the potassium salt of penicillin V, $C_{16}H_{17}KN_2O_5S$?

9. How many milligrams are in 0.017 mol of celecoxib, $C_{17}H_{14}F_3N_3O_2S$, the active ingredient of Celebrex?

10. How many moles are in 0.43 g of sodium phosphate?

Chapter 2
Answers to Additional Exercises

2.39   An element is composed of one type of atom—defined by its atomic number. Elements combine with other elements to form compounds.

2.41   Covalent compounds are formed between *nonmetal* atoms.

2.43   + and + and − and − are repulsive interactions.

2.45   a. Ca^{2+} b. Cr^{3+} and Cr^{6+} c. N^{3-} d. Ag^+

2.47   A cation is a positively charged atom: it has more protons than electrons. A cation is formed when an atom loses electrons. An anion is a negatively charged atom, which has more electrons than protons. An anion is formed when an atom gains electrons. Group 1A and 2A metals form cations.

2.49   Ions are not formed from group 8A elements because these elements have a full outer shell of electrons.

2.51 H^+ or H^-. If hydrogen gains an electron, it will have a full electron shell, similar to helium. If hydrogen loses an electron, it will not have any electrons in its outer shell.

2.53   a. Magnesium cation: 12 protons, 10 electrons. It has a +2 charge because the ion lost two electrons to have a full outermost electron shell.

b. Mercury cation: 80 protons, 78 electrons. It has a +2 charge because it has two more protons than electrons.

c. Chlorine anion or chloride: 17 protons, 18 electrons. It has a −1 charge because chlorine atom gained an electron to have a complete shell of electrons.

d. Fluorine anion or fluoride: 9 protons, 10 electrons. It has a −1 charge because fluorine atom gained an electron to have a complete shell of electrons.

e. Oxygen anion or oxide: 8 protons, 10 electrons. It has a −2 charge because oxygen gained 2 electrons to have a complete electron shell.

2.55   An ionic compound is formed by the mutual attraction of cations and anions. The ions are held together by electrostatic attraction.

2.57   When the ionic lattice of NaCl is dissolved in water, the lattice structure falls apart. The sodium ions (Na^+) and chloride ions (Cl^-) separate, and each ion is surrounded by water molecules.

2.59   a. potassium ion, K^+, and bromide, Br^-.

b. magnesium ion, Mg^{2+}, and chloride, Cl^-

     c. potassium ion, K^+, and iodide, I^-

     d. barium ion, Ba^{2+}, and chloride, Cl^-.

     e. sodium ion, Na^+, and fluoride, F^-.

2.61   a. LiI; lithium (cation), iodide (anion)

     b. RbF, rubidium (cation), fluoride (anion)

     c. $CaBr_2$ calcium (cation), bromide (anion)

     d. BaI_2, barium (cation), iodide (anion)

     e. FeS, iron (cation), sulfide (anion)

     f. Al_2O_3, aluminum (cation) oxygen (anion)

2.63   a. strontium oxide b. potassium iodide c. gallium oxide d. lithium fluoride e. sodium iodide f. iron(III) oxide

2.65 Na^+

2.67   a. Zn^{2+}, OH^- b. Cu^+, $CH_3CO_2^-$ c. Sn^{4+}, Cl^- d. V^{5+}, O^{2-} e. Cr^{6+}, O^{2-}

2.69   The ions in ionic compounds are held together by electrostatic forces, the attraction between oppositely charged objects. The atoms in covalent compounds are held together by covalent bonds. In a covalent bond, the two atoms share two valence electrons.

2.71   The diatomic elements are hydrogen, oxygen, nitrogen, fluorine, chlorine, bromine, and iodine.

2.73   a. $\cdot\overset{\displaystyle\cdot}{\underset{\displaystyle\cdot}{C}}\cdot$ b. $H\cdot$ c. $\overset{\displaystyle\cdot\cdot}{\underset{\displaystyle\cdot\cdot}{O}}\cdot$ d. $\overset{\displaystyle\cdot\cdot}{P}\cdot$

2.75   a. The carbon atom contains eight bonding electrons and zero nonbonding electrons.

     b. The nitrogen atom contains six bonding electrons and two nonbonding electrons.

     c. The carbon and nitrogen atoms have an octet of electrons; the hydrogen atom has a duet of electrons.

     d. There is a single bond between hydrogen and carbon. There is a triple bond between carbon and nitrogen.

     e. Six electrons are shared between the carbon atom and nitrogen atom in the triple bond.

2.77 $:\!\overset{\displaystyle\cdot\cdot}{\underset{\displaystyle\cdot\cdot}{Br}}\!-\!\overset{\displaystyle\cdot\cdot}{\underset{\displaystyle\cdot\cdot}{Br}}\!:$ This molecule is an element.

2.79 a. The total number of valence electrons are (6 × 1) + (2 × 4) = 14 electrons. There

$$\begin{array}{cc} H & H \\ | & | \\ H-C-C-H \\ | & | \\ H & H \end{array}$$

are no electrons to be distributed as non bonding pairs.

b. The total number of valence electrons are (3 × 1) + (1 × 5) = 8 electrons. There

are two electrons to be distributed as nonbonding pairs: 8 − (3 × 2) = 2 electrons.

$$H-\overset{\displaystyle ..}{\underset{\displaystyle |}{P}}-H$$
$$\overset{|}{H}$$

c. The total number of valence electrons are (4 × 7) + (1 × 4) = 32 electrons. There

are 24 electrons to be distributed as nonbonding pairs: 32 − (4 × 2) = 24

electrons.

$$:\overset{..}{\underset{..}{Cl}}:$$
$$:\overset{..}{\underset{..}{Cl}}-\overset{|}{\underset{|}{C}}-\overset{..}{\underset{..}{Cl}}:$$
$$:\overset{..}{\underset{..}{Cl}}:$$

d. The total number of valence electrons are (1 × 4) + (2 × 6) = 16 electrons. There

are 12 electrons to be distributed as nonbonding pairs: 16 − (2 × 2) = 12

electrons. $\overset{..}{\underset{..}{O}}=C=\overset{..}{\underset{..}{O}}$

e. The total number of valence electrons are (4 × 1) + (1 × 4) + (1 × 6) = 14

electrons. There are 4 electrons to be distributed as nonbonding pairs: 14 − (5 ×

$$\begin{array}{c} H \\ | \\ H-C-\overset{..}{\underset{..}{O}}-H \\ | \\ H \end{array}$$

2) = 4 electrons.

2.81 The phosphorus atoms have an expanded octet. Each atom has ten electrons.

2.83 SO_3 is sulfur trioxide. There is one sulfur atom, named after the element: sulfur.

There are three oxygen atoms, so the prefix "tri" is inserted before the element name

and the ending is changed to "ide."

2.85 PCl_3 is phosphorus trichloride. This molecule has three chlorine atoms, so the prefix

"tri" is inserted before the element name. PCl_5 is phosphorus pentachloride. This

molecule has five chlorine atoms, so the prefix "penta" is inserted before the element

name.

2.87 a. hydrogen carbonate (or bicarbonate) ion

b. acetate ion

c. hydroxide ion

2.89 a. NH_4^+

b. CO_3^{2-}

       c. $HPO_4{}^{2-}$

2.91   a.   The overall charge is −1.

       b.   Hydrogen carbonate is a polyatomic atom. It is an ion composed of several atoms joined by covalent bonds.

       c. $NaHCO_3$

       d. $Ca(HCO_3)_2$

2.93   a.   The overall charge is −2.

       b.   This ion is composed of several atoms joined by covalent bonds.

       c. K_2HPO_4

       d. $MgHPO_4$

2.95   a. Na_3PO_4

       b. NH_4Cl

       c. $Mg(OH)_2$

2.97 $Ca_3(PO_4)_2$

2.99   Avogadro's number is the number items in one mole of items. It is 6.02×10^{23}.

2.101 The mass of one mole of nickel atoms is 58.69 g.

2.103 A mole of zirconium is lighter than a mole of platinum. The molar mass of Zr is 91.22 g/mol while the molar mass of Pt is 195.1 g/mol.

2.105 $4\ C + 10\ H = (4 \times 12.01) + (10 \times 1.008) = 58.12$ amu.

2.107 For a compound, calculate the molar mass from its constituent masses, using the Periodic Table. Calculate the molar mass of calcium chloride, $CaCl_2$, from the atomic masses of its constituent atoms:

       $1\ Ca + 2\ Cl = (1 \times 40.08) + (2 \times 35.45) = 110.98$ g/mol.

       Express the molar mass as two possible conversion factors:

$$\frac{110.98\ \text{g}}{1\ \text{mol}} \quad \text{or} \quad \frac{1\ \text{mol}}{110.98\ \text{g}}$$

       Set up the calculation so that the supplied units cancel:

$$2.00\ \cancel{g} \times \frac{1\ \text{mol}}{110.98\ \cancel{g}} = 0.0180\ \text{mol}$$

2.109 Look up the molar mass of the element on the Periodic Table. The molar mass is 63.55 g/mol.

       Express the molar mass as two possible conversion factors:

$$\frac{63.55 \text{ g}}{1 \text{ mol}} \quad \text{or} \quad \frac{1 \text{ mol}}{63.55 \text{ g}}$$

Set up the calculation so that the supplied units cancel:

$$0.75 \cancel{g} \times \frac{1 \text{ mol}}{63.55 \cancel{g}} = 0.012 \text{ mol}$$

2.111   Look up the molar mass of the element on the Periodic Table. The molar mass is 69.72 g/mol.

Express the molar mass as two possible conversion factors:

$$\frac{69.72 \text{ g}}{1 \text{ mol}} \quad \text{or} \quad \frac{1 \text{ mol}}{69.72 \text{ g}}$$

Set up the calculation so that the supplied units cancel:

$$3.5 \cancel{mol} \times \frac{69.72 \text{ g}}{1 \cancel{mol}} = 244 \text{ g}$$

Rounded to 240 g

2.113   a. The molar mass of aluminum is 26.98 g/mol. b. The molar mass of hydrogen is 2 H = 2 × 1.008 = 2.016 g/mol. c. The molar mass of calcium is 40.08 g/mol. d. The molar mass of nitrogen is 2 N = 2 × 14.01 = 28.02 g/mol.

2.115   a. 2 Li + 1 C + 3 O = (2 × 6.941) + (1 × 12.01) + (3 × 16.00) = 73.89 g/mol.

b. 3 K + 1 P + 4 O = (3 × 39.10) + (1 × 30.97) + (4 × 16.00) = 212.3 g/mol.

2.117   Albumin should be present in a larger amount in a blood sample. The normal range for albumin is 3.9–5.0 g/ dL while the normal range for glucose is 64–128 mg/dL.

2.119   The molecular mass of urea (CH_4N_2O) is 1 C + 4 H + 2 N + 1 O = (1 × 12.01) + (4 × 1.008) + (2 × 14.01) + (1 × 16.00)= 60.06 amu. The molar mass of urea is 1 C + 4 H + 2 N + 1 O = (1 × 12.01) + (4 × 1.008) + (2 × 14.01) + (1 × 16.00) = 60.06 g/mol. Urea is a covalent compound. There are only covalent bonds between the atoms in the molecule.

In 1 dL of serum, there is a minimum of 7 mg of urea and a maximum of 20 mg of urea.

Express the molar mass as two possible conversion factors:

$$\frac{60.07 \text{ g}}{1 \text{ mol}} \quad \text{or} \quad \frac{1 \text{ mol}}{60.07 \text{ g}}$$

Milligrams also need to be converted to grams:

$$\frac{1 \times 10^{-3} \text{ g}}{1 \text{ mg}} \quad \text{or} \quad \frac{1 \text{ mg}}{1 \times 10^{-3} \text{ g}}$$

Set up the calculation so that the supplied units cancel:

$$7 \text{ mg} \times \frac{1 \times 10^{-3} \text{ g}}{1 \text{ mg}} \times \frac{1 \text{ mol}}{60.07 \text{ g}} = 1 \times 10^{-4} \text{ mol}$$

$$20 \text{ mg} \times \frac{1 \times 10^{-3} \text{ g}}{1 \text{ mg}} \times \frac{1 \text{ mol}}{60.07 \text{ g}} = 3 \times 10^{-4} \text{ mol}$$

Therefore, there are from 1×10^{-4} mol to 3×10^{-4} mol of urea in 1 dL of serum.

Chapter 3

Shapes of Molecules and Their Interactions

Chapter Summary

In this chapter, you have learned about the intermolecular forces of attraction between molecules. You have learned to distinguish the six basic shapes of simple molecules. The molecular shape and the individual bond dipoles determine the polarity of the molecule. The polarity of the molecule determines how the molecule interacts with other molecules. These intermolecular forces impact the physical properties of a compound and play an important role in the structure of biological molecules.

Section 3.1 Three-Dimensional Shapes of Molecules

Most molecules have a three-dimensional shape. The shape of a molecule containing one or two central atoms surrounded by two or more atoms is determined from the spatial arrangement of the atoms surrounding the central atom, which in turn is determined from the number and type of electrons on the central atom. The arrangement of electrons around the central atom is referred to as the molecule's electron geometry, while the arrangement of atoms around the central atom is the molecule's molecular shape.

In a Nutshell: Molecular Models

Molecular models are a common tool used for visualizing the three-dimensional shapes of molecules. Two types of molecular models are routinely used: the ball-and-stick model and the space-filling model. The ball-and-stick models depict molecules by representing atoms as balls and bonds as sticks and are used primarily to show bond angles. Space-filling models are used to show the relative space occupied by the atoms in a molecule. Occasionally a tube model will be used in place of a ball-and-stick model, particularly when the molecule is large. In a tube model both bonds and atoms appear as part of a tube, where end points represent the atoms.

In a Nutshell: Using Lewis Dot Structures to Predict Electron Geometry

The process for determining the molecular shape of a molecule can be described in three steps. Step 1) Write the Lewis dot structure for the molecule. Step 2) Determine the electron geometry from the Lewis dot structure. Step 3) Determine the molecular shape from the electron geometry.

The simplest theory for predicting electron geometry, known as valence shell electron pair repulsion theory (VSEPR), is based on the premise that electrons around a central atom adopt a geometry that places them as far apart from one another as possible, while still remaining attached to the central atom. To determine the electron geometry around the central atom, first count the groups of electrons around it. A group of electrons refers to one triple bond, one double bond, one single bond, or one nonbonding pair of electrons. The majority of molecules have two, three, or four electron groups around a central atom.

VSEPR theory gives us three common electron geometries that are derived from the number of electron groups around the central atom: four groups of electrons—tetrahedral; three groups of electrons—trigonal planar; two groups of electrons—linear. The linear geometry has the shape of a straight line, with both groups of electrons directed 180° from each other. The trigonal planar geometry is two-dimensional: Each of the three groups of electrons points to the three corners of an equilateral triangle, separated by 120°. The tetrahedral geometry is three-dimensional: Each of the four groups of electrons points to the four corners of a tetrahedron, creating angles of 109.5°.

Worked Example #1
Predict the electron geometry, molecular shape, and bond angles for following compounds.
- a. $CHCl_3$
- b. OCS (C is the central atom)

    a. *First, write the Lewis dot structure.*

       Second, determine the number of electron groups around the central atom. The carbon has four single bonds or four electrons groups around it.

       Third, determine the electron geometry from the number of electron groups.

Four electron groups indicate a tetrahedral geometry. The molecular shape is tetrahedral and the bond angles are 109.5°.

b. First, write the Lewis dot structure.

$$\ddot{\text{O}}=\text{C}=\ddot{\text{S}}$$

Second, determine the number of electron groups around the central atom. The carbon has two double bonds or two electron groups around it.

Third, determine the electron geometry from the number of electron groups. Two electron groups indicate a linear geometry. The molecular shape is linear and the bond angles are 180°.

Try It Yourself #1
Predict the electron geometry, molecular shape, and bond angles for following compounds.
a. CF_2Cl_2 40 32 8 24
b. HCN 18 10 8 2

a. Write the Lewis dot structure.

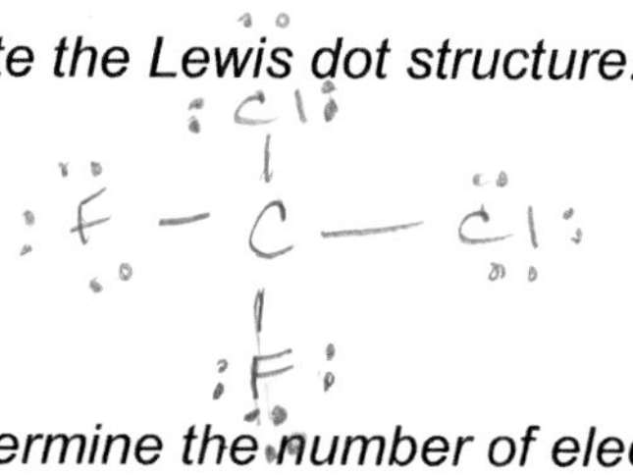

$$:\ddot{\text{F}} - \text{C} - \ddot{\text{Cl}}:$$

Determine the number of electron groups around the central atom.

Central atom: __C__

Number of electron groups: __4__

Determine the electron geometry.

Geometry: __tetrahedral__

b. Write the Lewis dot structure.

$$\text{H} - \text{C} \equiv \ddot{\text{N}}:$$

Determine the number of electron groups around the central atom.

Central atom: __C__

Number of electron groups: __2__

Determine the electron geometry.

Geometry: __linear__

In a Nutshell: Using Electron Geometries to Determine Molecular Shape

When the central atom in a molecule has only bonding electrons (and no nonbonding electrons), its molecular shape is the same as its electron geometry. A molecule with a central atom containing one or more nonbonding pairs of electrons will have a molecular shape that is different from its electron geometry, since shape is defined by the relative position of the atoms, and nonbonding electrons are not seen.

There are three different molecular shapes derived from a tetrahedral electron geometry depending on the number of nonbonding electrons on the central atom. Four bonding groups and no nonbonding groups form a tetrahedral molecular shape. Three bonding groups and one nonbonding group form a trigonal pyramidal shape. Two bonding groups and two nonbonding groups form a bent shape. The tetrahedral, trigonal pyramidal, and bent shapes all have bond angles of approximately 109.5°.

There are two molecular shapes derived from a trigonal planar geometry depending on the number of nonbonding electrons on the central atom. Three bonding groups and no nonbonding groups form a trigonal planar shape. Two bonding groups and one nonbonding group form a bent shape. The trigonal planar and bent shapes all have bond angles of approximately 120°.

When a central atom is surrounded by two groups of electrons, they will always be bonding groups; therefore, the molecule has a linear electron geometry and a linear molecular shape.

Worked Example #2

Predict the electron geometry, molecular shape, and bond angles for following compounds.

a. OCl_2

b. PCl_3

c. BCl_3 (an exception to the octet rule)

a. *Write the Lewis dot structure.*

$$:\overset{..}{\underset{..}{Cl}}-\overset{..}{\underset{..}{O}}-\overset{..}{\underset{..}{Cl}}:$$

Determine the number of electron groups around the central atom.
The central atom is oxygen and it has two nonbonding groups and two single bonds, so four electron groups around it.

Determine the electron geometry.
The electron geometry is tetrahedral.

Determine the number of nonbonding groups and bonding groups.
There are two nonbonding groups and two bonding groups; therefore, the molecular geometry is bent. The bond angles are approximately 109.5°.

b. *Write the Lewis dot structure.*

$$:\overset{..}{\underset{..}{Cl}}-\overset{..}{P}-\overset{..}{\underset{..}{Cl}}:$$
$$\underset{..}{\overset{}{:}Cl:}$$

Determine the number of electron groups around the central atom.
Phosphorus is the central atom; there are three single bonds and one nonbonding pair; therefore, there are four electron groups around phosphorus.

Determine the electron geometry.
The electron geometry is tetrahedral.

Determine the number of nonbonding groups and the number of bonding groups.
There are one nonbonding group and three bonding groups; therefore, the molecular geometry is trigonal pyramidal. The bond angles are approximately 109.5°.

c. *Write the Lewis dot structure.*

$$:\overset{..}{\underset{..}{Cl}}-B-\overset{..}{\underset{..}{Cl}}:$$
$$\underset{..}{:Cl:}$$

Determine the number of electron groups around the central atom.
The central atom is boron and there are three single bonds around it, therefore, there are three electron groups around it.

Determine the electron geometry.
The electron geometry is trigonal planar.

Determine the number of nonbonding groups and the number of bonding groups.
There are three bonding groups and no nonbonding groups; therefore, the molecular geometry is trigonal planar. The bond angles are approximately 120°.

Try It Yourself #2

Predict the electron geometry, molecular shape, and bond angles for following compounds.

a. PH_3 14 8 6 2
b. SF_2 24 20 4 16

a. Write the Lewis dot structure.

$$H - \overset{\bullet \bullet}{P} - H$$
$$|$$
$$H$$

Determine the number of electron groups around the central atom.
Central atom: __P__
Number of electron groups: __4__
Determine the electron geometry.
Geometry: __trigonal planar__
Determine the molecular geometry.
Number of bonding groups: __3__
Number of nonbonding groups: __1__
Geometry: __tetrahedral__

b. Write the Lewis dot structure.

$$:\overset{\bullet \bullet}{F} - \overset{\bullet \bullet}{S} - \overset{\bullet \bullet}{F}:$$

Determine the number of electron groups around the central atom.

Central atom: S

Number of electron groups: 4

Determine the electron geometry.

Geometry: linear

Determine the molecular geometry.

Number of bonding groups: 2

Number of nonbonding groups: 2

Geometry: tetrahedral

In a Nutshell: Shapes of Larger Molecules

The six different molecular shapes can also be used to evaluate the shapes of larger molecules having more than one central atom. The shape of a larger molecule can be viewed as a combination of the shapes of the individual atom centers.

Worked Example #3

Predict the shape of ethanol, CH_3CH_2OH.

Write the Lewis dot structure.

$$H-\overset{\overset{\displaystyle H}{|}}{\underset{\underset{\displaystyle H}{|}}{C}}-\overset{\overset{\displaystyle H}{|}}{\underset{\underset{\displaystyle H}{|}}{C}}-\overset{\cdot\cdot}{\underset{\cdot\cdot}{O}}-H$$

Evaluate the geometry around each central atom.

The two carbons have four bonding groups; therefore, they are tetrahedral. The oxygen atom has two bonding groups and two nonbonding groups; therefore, it is bent. The overall structure has a zigzag shape.

Try It Yourself #3

Predict the shape of hydrogen peroxide, H_2O_2.

Write the Lewis dot structure.

Central Atoms: _____

Number of bonding groups around each central atom: _____

Number of nonbonding groups around each central atom: _____

Geometry of each central atom: _____

Practice Problems for Three-Dimensional Shapes of Molecules

1. Predict the electron geometry, molecular shape, and bond angles for following compounds.

    a. H_2CO_3

    b. $COCl_2$

    c. CS_2

2. Predict the shapes of the central atoms in the following compounds.

    a. N_2H_4

    b. CH_3SH

    c. C_2Cl_4

Section 3.2 Molecular Polarity

Molecular shape is one of the factors that determine whether a molecule is polar or nonpolar. A polar molecule has a separation of charge or a dipole. A nonpolar molecule has no separation of charge. The polarity of a molecule is determined by the types of covalent bonds in the molecule and the molecular shape of the molecule.

In a Nutshell: Electronegativity

A covalent bond is polar if the two atoms sharing electrons have a significantly different electronegativity. Electronegativity is the measure of an atom's ability to draw electrons toward itself in a covalent bond. The most electronegative element is fluorine. Within a group of elements, electronegativity increases as you go from the bottom of a group to the top of a group. Within a period of elements, electronegativity increases as you go from left to right.

In a Nutshell: Bond Dipoles

If two identical atoms form a covalent bond, then the bond is a nonpolar covalent bond. The electron density is evenly distributed around both nuclei. When covalent bonds are formed between atoms with comparable electronegativity values, these bonds are also nonpolar. When a covalent bond is formed between two atoms with significantly different electronegativities, the bonding electrons spend a greater amount of time around the more electronegative atom. The bond will have a separation of charge, or a dipole. This type of bond is a polar covalent bond.

The Greek symbols $\delta+$ and $\delta-$ are used to indicate partial positive and partial negative charge on an atom. A dipole arrow can be placed alongside the bond with the tail of the arrow shown next to the less electronegative atom and the head of the arrow pointing toward the more electronegative atom. An electron density diagram is a space filling model that indicates regions of higher and lower concentrations of electrons by color.

Worked Example #4

Identify the more electronegative atom in each pair.
   a. sodium and rubidium
   b. chlorine and silicon

   a. *Figure 3-10 shows that sodium is above rubidium in the same group; therefore, sodium is the more electronegative atom.*

b. *Figure 3-10 shows that chlorine is to the right of silicon in the same period; therefore, chlorine is more electronegative.*

Try It Yourself #4

Identify the more electronegative atom in each pair.

   a. sulfur and copper
   b. platinum and bromine

Tools: Figure 3-10

   a. *The more electronegative atom is _____*
   b. *The more electronegative atom is _____*

Worked Example #5

Indicate which of the following represent polar covalent bonds by showing a bond dipole arrow.

   a. P—F
   b. O—H
   c. N—O

   a. P–F *Figure 3-10 shows that fluorine is more electronegative than phosphorus.*
   b. O–H *Figure 3-10 shows that oxygen is more electronegative than hydrogen.*
   c. N—O *is a nonpolar covalent bond. They have similar electronegativities.*

Try It Yourself #5

Indicate which of the following represent polar covalent bonds by showing a bond dipole arrow.

   a. S—O
   b. C—Br
   c. Si—C

Tools: Figure 3-10

    a. S—O

    b. C—Br

    c. Si—C

In a Nutshell: Polar Molecules

When a molecule contains only one polar covalent bond, it is a polar molecule. For molecules with more than one covalent bond, a molecule will be nonpolar in the absence of any polar covalent bonds. If a molecule contains more than one covalent bond, its molecular shape, as well as the individual bond dipoles, must be considered to determine if the molecule is polar. If the bond dipoles counterbalance each other, the molecule will be nonpolar. In general, when the covalent bonds around a central atom are identical and the electron geometry is the same as the molecular geometry (tetrahedral, trigonal planar and linear), the bond dipoles will cancel and the molecule is nonpolar.

Worked Example #6

Which of the following molecules are nonpolar? Explain why.

    a. O_2

    b. CF_4

    c. CH_3Cl

    d. HBr

    a. *O_2 is nonpolar. The electrons in the covalent bond are shared equally by the identical atoms that form the covalent bond.*

    b. *Carbon tetrafluoride is nonpolar. CF_4 is tetrahedral and has four C—F bonds which are polar covalent bonds. The dipoles are equivalent and are directed into the four opposite corners of the tetrahedron; therefore, the dipoles cancel each other, causing the molecule to be nonpolar.*

    c. *CH_3Cl is a polar molecule. There is one polar covalent bond (C—Cl) in the molecule.*

   d. H—Br is a polar molecule.

Try It Yourself #6

Which of the following molecules are nonpolar? Explain why.

   a. CF_3Cl

   b. NBr_3

   c. BCl_3

   d. C_4H_{10}

Solution

_____ C_4H_{10} _____ are nonpolar.

Practice Problems for Molecular Polarity

1. Which of the following elements is less electronegative?

   a. fluorine and hydrogen

   b. arsenic and nitrogen

   c. sodium and bromine

2. Indicate which of the following represent a polar covalent bond by showing a bond dipole arrow.

   a. C—H

   b. N—H

   c. P—Cl

   d. C—O

3. Which of the following molecules are nonpolar?

   a. CH_2ClF

b. C_2H_4

c. PF_3

d. HCN

Section 3.3 Intermolecular Forces of Attraction

Molecules interact with each other through intermolecular forces of attraction—those that exist between molecules rather than within molecules. Intermolecular forces are much weaker than covalent bonding forces. Intermolecular forces arise from electrostatic interactions between partial charges resulting from bond dipoles in one molecule being attracted to the opposite partial charges in other molecules. A dipole can be either a permanent dipole or a temporary dipole. The three basic types of intermolecular forces of attraction between molecules listed in order of increasing strength are dispersion forces, dipole-dipole forces, and hydrogen-bonding forces.

In a Nutshell: Dispersion Forces

Dispersion forces are the sole force of attraction between nonpolar molecules. In nonpolar molecules, the electrons can shift temporarily toward one end of the molecule, creating a temporary dipole. A temporary dipole induces a corresponding dipole in adjacent molecules, causing them to shift in a way that brings opposite partial charges together. Since the induced dipoles are short-lived, dispersion forces are the weakest of the three types of intermolecular forces of attraction. The strength of a compound's dispersion forces determines many of its physical properties.

In a Nutshell: Dipole-Dipole Forces

In addition to dispersion forces, polar molecules are capable of interacting through dipole-dipole forces, which are attractions between the permanent dipoles of polar molecules. The molecules arrange themselves in a way that brings together opposite charges. Because the dipoles are permanent, dipole-dipole forces are much stronger than dispersion forces.

In a Nutshell: Hydrogen Bonding Forces

Hydrogen bonding is the strongest intermolecular force of attraction between molecules. It can exist when a molecule contains an H—F bond, an H—O, or an H—N bond. In these bonds, because of the differences in electronegativity, the hydrogen atom has the greatest partial positive charge of any atom in a covalent bond, and F, O, and N have the greatest partial negative charge of any atom in a covalent bond. Therefore, a molecule containing an O—H or N—H bond should exhibit the strongest force of attraction between the positive pole—hydrogen—of the bond and a nitrogen or oxygen atom in another molecule—the negative pole. The presence of hydrogen bonding impacts the boiling points of molecules.

Worked Example #7

Which type of intermolecular force of attraction is formed between temporary dipoles? What types of molecules have temporary dipoles?

Dispersion forces are formed between temporary dipoles of different molecules. All molecules and elements have temporary dipoles.

Try It Yourself #7

How do dipole-dipole forces differ from dispersion forces? How are they the same?

Solution

Worked Example #8

What type of intermolecular force will be exerted by the following molecules?
   a. CH_3OH (methanol)
   b. C_4H_8

   a. *Follow the flow chart in Figure 3-13. Are there polar molecules present? Yes, CH_3OH is polar. It has a polar covalent bond, O—H. Are H—O or H—N bonds present? Yes, therefore it has hydrogen bonding.*

b. *Follow the flow chart in Figure 3-13. Are there polar molecules present? No, C_4H_8 has only C—H bonds, which are nonpolar covalent bonds. Therefore, dispersion forces are present.*

Try It Yourself #8

What type of intermolecular force will be exerted by the following molecules?

    a. CH_3Cl

    b. CH_3NH_2

Solution

Tools: Figure 3-13

    a. *CH_3Cl has _____*

    b. *CH_3NH_2 has _____*

Worked Example #9

Which molecule in each pair exhibits stronger intermolecular forces of attraction with like molecules? Explain your choice. State the type of intermolecular force.

    a. C_4H_{10} or $C_{10}H_{22}$

    b. NH_3 or PH_3

    a. *$C_{10}H_{22}$ exerts stronger dispersion forces than C_4H_{10} because there are more atoms in the molecule and therefore more electrons. Dispersion forces increase with the number of electrons in a molecule.*

    b. *NH_3 exhibits the stronger force because it interacts with other NH_3 molecules through hydrogen bonding, the strongest force of attraction. PH_3 takes part in dipole-dipole interactions because it has a permanent dipole.*

Try It Yourself #9

Which molecule in each pair exhibits stronger intermolecular forces of attraction with like molecules? Explain your choice. State the type of intermolecular force.

    a. CH_4 or H_2O

b.

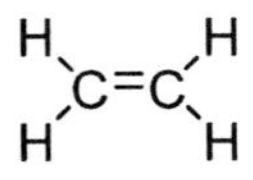

or

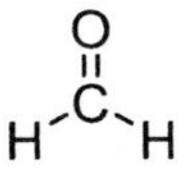

Solution

a. _____ exhibits stronger intermolecular forces of attraction. The

reason for the stronger intermolecular forces of attraction is

_____ The intermolecular force is

_____

b. _____ exhibits stronger intermolecular forces of attraction. The

reason for the stronger intermolecular forces of attraction is

_____ The intermolecular force is

_____

In a Nutshell: Hydrogen Bonding in DNA

DNA stores all your genetic information. It is a huge molecule made up of two strands
wound together in the shape of a helix. Hydrogen bonds hold the two strands of DNA
together throughout the length of the helix.

Worked Example #10

Draw the hydrogen bonds that are formed between C and G of DNA below. Label the
atoms in the hydrogen bonds as δ^+ or δ^-, appropriately.

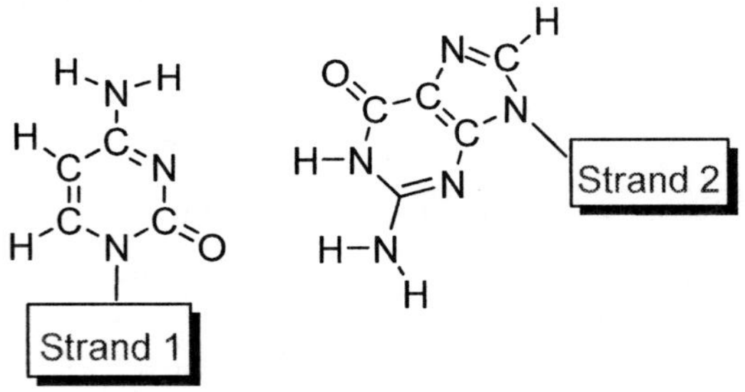

Solution

The hydrogen bonds are shown as dashed lines below. There are three: two between O
and N—H and one between N and N—H.

Try It Yourself #10

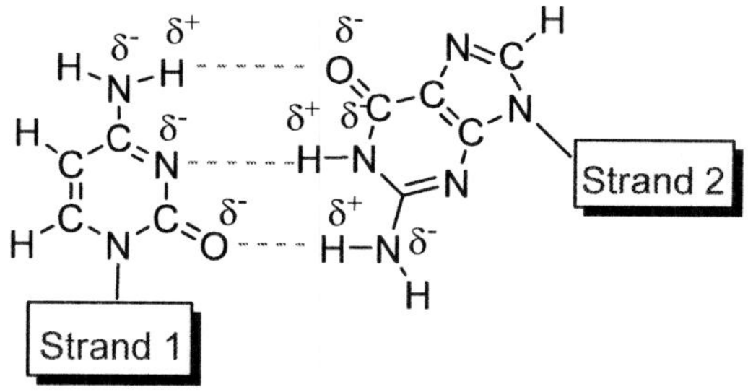

Draw the hydrogen bonds that are formed between C and G of DNA on the following page. Label the atoms in the hydrogen bonds as δ^+ or δ^-, appropriately.

Solution

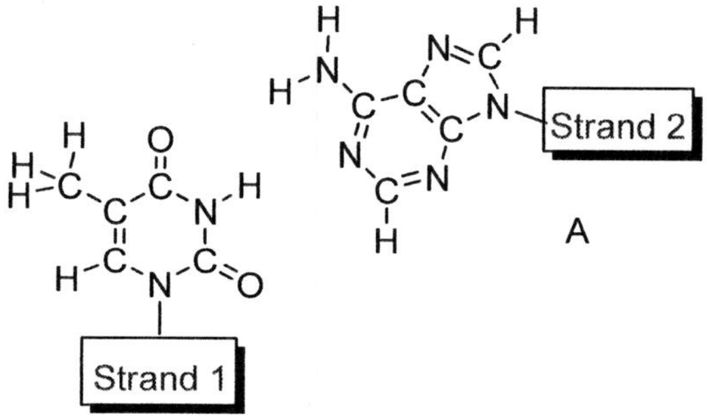

Practice Problems for Intermolecular Forces of Attraction

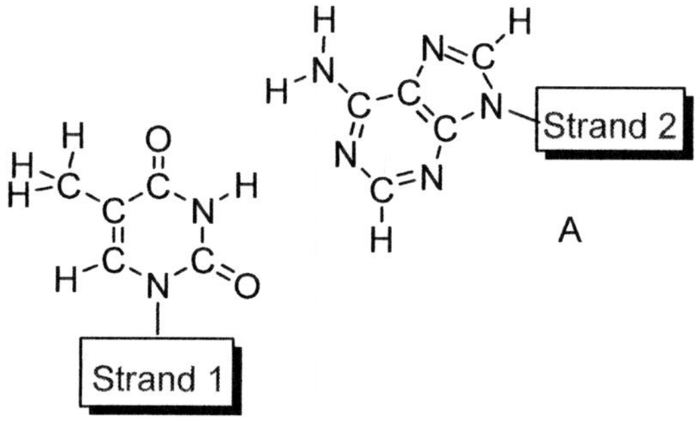

1. What type of intermolecular force will be exhibited by the following molecules?

    a. CH_3F

  b. C_2H_6

  c. CH_3CH_2OH

  d. 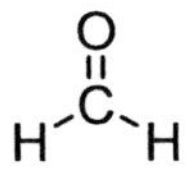 $H\overset{\overset{\displaystyle O}{\parallel}}{\underset{}{C}}H$

2. Draw the hydrogen bonds that can be formed between three water molecules.

3. Which molecule in each pair is capable of hydrogen bonding?
   a. CHF_3 or CH_3OH

   b. CH_3OH or CH_3OCH_3

Chapter 3 Quiz

1. Predict the electron geometry, molecular shape, and bond angles for the following compounds.

    a. $CHClF_2$

    b. CF_4

    c. C_2H_2

    d. C_2H_4

    e. H_2CO

    f. $CH_3CH_2CH_2OH$

2. Which molecules in Question 1 are polar molecules?

3. Which molecules in Question 1 are three-dimensional?

4. Which molecules in Question 1 exert only dispersion forces as an intermolecular force of attraction? Explain.

5. Which molecules in Question 1 have hydrogen bonding?

6. Which molecules in Question 1 exert dipole-dipole interactions?

7. Predict the electron geometry, molecular shape, and bond angles for the following compounds.
    a. CH_3CN

    b. $SiCl_4$

    c. C_3H_6

    d. H_2S

8. Which molecules in Question 7 are polar molecules?

9. Which molecules in Question 7 are three-dimensional?

10. What types of intermolecular forces of attraction are exerted by the molecules in Question 7?

Chapter 3
Answers to Additional Exercises

3.21 a.

H—C—Cl structure with H above carbon and Cl below

There are four bonding groups around the carbon. The electron geometry is tetrahedral. The molecular shape is tetrahedral. The bond angles are 109.5°.

b. I—N—I structure with I below nitrogen

There are three bonding groups and one nonbonding group around the nitrogen. The electron geometry is tetrahedral. The molecular shape is trigonal pyramidal. The bond angles are approximately 109.5°.

c. :O—Se=O

There are three bonding groups and one nonbonding group around the selenium. The electron geometry is trigonal planar. The molecular shape is bent. The bond angles are 120°.

3.23 The molecule is linear if there are only two electron groups surrounding the central atom.

3.25 The electron geometry for four electron groups around a central atom is tetrahedral. The possible molecular shapes are tetrahedral, trigonal pyramidal, and bent. The bond angles for all three molecular shapes are 109.5°.

3.27 The trigonal pyramidal shape is three-dimensional; the bent and trigonal planar shapes are two dimensional.

3.29 The bond angles are 180° in a linear electron geometry.

3.31 The bond angles are 109.5° in a tetrahedral geometry.

3.33 a. The molecular shape is linear. The bond angle around the carbon atom is 180°.

     b. The molecular shape is trigonal planar. The bond angles around the carbon atom are 120°.

3.35 The geometry of this molecule is tetrahedral; there are four bonding groups and no nonbonding groups. The O—S—O bond angle is 109.5°.

3.37 The bond angle decreases around the central atom as the number of electron groups surrounding the central atom because there are more groups to distribute in the same amount of space. The largest bond angle is 180°; the smallest bond angle is typically 109.5°.

3.39   a.

H   H
 \ /
  C=C
 / \
H   H

The molecular shape around each carbon atom is trigonal planar; there are three bonding groups around each carbon atom. The bond angles are 120°.

b.

    :F:
     |
:F−Si−F:
     |
    :F:

The molecular shape is tetrahedral; there are four bonding groups. The bond angles are 109.5°.

c.

  H H
  | |
H−C−C−H
  | |
  H H

The molecular shape around each carbon atom is tetrahedral; there are four bonding groups and zero nonbonding groups. The bond angles are 109.5°.

d.   H−C≡C−H The molecular shape around each carbon atom is linear; there are two bonding groups. The bond angles are 180°.

3.41   a.

      H   H
      \   /
   H   C
    \ / \
     C   O−H
    /|
   H H

b.   The bond angle around each carbon is 109.5°.

c.   The bond angle around the oxygen atom is approximately 109.5°.

d.   The molecular shape around each carbon atom is tetrahedral.

e.   The molecular shape around the oxygen atom is bent.

3.43   The geometry is trigonal planar when there are three bonding groups and no nonbonding groups surrounding a central atom. The bond angles are 120°. The geometry is trigonal pyramidal when there are three bonding groups and one nonbonding group surrounding a central atom. The bond angles are 109.5°.

3.45   The geometry is linear for two atoms because two points define a line and no other geometry is possible.

3.47   **II. Observations of Formaldehyde Model**

2.   :O:
         ‖
       H−C−H

4.   There are three electron groups surrounding the central carbon atom.

5.   There are three groups of bonding electrons.

6.   There are no nonbonding electrons on the central atom.

7.   The electron geometry is trigonal planar.

8.   The molecular shape of formaldehyde is trigonal planar.

9.   This molecule has a two-dimensional, flat shape.

10. The H—C—O bond angle is 120°. All the bond angles are the same.

11. The trigonal planar shape allows all three groups of electrons to be spaced the farthest possible distance from each other. In the T-shape, two groups are farther apart, but two groups are closer together.

3.49   Ball-and-stick models show the bond angles present in a molecule. When the size of the atoms is important, a space-filling model is used.

3.51   Electronegativity is a measure of the ability of an atom to attract electrons to itself in a covalent bond.

3.53   a. Oxygen is the more electronegative element. Oxygen lies farther to the right of carbon within the same period.

b. Oxygen is the more electronegative element. Oxygen and sulfur are in the same group, and oxygen lies above sulfur.

c. Fluorine is the more electronegative element. Fluorine lies farther to the right of lithium within the same period.

3.55   Electronegativity increases as you move up in a group of elements.

3.57   The halogens are the most electronegative family of elements.

3.59   If the electronegativities of the two atoms in a covalent bond are similar, then the bond will be nonpolar. If the electronegativities are different, then the covalent bond will be a polar covalent bond.

3.61   a. Water is polar.

b. Ethanol is polar.

c. C_2H_4 is nonpolar. Carbon and hydrogen have similar electronegativities.

3.63   a. I_2 is nonpolar. The two atoms in the covalent bond are the same.

b. CH_4 is nonpolar. Carbon and hydrogen have similar electronegativities.

c. H—Br is polar. Hydrogen and bromine have different electronegativities.

3.65   A covalent bonding force occurs between two atoms within the same molecule and is much stronger than an intermolecular force of attraction. An intermolecular force of attraction occurs between atoms on different molecules.

3.67   The three types of intermolecular forces of attraction are: dispersion forces, dipole-dipole interactions, and hydrogen bonds. The strongest forces of attraction are hydrogen bonds. The weakest forces of attraction are dispersion forces.

3.69   Dispersion forces are the weakest of the intermolecular forces because they are due to the interaction between temporary dipoles.

3.71
```
H  Br H  Br
|  |  |  |
Br H  Br H

H  Br H  Br
|  |  |  |
Br H  Br H
```
The H-Br molecules exhibit dipole-dipole interactions. They are lined up so that the partial positive charge (H) lies next the partial negative charge (Br) of an adjacent molecule.

3.73   a. C_5H_{12} is nonpolar, so it should only exert dispersion forces.

b. Acetone is polar, so it should exert dispersion forces and dipole-dipole interactions.

c. Water is polar and has O—H bonds, so it should exert hydrogen bonding, as well as dispersion forces, and dipole-dipole interactions.

3.75   Ice floats on water because hydrogen bonding causes the molecules to occupy a greater volume in the solid state than in the liquid state, so the mass per volume is greater for the liquid state than the solid state.

3.77   The dashed lines represent the hydrogen bond between an oxygen atom of one molecule and a hydrogen atom of another molecule.

3.79   Hydrogen bonding gives DNA a helical shape.

3.81   Estrogen binds to the estrogen receptor and activates several genes. This gene activation also stimulates the proliferation of breast cancer cells.

3.83   The estrogen receptor is a large protein molecule. Estradiol fits the estrogen binding site perfectly because it has a complimentary shape to the binding site—a cavity within the receptor.

3.85   When Tamoxifen binds to the receptor, it changes the shape of the receptor, preventing gene activation from occurring.

Chapter 4

Solids, Liquids, and Gases

Chapter Summary

In this chapter, you have examined the factors that determine the physical states of matter and were introduced to the central role that energy plays in determining the physical state. You learned about pressure and how it relates to blood pressure. You studied gases more in depth and have learned that having a basic knowledge of the special behavior of gases helps you understand important medical issues such as breathing, decompression sickness, and anesthesia.

Section 4.1 States of Matter

All matter exists in one of three physical states or phases: solid, liquid, or gas. A solid has a definite shape and volume, which is independent of its container. A liquid has a definite volume but does not have a definite shape; it conforms to the shape of the container. A gas has neither a definite volume nor a definite shape. A volume of gas expands to fill its entire container.

In a Nutshell: Kinetic and Potential Energy

Energy is the capacity to do work, where work is the act of moving an object. There are two fundamental forms of energy: kinetic energy and potential energy. Kinetic energy is the energy of motion, the energy a substance possesses as a result of the motion of its molecules or atoms. Faster-moving molecules have greater kinetic energy than slower ones. Potential energy is stored energy, the energy a substance possesses as a result of position, composition, and condition of its atoms. Heat is kinetic energy that is transferred from one object to another due to a difference in temperature. Heat energy always flows from a hot object to a cold object. While heat is a form of energy, temperature is a measure of the average kinetic energy of the molecules, ions, or atoms that make up a substance.

In a Nutshell: Temperature Scales

Temperature can be measured using a thermometer and can be reported in one of three temperature scales: Celsius, Fahrenheit, and Kelvin. The Kelvin scale is important in many applications in science because it represents an absolute scale, one that assigns 0 K to the condition where all molecular motion has stopped. The Celsius scale is a relative scale, based on the freezing and boiling points of water. Equations are needed to convert between the temperature scales.

To convert from °C to °F, use the following equation:

$$°F = \left(\frac{9}{5}°C\right) + 32$$

To convert from °F to °C, use the following equation:

$$°C = \frac{5}{9}\left(°F - 32\right)$$

To convert from °C to K, use the following equation:

$$K = °C + 273$$

To convert from K to °C, use the following equation:

$$°C = K - 273$$

Worked Example #1

It's snowing in New York City and the temperature is 28 °F. Your friend is visiting from Europe and would like to know what the temperature 28 °F corresponds to in °C. What do you tell him? What is the Kelvin temperature on this winter day?

Convert °F to °C by using the equation that has °C isolated on one side of the equal sign:

$$°C = \frac{5}{9}\left(°F - 32\right)$$

Substitute the given value for °F into the equation and solve for °C:

$$°C = \frac{5}{9}\left(28 - 32\right) = -2.0 \ °C$$

A temperature of 28 °F corresponds to a temperature of −2.0 °C.

To find this temperature in Kelvin, use the equation that converts °C to K:

$$K = °C + 273$$

Substitute the given value for °C into the equation and solve for K:

$$K = -2 + 273 = 271 \ K$$

Thus, 28 °F = −2 °C = 271 K.

Try It Yourself #1

It's a spring day in Paris, France, and the temperature is 19 °C. What is this temperature in °F and in K?

Convert °C to °F by using the equation that has °F isolated on one side of the equal sign. Substitute the given value for °C into the equation and solve for °C.

To find this temperature in Kelvin, use the equation that converts °C to K. Substitute the given value for °C into the equation and solve for K.

In a Nutshell: A Kinetic-Molecular View of the States of Matter

Intermolecular forces of attraction draw molecules together and are a form of potential energy. Kinetic energy tends to pull molecules apart. These two opposing forces determine the physical state of a substance.

In the gas phase, atoms or molecules have high kinetic energy. Atoms and molecules are moving faster than such particles in the solid or liquid phases and they are much farther apart. Since the particles are so far apart, the intermolecular forces of attraction are minimal or nonexistent.

In the liquid phase, the particles are much closer together; they are in contact with other molecules moving randomly and tumbling over one another. Intermolecular forces pull the particles together.

In the solid phase, atoms or molecules exist in a regular ordered pattern. Molecules in the solid phase have less kinetic energy than liquids, so they remain in a fixed position with mainly vibrational, jiggling motions. Intermolecular forces dominate kinetic forces in the solid phase.

Worked Example #2

Indicate whether each of the following examples is a demonstration of potential energy or kinetic energy:

    a. molecules colliding with the walls of their container

    b. dipole-dipole interaction

    a. The molecules are in motion, so it is a demonstration of kinetic energy.

    b. Dipole-dipole interactions are a form of potential energy; there is stored energy.

Try It Yourself #2

Indicate whether each of the following examples is a demonstration of potential energy or kinetic energy:

    a. water in the lake behind a dam

    b. water falling over a dam

    a. Is there stored energy or is there motion?

    b. Is there stored energy or is there motion?

Worked Example #3

Match the following descriptions to the state that describes it: solid, liquid, or gas.

    a. Intermolecular forces are important in these two states.

    b. Molecules in this state have the least amount of kinetic energy.

    c. This state has a definite volume, but it conforms to the shape of the container.

    a. Intermolecular forces are important in the solid state and the liquid state. Molecules in the gas state are spaced too far apart to exert intermolecular forces.

    b. In the solid state, the molecules are in a fixed position and do not move, except for jiggling motions.

    c. Liquids have a definite volume and conform to the shape of the container.

Try It Yourself #3

Match the following descriptions to the state that it describes: solid, liquid, or gas.

    a. This state has neither a definite volume nor a definite shape.

    b. Dispersion forces are more important than kinetic forces in this state.

    c. Kinetic forces and intermolecular forces are both important in this state.

    a. State described: _____

    b. State described: _____

    c. State described: _____

Practice Problems for States of Matter

1. Which of the following would conform to shape of a glass bottle: liquid water, steam, or ice?

2. A thermometer reads 48 °F. What is this temperature in Celsius and in Kelvin?

3. Dry ice has a temperature of 194.5 K. Convert this temperature to Fahrenheit and Celsius.

4. Indicate whether the following examples is a demonstration of potential energy or kinetic energy:

    a. a bouncing ball

    b. dispersion forces

    c. an airplane flying

Section 4.2 Changes of State

A change in state is the physical process of going from one state to another. Changes of state are physical changes, not chemical changes, because the chemical bonds in the molecules do not change.

A change of state from the solid to the liquid phase is known as melting; the reverse change from liquid to solid is known as freezing. A change of state from the liquid to the gas phase is known as vaporization, while the reverse change from gas to liquid is known as condensation. A solid can go directly to the gas phase; a process known as sublimation. The reverse process changing from the gas phase to the solid phase is known as deposition.

In a Nutshell: Energy and Changes of State

Intermolecular forces are disrupted or reformed during a change of state, particularly if the substance is changing to or from the gas phase. In a change of state, energy must be added or removed from the substance. Energy must be added to achieve melting, vaporization or sublimation. Energy must be removed to achieve freezing, condensation, or deposition. The heat of vaporization is the amount of energy that must be added to a liquid to transform it into a gas. The heat of fusion is the amount of energy that must be added to turn a solid into a liquid.

In a Nutshell: Melting and Boiling Points

At a specific temperature unique for every substance, known as the melting point, a substance changes from the solid state to the liquid state. The boiling point is a specific temperature unique for every substance, where the substance changes from a liquid to a gas. The melting point and the boiling point depend on the strength and number of intermolecular forces of attraction between the molecules of that substance. When the chemical structure of a substance allows it to form hydrogen bonds, a greater amount of energy must be supplied to change the substance from a liquid to a gas.

Worked Example #4

Identify the change of state and the term that describes the change of state.

    a. formation of ice cubes in the freezer

    b. seeing your breath on a cold winter day

c. a puddle of water disappearing on a hot summer day

a. *The formation of ice cubes is water changing from the liquid state to the solid state. This process is known as freezing.*

b. *Seeing your breath on a cold winter day is seeing water vapor (a gas) change into liquid water. This process is known as condensation.*

c. *When a puddle of water disappears, the liquid is changing into a gas. This process is known as evaporation.*

Try It Yourself #4

Identify the change of state and the term that describes the change of state.

a. a solid block of metal heated so that it turns into a pool of molten metal

b. the mist coming from a humidifier

c. snow forming from water vapor in clouds

a. *Change in state:_____*

Term to describe change in state:_____

b. *Change in state:_____*

Term to describe change in state:_____

c. *Change in state:_____*

Term to describe change in state:_____

Practice Problems for Changes of State

1. Which changes in state involve transfer of heat from the surroundings?

2. Which molecule would you expect to have a higher boiling point: $CHCl_3$ or CH_3OH?

3. Identify the change of state and the term that describes the change of state.

  a. steam coming from your hot cup of coffee

  b. frost forming from water vapor in the air

  c. rain forming from water vapor in clouds

Extension Topic 4-1: Specific Heat

Specific heat is the amount of heat required to raise the temperature of 1 g of substance by 1 °C. The specific heat is a physical property. It can be calculated using the following equation:

$$\text{Specific heat} = \frac{\text{Amount of heat } (q)}{\text{Mass} \times \text{change in temperature } (\Delta T)}$$

Heat refers to the amount of heat transferred to the substance and the change in temperature is the difference in temperature before and after heat has been transferred. Heat is usually measured in calories and mass in grams. A substance with a large specific heat requires a large amount of heat for a small increase in temperature.

Worked Example #5

Using Table 4-2, calculate the number of calories that must be added to warm 17.3 g of each of the following substances below from 15 °C to 28 °C. Which substance requires the most heat?

  a. sand
  b. water

First, rearrange the equation for specific heat to solve for q, the amount of heat (the number of calories).

$$\text{Amount of heat } (q) = \text{specific heat } (\frac{\text{cal}}{\text{g °C}}) \times \text{mass (g)} \times \Delta T \text{ (°C)}$$

Second, calculate the temperature difference:

  ΔT = 28 – 15 = 13 °C

Third, substitute the supplied values into the equation and solve for the amount of heat.

a. *For sand:* Amount of heat $= 0.16 \dfrac{\text{cal}}{\cancel{\text{g}} \ \cancel{°C}} \times 17.2 \ \cancel{\text{g}} \times 13 \cancel{°C} = 36$ cal

b. *For water:* Amount of heat $= 1.00 \dfrac{\text{cal}}{\cancel{\text{g}} \ \cancel{°C}} \times 17.2 \ \cancel{\text{g}} \times 13 \cancel{°C} = 2.2 \times 10^2$ cal

Water requires the greater amount of heat to raise the temperature of 17.3 g by 13 °C.

Try It Yourself #5

Using Table 4-2, calculate the number of calories that must be added to warm 56 g of each of the following substances below from 31 °C to 68 °C. Which substance requires the most heat?

    a. paraffin wax

    b. lead

First, rearrange the equation for specific heat to solve for q, the amount of heat (the number of calories).

Second, calculate the temperature difference.

Third, substitute the supplied values into the equation and solve for the amount of heat.

Practice Problems for Specific Heat

    1. Calculate the number of calories required to raise 31 g of each of the following substances from 3 °C to 29 °C.

        a. ambient air

    b. ethanol

    c. aluminum

    d. copper

2. An unknown liquid weighing 25 g is water or ethanol. In going from a temperature of 28 °C to 57 °C, the liquid absorbed 420 calories. Use Table 4-2 to determine the identity of the liquid.

Section 4.3 Pressure

Pressure, P, is the measure of amount of force applied over a given area:

$$Pressure = \frac{Force}{Area}$$

Pressure is an important factor that influences the states of matter, in particular, the gas state.

In a Nutshell: Pressure Units

Pressure can be measured and reported in a variety of units. The most common unit used in medicine is mmHg. Atmospheres are the unit used to measure atmospheric pressure. Atmospheric pressure is the pressure exerted by the weight of air at any given place (area) on the Earth. The other common units for pressure are pounds per square inch (psi), torr, and Pascal.

Worked Example #6

A tire on an automobile is filled to 35 psi. What is this pressure equivalent to in mmHg?

Identify the conversion.

Use Table 4-3 to find the conversion between psi and mmHg: 14.70 psi = 760 mmHg.

Express the conversion as a conversion factor:

$$\frac{14.70 \text{ psi}}{760 \text{ mmHg}} \quad \text{or} \quad \frac{760 \text{ mmHg}}{14.70 \text{ psi}}$$

Set up the equation so that the supplied units cancel:

$$35 \text{ \sout{psi}} \times \frac{760 \text{ mmHg}}{14.70 \text{ \sout{psi}}} = 1.8 \times 10^3 \text{ mmHg}$$

Try It Yourself #6

A patient has an intraocular eye pressure of 12 mmHg. What is this pressure in Pa?

Tools: Table 4-3

Identify the conversion.

Express the conversion as a conversion factor.

Set up the equation so that the supplied units cancel.

In a Nutshell: Blood Pressure

Blood pressure is the pressure exerted by blood on the walls of blood vessels. The heart acts like a pump, contracting and relaxing, causing blood to flow throughout the body. When the heart muscle contracts, blood pressure against the arteries reaches a maximum as blood is forced through the arteries. The blood pressure maximum is known as the systolic pressure. When the heart muscle relaxes, blood pressure drops and the minimum pressure is known as the diastolic pressure. A device called a sphygmomanometer is used to measure blood pressure.

Worked Example #7

A patient's blood pressure is 110/68.

    a. Convert the systolic pressure to units of atm.

    b. Convert the diastolic pressure to units of psi.

The systolic pressure is 110 and the diastolic pressure is 68.

    a. Identify the conversion.

        Use Table 4-3 to find the conversion between mmHg and atm: 760 mmHg = 1 atm.

        Express the conversion as a conversion factor:

$$\frac{1\ atm}{760\ mmHg} \quad or \quad \frac{760\ mmHg}{1\ atm}$$

        Set up the calculation so that the supplied units cancel:

$$110\ \cancel{mmHg} \times \frac{1\ atm}{760\ \cancel{mmHg}} = 0.14\ atm$$

    b. Identify the conversion.

        Use Table 4-3 to find the conversion between mmHg and psi: 760 mmHg = 14.70 psi.

        Express the conversion as a conversion factor:

$$\frac{14.70\ psi}{760\ mmHg} \quad or \quad \frac{760\ mmHg}{14.70\ psi}$$

        Set up the calculation so that the supplied units cancel:

$$68\ \cancel{mmHg} \times \frac{14.70\ psi}{760\ \cancel{mmHg}} = 1.3\ psi$$

Try It Yourself #7

A patient's blood pressure is 147/105.

    a. Does the patient have normal blood pressure? If not, what is his condition?

    b. Convert the systolic pressure to units of torr.

    c. Convert the diastolic pressure to units of atm.

Tools: Table 4-3 and Table 4-4

    a. The patient has_____

    b–c. Identify the systolic pressure and the diastolic pressure.

        Identify the conversion.

        Express the conversion as a conversion factor.

        Set up the calculation so that the supplied units cancel.

In a Nutshell: Vapor Pressure

At the surface of a liquid, some of the liquid particles enter the gas phase while simultaneously some of the gas phase particles enter the liquid phase. The movement of the particles between the liquid and gas phases is a result of the vapor pressure of the liquid substance. Vapor pressure is the pressure exerted by molecules in the gas phase in contact with molecules in the liquid phase at a given temperature. Liquids with a high vapor pressure at room temperature enter the gas phase readily and are considered volatile liquids.

The vapor pressure of a liquid increases with temperature. When the temperature is high enough that the vapor pressure of the substance equals the atmospheric pressure, the liquid boils. The boiling point of a liquid is the temperature at which the vapor pressure of the liquid equals the atmospheric pressure. The lower the boiling point of a substance, the higher the vapor pressure will be.

Worked Example #8

Diethyl ether was has been used as an anesthetic. It has a vapor pressure of 400 mmHg at 17.9 °C. Is this compound volatile? Does it have a high or a low boiling point?

The vapor pressure is high; therefore, the compound is volatile and the boiling point should be low.

Try It Yourself #8

The vapor pressure of ethyl acetate (commonly used in nail polish remover) is 76 mmHg at 20 °C. The vapor pressure of phenol is 0.36 mmHg at 20 °C. Which would you expect to have the lower boiling point?

Compound with higher vapor pressure: _____

Compound with lower boiling point: _____

Practice Problems for Pressure

1. A patient has an intraocular eye pressure of 2.0×10^3 Pa. If the intraocular eye pressure is greater than 21.5 mmHg, then the patient has a greater risk for developing glaucoma. What is the patient's intraocular eye pressure in mmHg? Does this patient have a risk for developing glaucoma?

2. A patient had a systolic pressure of 124 torr and a diastolic pressure of 0.089 atm. Does this patient have hypertension? What is the patient's blood pressure in mmHg?

3. The vapor pressure of cinnamaldehyde (it gives cinnamon its flavor and smell) is 5.8 × 10^{-4} psi at 25 °C. What is the vapor pressure in mmHg? Do you expect cinnamaldyde to have a high or low boiling point?

Section 4.4 Gases

In a Nutshell: Kinetic Molecular Theory

A set of gas laws has been identified to describe the behavior of gases in terms of four variables: pressure (P), volume (V), number of moles (n), and temperature (T). Since atoms and molecules are in constant motion, molecules and atoms in the gas phase are continuously colliding with the walls of their container creating pressure. The characteristics of a gas can be described by the kinetic molecular view of gases as follows:

The particles of a gas are in constant, random motion.

The total volume of all the gas particles in a container is negligible compared to the volume of the container.

The attractive forces among the particles of a gas are negligible.

The temperature of a gas depends on the average kinetic energy of the gas particles.

Worked Example #9

Use kinetic molecular theory of gases to explain the following:

a. Dispersion forces are not important in the gas phase.

b. Perfume dropped in one part a room is quickly detected at the other end of the room.

a. *Molecules in the gas phase do not have intermolecular forces of attraction; therefore, they do not exert dispersion forces.*

b. *The particles of a gas do not have intermolecular forces of attraction; therefore gas particles move at high speeds in all directions, filling the volume of the container they are in.*

Try It Yourself #9

Use kinetic molecular theory of gases to explain the following:

a. A balloon filled with helium on a winter day would be smaller than a balloon filled with the same amount of helium on a summer day.

b. Gas molecules can be compressed to fit into a smaller volume.

Solutions

a.

b.

In a Nutshell: STP and the Molar Volume of a Gas

In chemistry, the properties of a gas are described under a standard set of reference conditions. These conditions are known as STP, Standard Temperature and Pressure (273 K and 760 mmHg). Under the conditions of STP, one mole of any gas occupies a volume of 22.4 L, the molar volume of a gas. The identity or the mass of the gas does not matter.

Worked Example #10

What volume will be occupied by 10.2 moles of helium at STP? Would the volume be the same for 10.2 moles of xenon?

Express conversions as conversion factors:

$$\frac{22.4 \text{ L}}{1 \text{ mol}} \quad \text{or} \quad \frac{1 \text{ mol}}{22.4 \text{ L}}$$

Set up the calculation so that the supplied units cancel:

$$10.2 \text{ mol} \times \frac{22.4 \text{ L}}{1 \text{ mol}} = 228 \text{ L}$$

Therefore, 228 L would be occupied by 10.2 moles of helium. Yes, the volume would be the same for 10.2 moles of xenon because the identity of the gas is not important.

Try It Yourself #10

What volume will be occupied by 2.7 moles of krypton at STP?

Express the conversion as conversion factors.

Set up the calculation so that the supplied units cancel.

In a Nutshell: Pressure-Volume Relationship of Gases

As the volume of gas decreases, the pressure of the gas increases; and vice versa, given constant T and n. Pressure and volume are inversely related to each other, as one goes up, the other goes down. This relationship is known as Boyle's law and can be written as follows:

$$P_1 V_1 = P_2 V_2 \qquad (n \text{ and } T \text{ are constant})$$

where the subscripts 1 and 2 refer to the initial and final values for P and V. The pressure-volume relationship of gases is demonstrated every time you inhale and exhale. Upon inhalation, the pressure of the lungs decreases as the volume increases, and upon exhalation, the pressure of the lungs increases and the volume decreases.

Worked Example #11

The pressure gauge on a patient's full 13.9 L oxygen tank reads 10.2 atm. At constant temperature, how many liters of oxygen can the patient's entire tank hold at a pressure of 25.1 atm?

Since both P and V are changing, while n and T are constant, we can use Boyle's law:
$P_1V_1 = P_2V_2$.

Define the variables and select the variable to solve for.
The problem indicates that P_1 = 10.2 atm, V_1 = 13.9 L, and P_2 = 25.1 atm; therefore, you need to solve for V_2 the final volume.

Algebraically isolate the unknown variable on one side of the equation.
Use algebra to manipulate the equation so that V_2 is isolated:

$$V_2 = P_1 \times \frac{V_1}{P_2}$$

Substitute the known values into the equation and solve for the unknown variable:

$$V_2 = 10.2 \text{ atm} \times \frac{13.9 \text{ L}}{25.1 \text{ atm}} = 5.65 \text{ L}$$

Try It Yourself #11

An air bubble forms at the bottom of a lake where the total pressure is 4.35 atm. At this pressure, the bubble has a volume of 7.32 mL. What is the volume of the bubble when the bubble rises to the surface, where the pressure is 1.00 atm?

The variables that are changing are: _____
The equation to use is: _____
Define the variables and select the variable to solve for.

Algebraically isolate the unknown variable on one side of the equation.

Substitute the known variables into the equation and solve for the unknown variable.

In a Nutshell: Pressure-Temperature Relationship of Gases

As the temperature of a gas increases, the pressure of the gas increases when the moles of gas and the volume are constant. The pressure of a gas is directly proportional to the temperature of a gas: As one goes up, the other one goes up. The relationship between temperature and pressure is known as Gay-Lussac's law and can be written as follows, where 1 and 2 refer to the initial and final values for P and T:

$$\frac{P_1}{T_1} = \frac{P_2}{T_2}$$

(n and V are constant, T must be in Kelvin)   T↑ P↑

When performing calculations using this or any other gas law equation, the Kelvin scale is used because it is an absolute scale.

Worked Example #12

The contents of an aerosol can have a pressure of 150 psi at 54.4 °C. If the contents of the can are heated to 82.2 °C, the can will explode. What is the pressure of the gas at this temperature?

Since P and T are changing and the volume and the number of moles of gas in the can are constant, we can use Gay-Lussac's law:

$$\frac{P_1}{T_1} = \frac{P_2}{T_2}$$

Define the variables and select the variable to solve for.

The problems indicates that P_1 = *150 psi,* T_1 = *54.4* °*C, and* T_2 = *82.2* °*C; therefore, you need to solve for* P_2.

Algebraically isolate the unknown variable on one side of the equation.

Use algebra to manipulate the equation so that P_2 *is isolated:*

$$P_2 = T_2 \times \frac{P_1}{T_1}$$

Substitute the known values into the equation and solve for the unknown variable.

REMEMBER, YOU MUST FIRST CONVERT TEMPERATURES TO KELVIN!

$$T_1 = 54.4 \text{ °C} + 273 = 327.4 \text{ K}$$
$$T_2 = 82.2 \text{ °C} + 273 = 355.2 \text{ K}$$

$$P_2 = 355.2 \text{ \cancel{K}} \times \frac{150 \text{ psi}}{327.4 \text{ \cancel{K}}} = 163 \text{ psi, rounded to 160 pst}$$

Try It Yourself #12

Your tire pressure at 28 °C reads 34.2 psi. After driving around for a few hours, your tire pressure reads 36.3 psi. What is the temperature of air in your tires?

The variables that are changing are: _____

The equation to use is: _____

Define the variables and select the variable to solve for.

Algebraically isolate the unknown variable on one side of the equation.

Substitute the known variables into the equation and solve for the unknown variable.

REMEMBER, YOU MUST FIRST CONVERT TEMPERATURES TO KELVIN!

In a Nutshell: Volume-Temperature Relationship of Gases

As the temperature of a gas increases, the volume of a gas increases if the number of moles of gas and the pressure of the gas remain constant. The volume of a gas is directly proportional to the temperature of a gas: As one goes up, the other one goes up. The relationship between temperature and volume is known as Charles' law and can be written as follows, where 1 and 2 refer to the initial and final values for *V* and *T*:

$$\frac{V_1}{T_1} = \frac{V_2}{T_2}$$ (*n* and *P* are constant, *T* must be in Kelvin)   T↑ V↑

When performing calculations using this or any other gas law equation, the Kelvin scale is used because it is an absolute scale.

Worked Example #13

Oxygen is warmed from 27 °C to 58 °C. The original volume of the gas was 19.8 L. What is the final volume of the gas assuming *P* and *n* have remained constant?

Since V *and* T *are changing and* P *and* n *are constant we can use Charles' law:*

$$\frac{V_1}{T_1} = \frac{V_2}{T_2}$$

Define the variables and select the variable to solve for.

The problem indicates T_1 = 27 °C, V_1 = 19.8 L, *and* T_2 = 58 °C; *therefore you need to solve for* V_2.

Algebraically isolate the unknown variable on one side of the equation.

$$V_2 = T_2 \times \frac{V_1}{T_1}$$

Substitute the known variables into the equation and solve for the unknown variable.
REMEMBER, YOU MUST FIRST CONVERT TEMPERATURES TO KELVIN!

$$T_1 = 27 \text{ °C} + 273 = 300 \text{ K}$$

$$T_2 = 58 \text{ °C} + 273 = 331 \text{ K}$$

$$V_2 = 331 \text{ K} \times \frac{19.8 \text{ L}}{300 \text{ K}} = 21.8 \text{ L}$$

Try It Yourself #13

A balloon taken from the freezer has a temperature of 2 °C and a volume of 1.63 L. The balloon is warmed up and the final volume is 1.80 L. Assume that the pressure and the number of moles of gas in the balloon are constant, what is the final temperature of the gas in the balloon?

The variables that are changing are: _____

The equation to use is: _____

Define the variables and select the variable to solve for.

Algebraically isolate the unknown variable on one side of the equation.

Substitute the known variables into the equation and solve for the unknown variable. REMEMBER, YOU MUST FIRST CONVERT TEMPERATURES TO KELVIN!

In a Nutshell: Combined Gas Law

A single relationship can be formed by combining the three previous relationships: pressure-volume, pressure-temperature, and volume-temperature. As long as the number of moles is held constant, you can predict how changing three variables P, V, or T, affects an unknown variable. This combined relationship is known as the combined gas law and can be written as follows:

$$\frac{P_1 V_1}{T_1} = \frac{P_2 V_2}{T_2} \quad (n \text{ is constant})$$

Remember, all temperatures must be converted to Kelvin!

Worked Example #14

A sample of nitrogen gas has a volume of 3.86 L at a pressure of 345 torr and a temperature of 28 °C. Calculate the volume of the gas, when the temperature is 35 °C and the pressure is 154 torr.

Since P, V, *and* T *are changing we need to use the combined gas law:*

$$\frac{P_1 V_1}{T_1} = \frac{P_2 V_2}{T_2}$$

Define the variables and select the variable to solve for.

The problem indicates that V_1 = 3.86 L, T_1 = 28 °C, P_1 = 345 torr, T_2 = 35 °C, P_2 = 154 torr; *therefore, you need to solve for* V_2.

Algebraically isolate the unknown variable on one side of the equation:

$$V_2 = \frac{P_1 V_1 T_2}{P_2 T_1}$$

Substitute the known variables into the equation and solve for the unknown variable.

REMEMBER, YOU MUST FIRST CONVERT TEMPERATURES TO KELVIN!

$$T_1 = 28 \text{ °C} + 273 = 301 \text{ K}$$
$$T_2 = 35 \text{ °C} + 273 = 308 \text{ K}$$

$$V_2 = \frac{345 \ \cancel{Torr} \times 3.86 \text{ L} \times 308 \ \cancel{K}}{154 \ \cancel{Torr} \times 301 \ \cancel{K}} = 8.85 \text{ L}$$

Try It Yourself #14

A sample of argon gas has a volume of 5.46 L at a pressure of 2.64 atm and a temperature of 24 °C. Calculate the temperature of the gas in Celsius when the pressure is 4.75 atm and a volume of 1.22 L.

The variables that are changing are: _____

The equation to use is: _____

Define the variables and select the variable to solve for.

Algebraically isolate the unknown variable on one side of the equation.

Substitute the known variables into the equation and solve for the unknown variable. REMEMBER, YOU MUST FIRST CONVERT TEMPERATURES TO KELVIN!

In a Nutshell: Partial Pressure and Gas Mixtures

Dalton's law states that, if you have a mixture of gases, each gas in the mixture will exert a pressure independent of the other gases present, and each gas will behave as it alone occupied the total volume. The pressure exerted by a gas in a mixture is known as the partial pressure, P_N. Dalton's law states that the sum of the partial pressures of each gas present in the mixture equals the total pressure (P_{TOT}) exerted by the mixtures of gases:

$$P_{TOT} = P_1 + P_2 + P_3 + ... \ P_N$$

where P_1, P_2, P_3,.....P_N represent the partial pressures of each gas in the mixture.

Worked Example #15

SCUBA divers often use a mixture of nitrogen and gas in their air tanks. An air tank has a total pressure of 4.67 atm and contains a mixture of oxygen and nitrogen. If the partial pressure of nitrogen is 3.18 atm, what is the partial pressure of oxygen?

Since partial pressures are provided, use the following equation:

$$P_{TOT} = P_{oxygen} + P_{nitrogen}$$

Define the variables and select the variable to solve for.
The problem indicates that P_{TOT} = 4.67 atm and $P_{nitrogen}$ = 3.18 atm; therefore we need find P_{oxygen}.

Algebraically isolate the unknown variable on one side of the equation:

$$P_{oxygen} = P_{TOT} - P_{nitrogen}$$

Substitute the known variables into the equation and solve for the unknown variable:

$$P_{oxygen} = P_{TOT} - P_{nitrogen} = 4.67 \text{ atm} - 3.18 \text{ atm} = 1.49 \text{ atm}$$

Try It Yourself #15

A SCUBA diver is going to dive to a depth of 330 ft. He will be using an air tank filled with a mixture of helium, nitrogen, and oxygen. The total pressure of the gas mixture is 8.45 atm. If the partial pressure of helium is 5.91 atm and the partial pressure of nitrogen is 1.69 atm, what is the partial pressure of oxygen in the tank?

The variables that are changing are: _____

The equation to use is: _____

Define the variables and select the variable to solve for.

Algebraically isolate the unknown variable on one side of the equation.

Substitute the known variables into the equation and solve for the unknown variable.

In a Nutshell: Henry's Law

Concentration is a measure of how much of a given substance is mixed with another substance to form a solution. A solution is a uniform mixture of substances. The amount of gas dissolved in a solution is directly proportional to the partial pressure of that gas above the solution. As the pressure of the gas above the solution increases, the concentration of the gas in the solution increases. This relationship is known as Henry's law and can be described by the following equation:

Pressure = k × concentration

$P = kC$

where P is the pressure of the gas above the solution, k is a constant, which depends on the gas, and C is the concentration of the gas in the solution.

Henry's law is useful in the field of anesthesiology when gaseous anesthetics are inhaled. Anesthetics have different Henry's constants, k, and therefore are present in different concentrations when dissolved in blood at given pressure. The smaller the Henry's constant of an anesthetic, the higher the concentration of anesthetic dissolved in blood.

Worked Example #16

Where would you expect a glass of soda to have more bubbles: at the beach or in the mountains?

Henry's law shows that there is a direct relationship between the pressure above a solution and the concentration of gas in a solution. Therefore, at lower altitudes and higher atmospheric pressure, you would have a corresponding higher concentration of carbon dioxide (the bubbles) in the solution. Thus, there would be more bubbles in the glass of soda on the beach. In the mountains, the CO_2 would escape more quickly once the bottle is opened and depressurized causing the soda to go "flat" sooner.

Try It Yourself #16

Desflurane and sevoflurane have large Henry's constant compared to diethyl ether. Which anesthetic would allow a patient to regain conscious more quickly? Explain.

Solution:

Practice Problems for Gases

1. A sample of xenon has a volume of 12.3 L at a pressure of 672 mmHg and a temperature of 15 °C. Calculate the pressure of the gas when the temperature is 47 °C and the volume is 20.6 L.

2. A balloon is filled with helium and it has a volume of 17.8 L at 20 °C. What is the volume of this balloon when the temperature changes to 56 °C (and the pressure and the number of moles of gas remain constant)?

3. A mixture of two gases, argon and oxygen, is sometimes used in SCUBA diving. If the total pressure in air tank containing a mixture of argon and oxygen is 12.2 atm and the partial pressure of oxygen is 2.4 atm, what is the partial pressure of argon?

4. A sample of oxygen has a pressure of 746 Torr and a volume of 30.8 L, what is the pressure of the gas if the volume of the gas changes to 50.3 L (assume that the number of moles and the temperature remain constant)?

5. A sample of nitrogen has a pressure of 30.2 psi at 26 °C. Calculate the temperature, in Celsius, of the gas when the pressure is 15.9 psi. Assume that the number of moles of gas and the volume of the gas stays constant.

Chapter 4 Quiz

1. A thermometer in Europe reads 32 °C. Is it a winter day or a summer day? What temperature does this correspond to in degrees Fahrenheit and Kelvin?

2. A tank that contains a mixture of carbon dioxide, nitrous oxide, and oxygen has a total pressure of 5.63 atm. The partial pressure of carbon dioxide is 0.56 atm, and the partial pressure of oxygen is 3.66 atm. What is the partial pressure of nitrous oxide?

3. Using the kinetic theory of gases, explain why a gas will conform to the shape of the container.

4. A sample of argon has a volume of 5.3 L at a pressure of 270 torr and a temperature of 25 °C. Calculate the volume of the gas when the temperature is 47 °C and the pressure is 105 mmHg.

5. A sample of carbon dioxide has a pressure of 0.67 atm at 12 °C. Calculate the temperature, in Celsius, of the gas when the pressure is 0.012 atm. Assume that the number of moles of gas and the volume of the gas stays constant.

6. A balloon is filled with helium and it has a volume of 345 mL at 18 °C. What is the temperature, in Celsius, of the helium in this balloon when the volume changes to 1098 mL (and the pressure and the number of moles of gas remain constant)?

7. Which has more bubbles a glass of seltzer water in Santa Fe, New Mexico (elevation 7000 ft) or a glass of seltzer water in New Orleans, Louisana (elevation −8 ft)?

8. A sample of nitrogen has a pressure of 12.1 psi and a volume of 53.8 L. What is the volume of the gas if the pressure of the gas changes to 3.2 psi (assume that the number of moles and the temperature remain constant)?

9. A patient had her blood pressure taken. The systolic pressure was 2.32 psi. The diastolic pressure was 0.089 atm. What is this patient's blood pressure in mmHg? Does she have hypertension?

10. If a compound has a low vapor pressure, do you expect it to have a low or high boiling point? Explain your answer.

Chapter 4
Answers to Additional Exercises

4.53   The three physical states of matter are the solid state, the liquid state, and the gas state.

4.55   The solid state and the liquid state have fixed volumes compared to the container. A gas will expand to fill the container completely.

4.57   A rock rolling down a hill is in motion, so it has kinetic energy.

4.59   a. Kinetic energy (the biker is in motion) b. Potential energy (the hiker has stored energy) c. Kinetic energy (the helium atoms are in motion) d. Potential energy (the chemical bonds in the molecules of wax have stored energy)

4.61   Heavy molecules have more kinetic energy, if they have the same average speed.

4.63   In the solid state, molecules have the least amount of kinetic energy.

4.65   Steam molecules have the most kinetic energy because the water molecules are in the gas phase.

4.67 $°F = \left(\dfrac{9}{5}°C\right) + 32 = \left(\dfrac{9}{5} \times 31\right) + 32 = 88 \ °F$

$K = °C + 273 = 31 + 273 = 304 \ K$

You would be wearing summer clothes.

4.69 $°C = \dfrac{5}{9}(°F - 32) = \dfrac{5}{9}(98.6 - 32) = 37.0 \ °C$

4.71 $°C = K - 273 = 2.7 - 273 = -270 \ °C$

$°F = \left(\dfrac{9}{5}°C\right) + 32 = \left(\dfrac{9}{5} \times (-270)\right) + 32 = -454 \ °F$; rounded to -450 °F

4.73   Intermolecular forces are greatest in the solid state.

4.75   A block of aluminum would sink in a container of molten aluminum; the atoms are much more closely packed together in the solid state.

4.77   Intermolecular forces of attraction are affected when a change of state occurs.

4.79   Energy needs to be removed to achieve freezing, condensation, or deposition.

4.81   a. Liquid → gas b. Liquid → solid c. Liquid → solid d. Solid → gas e. Gas → liquid

4.83   Steam causes burns because of the change of state that occurs when steam comes in contact with your skin. Steam condenses when it comes in contact with the skin, which requires heat to be removed from the steam by an amount equivalent to the heat of vaporization of water. The heat is removed from the steam and transferred to your skin. Additional heat is transferred as the liquid water cools from 100 °C to 37 °C.

4.85   Ethanol can form hydrogen bonds, while carbon dioxide cannot. In order for ethanol to enter the gas phase, more energy must be supplied to break the hydrogen bonds holding the ethanol molecules together.

4.87 $\Delta T = 49 - 22 = 27$ °C

a. For brick: Amount of heat $= 0.20 \dfrac{\text{cal}}{\text{g} \cdot \text{°C}} \times 10.9 \text{ g} \times 27 \text{ °C} = 59$ cal

b. For ethanol: Amount of heat $= 0.58 \dfrac{\text{cal}}{\text{g} \cdot \text{°C}} \times 10.9 \text{ g} \times 27 \text{ °C} = 170$ cal

c. For wood: Amount of heat $= 0.10 \dfrac{\text{cal}}{\text{g} \cdot \text{°C}} \times 10.9 \text{ g} \times 27 \text{ °C} = 29$ cal

Ethanol requires the greatest input of heat to warm because ethanol has the largest heat capacity, 0.58 cal/g • °C..

4.89   a. Amount of heat transferred by water cooling:

$\Delta T = 100 - 37 = 63$ °C

Amount of heat $= 1.00 \dfrac{\text{cal}}{\text{g} \cdot \text{°C}} \times 25 \text{ g} \times 63 \text{ °C} = 1{,}575$ cal , rounded to 1600 cal.

b. Amount of heat transferred to condense steam:

Amount of heat = heat of vaporization (cal/g) × mass(g)

Amount of heat = $\dfrac{540 \text{ cal}}{1 \text{ g}} \times 25.0 \text{ g} = 13{,}500$ cal , rounded to 14,000 cal

The amount of energy released to condense 25.0 g of steam is 14,000 cal. The total amount of heat released is 14,000 cal + 1600 cal = 15,600 cal

4.91   Atmospheres are a unit of pressure. Atmospheric pressure is the pressure caused by the molecules of air pressing down on us as a result of gravity.

4.93   a. Identify the conversion:

760 mmHg = 14.70 psi

Express the conversion as a conversion factor:

$\dfrac{14.70 \text{ psi}}{760 \text{ mmHg}}$ or $\dfrac{760 \text{ mmHg}}{14.70 \text{ psi}}$

Set up the calculation so that the supplied units cancel:

$$160 \; \cancel{\text{mmHg}} \times \frac{14.70 \text{ psi}}{760 \; \cancel{\text{mmHg}}} = 3.1 \text{ psi}$$

b. Identify the conversion:

760 mmHg = 760 torr

Express the conversion as a conversion factor:

$$\frac{760 \text{ torr}}{760 \text{ mmHg}} \quad \text{or} \quad \frac{760 \text{ mmHg}}{760 \text{ torr}}$$

Set up the calculation so that the supplied units cancel:

$$110 \; \cancel{\text{mmHg}} \times \frac{760 \text{ torr}}{760 \; \cancel{\text{mmHg}}} = 110 \text{ torr}$$

4.95   Identify the conversion:

760 mmHg = 14.70 psi

Express the conversion as a conversion factor:

$$\frac{14.70 \text{ psi}}{760 \text{ mmHg}} \quad \text{or} \quad \frac{760 \text{ mmHg}}{14.70 \text{ psi}}$$

Set up the calculation so that the supplied units cancel:

$$30 \; \cancel{\text{mmHg}} \times \frac{14.70 \text{ psi}}{760 \; \cancel{\text{mmHg}}} = 0.6 \text{ psi}$$

4.97   Systolic pressure

Identify the conversion:

760 mmHg = 1 atm

Express the conversion as a conversion factor:

$$\frac{760 \text{ mmHg}}{1 \text{ atm}} \quad \text{or} \quad \frac{1 \text{ atm}}{760 \text{ mmHg}}$$

Set up the calculation so that the supplied units cancel:

$$0.23 \; \cancel{\text{atm}} \times \frac{760 \text{ mmHg}}{1 \; \cancel{\text{atm}}} = 170 \text{ mmHg}$$

Diastolic pressure

Identify the conversion

760 mmHg = 14.70 psi

Express the conversion as a conversion factor:

$$\frac{14.70 \text{ psi}}{760 \text{ mmHg}} \quad \text{or} \quad \frac{760 \text{ mmHg}}{14.70 \text{ psi}}$$

Set up the calculation so that the supplied units cancel:

$$2.17 \; \cancel{psi} \times \frac{760 \; mmHg}{14.70 \; \cancel{psi}} = 112 \; mmHg$$

The patient's blood pressure is 170/112. This patient has hypertension.

4.99   Acetone has a higher boiling point than methylene chloride. The vapor pressure of acetone is lower.

4.101   The boiling point of water should be lower at the base camp for Mt. Everest because the atmospheric pressure is significantly lower there.

4.103   One mole of carbon dioxide occupies 22.4 L at STP.

4.105   Express the conversion as a conversion factor:

$$\frac{1 \; mol}{22.4 \; L} \; or \; \frac{22.4 \; L}{1 \; mol}$$

Set up the calculation so that the supplied units cancel:

$$4.1 \; \cancel{mol} \times \frac{22.4 \; L}{1 \; \cancel{mol}} = 92 \; L$$

4.107   Express the conversion as a conversion factor:

$$\frac{1 \; mol}{22.4 \; L} \; or \; \frac{22.4 \; L}{1 \; mol}$$

Set up the calculation so that the supplied units cancel:

$$220 \; \cancel{L} \times \frac{1 \; mol}{22.4 \; \cancel{L}} = 9.8 \; mol$$

4.109   At the lower elevation in Denver, the atmospheric pressure is higher compared to the mountains where you capped the bottle. The pressure of a gas is inversely proportional to the volume of the gas. As the atmospheric pressure increases as you come down from the mountains, the volume of gas in the water bottle decreases.

4.111   Since both P and V are changing, while n and T are constant, we can use Boyle's law: $P_1V_1 = P_2V_2$.

Define the variables and select the variable to solve for.
The problem indicates that P_1 = 142.8 atm, V_1 = 2.24 L, and P_2 = 0.84 atm; therefore, you need to solve for V_2 the final volume.

Algebraically isolate the unknown variable on one side of the equation.
Use algebra to manipulate the equation, so that V_2 is isolated:

$$V_2 = P_1 \times \frac{V_1}{P_2}$$

Substitute the known values into the equation and solve for the unknown variable:

$$V_2 = 142.8 \; \cancel{atm} \times \frac{2.24 \text{ L}}{0.84 \; \cancel{atm}} = 380 \text{ L}$$

4.113   Since both P and V are changing, while n and T are constant, we can use Boyle's law: $P_1V_1 = P_2V_2$.

Define the variables and select the variable to solve for.

The problem indicates that $P_1 = 1$ atm, $V_1 = 3.6$ L, and $P_2 = 5.9$ atm; therefore, you need to solve for V_2 the final volume.

Algebraically isolate the unknown variable on one side of the equation.

Use algebra to manipulate the equation, so that V_2 is isolated:

$$V_2 = P_1 \times \frac{V_1}{P_2}$$

Substitute the known values into the equation and solve for the unknown variable:

$$V_2 = 1 \; \cancel{atm} \times \frac{3.6 \text{ L}}{5.9 \; \cancel{atm}} = 0.6 \text{ L}$$

4.115   Upon exhalation, the pressure of the lungs *increases* as the volume of the lungs *decreases*.

4.117   Gay-Lussac's law states that the pressure of a gas is directly proportional to the temperature of a gas. If you heat the can of unopened beans directly on the stove, the temperature of the gas inside the can will increase. As the temperature of the gas increases, the pressure of the gas will increase, which could cause the can of beans to explode.

4.119   Temperatures are higher in the summer, and pressure is directly proportional to temperature, so the pressure inside the tires will be higher in summer.

4.121   Since both P and T are changing, while n and V are constant, we can use Gay-Lussac's law: $\dfrac{P_1}{T_1} = \dfrac{P_2}{T_2}$.

Define the variables and select the variable to solve for.

The problem indicates that P_1 = 30.1 psi, T_1 = 20.0 °C, and T_2 = 35.5 °C; therefore, you need to solve for P_2, the final pressure.

Algebraically isolate the unknown variable on one side of the equation.
Use algebra to manipulate the equation, so that P_2 is isolated:

$$P_2 = T_2 \times \frac{P_1}{T_1}$$

Substitute the known values into the equation and solve for the unknown variable.
REMEMBER, YOU MUST FIRST CONVERT TEMPERATURES TO KELVIN!
T_1 = 20.0 °C + 273 = 293 K
T_2 = 35.5 °C + 273 = 309 K

$$P_2 = 309 \cancel{K} \times \frac{30.1 \text{ psi}}{293 \cancel{K}} = 31.7 \text{ psi}$$

4.123   Charles' law states that temperature and volume are directly proportional to each other. As you heat the cake in the oven, the carbon dioxide in the cake heats up; therefore, the volume of the carbon dioxide increases and the cake rises.

4.125   Since both V and T are changing, while n and P are constant, we can use Charles'

law: $\frac{V_1}{T_1} = \frac{V_2}{T_2}$.

Define the variables and select the variable to solve for.
The problem indicates that V_1 = 8.7 L, T_1 = 12 °C, and T_2 = 34 °C; therefore, you need to solve for V_2, the final volume.

Algebraically isolate the unknown variable on one side of the equation.
Use algebra to manipulate the equation, so that V_2 is isolated:

$$V_2 = T_2 \times \frac{V_1}{T_1}$$

Substitute the known values into the equation and solve for the unknown variable.
REMEMBER, YOU MUST FIRST CONVERT TEMPERATURES TO KELVIN!
T_1 = 12 °C + 273 = 285 K
T_2 = 34 °C + 273 = 307 K

$$V_2 = 307 \cancel{K} \times \frac{8.7 \text{ L}}{285 \cancel{K}} = 9.4 \text{ L}$$

4.127   As the air inside of the balloon is heated up, the volume of the air increases (Charles' law). The number of molecules of air in the balloon stays the same, but the density (mass/volume) of the heated air decreases. The less dense air inside the balloon will cause the balloon to float.

4.129   Since V, P, and T are changing, while n is constant, we can use the combined gas law: $\dfrac{P_1V_1}{T_1} = \dfrac{P_2V_2}{T_2}$.

Define the variables and select the variable to solve for.

The problem indicates that $V_1 = 2.25$ L, $T_1 = 12.0\ °C$, $P_1 = 2.7$ atm, $P_2 = 100.$ kPa and $T_2 = 0.00\ °C$; therefore, you need to solve for V_2, the final volume.

Algebraically isolate the unknown variable on one side of the equation.

Use algebra to manipulate the equation, so that V_2 is isolated:

$$V_2 = \frac{T_2 \times P_1 \times V_1}{P_2 \times T_1}$$

Substitute the known values into the equation and solve for the unknown variable.

REMEMBER, YOU MUST FIRST CONVERT TEMPERATURES TO KELVIN!

$T_1 = 12.0\ °C + 273 = 285$ K

$T_2 = 0.00\ °C + 273 = 273$ K

The pressure units need to be the same, so kPa needs to be converted to atm:

1 atm = 101.3 kPa

$$100.\ \text{kPa} \times \frac{1\ \text{atm}}{101.3\ \text{kPa}} = 0.987\ \text{atm}$$

$$V_2 = \frac{273\ \cancel{K} \times 2.7\ \cancel{\text{atm}} \times 2.25\ L}{0.987\ \cancel{\text{atm}} \times 285\ \cancel{K}} = 5.9\ L$$

4.131   Dalton's law states that each gas in a mixture of gases will exert a pressure independent of the other gases present; and each gas will behave as if it alone occupied the total volume. The sum of the partial pressures of each gas present is equal to the total pressure.

4.133   Since partial pressures are provided, use the following equation:

$$P_{TOT} = P_{neon} + P_{argon}$$

Define the variables and select the variable to solve for.

The problem indicates that P_{TOT} = 2.42 atm and P_{neon} = 1.81 atm; therefore we need find P_{argon}.

Algebraically isolate the unknown variable on one side of the equation:

$$P_{argon} = P_{TOT} - P_{neon}$$

Substitute the known variables into the equation and solve for the unknown variable:

$$P_{argon} = P_{TOT} - P_{neon} = 2.42 \text{ atm} - 1.81 \text{ atm} = 0.61 \text{ atm}$$

4.135   The person who is SCUBA diving is under greater pressure and therefore will have a higher concentration of oxygen in their blood, because concentration is proportional to pressure.

4.137   Desflurane has a larger Henry's constant than diethyl ether. The concentration of desflurane will be lower in the blood; therefore, the patient would regain consciousness more quickly.

4.139   The increase in pressure within the chamber causes the nitrogen bubbles to re-dissolve. The redissolved nitrogen circulates to the lungs where it can safely be exhaled. Henry's law shows that there is a direct relationship between the pressure and the concentration of a gas in solution. As the pressure increases, the concentration of nitrogen in the blood should increase.

4.141   Carbon monoxide is poisonous because it binds to hemoglobin more strongly than oxygen. When carbon monoxide enters the blood stream, it binds with hemoglobin replacing the oxygen, and the level of oxygen available to tissues drops to dangerous levels.

4.143   a.   Since both P and V are changing, while n and T are constant, we can use Boyle's law: $P_1V_1 = P_2V_2$.

Define the variables and select the variable to solve for.

The problem indicates that P_1 = 2.81 atm, V_1 = 0.021 mL, and P_2 = 1.00 atm; therefore, you need to solve for V_2, the final volume.

Algebraically isolate the unknown variable on one side of the equation.

Use algebra to manipulate the equation, so that V_2 is isolated:

$$V_2 = P_1 \times \frac{V_1}{P_2}$$

Substitute the known values into the equation and solve for the unknown variable:

$$V_2 = 2.81 \; \cancel{atm} \times \frac{0.021 \; mL}{1.00 \; \cancel{atm}} = 0.059 \; mL$$

b. Since both P and V are changing, while n and T are constant, we can use Boyle's law: $P_1 V_1 = P_2 V_2$.

Define the variables and select the variable to solve for.

The problem indicates that P_1 = 1.00 atm, V_1 = 0.059 mL, and P_2 = 2.25 atm; therefore, you need to solve for V_2, the final volume.

Algebraically isolate the unknown variable on one side of the equation.

Use algebra to manipulate the equation, so that V_2 is isolated:

$$V_2 = P_1 \times \frac{V_1}{P_2}$$

Substitute the known values into the equation and solve for the unknown variable:

$$V_2 = 1.00 \; \cancel{atm} \times \frac{0.059 \; mL}{2.25 \; \cancel{atm}} = 0.026 \; mL$$

4.145 We need to determine the volume of the gas at STP, and then use that volume to determine the number of moles of carbon dioxide. Since V, P, and T are changing, while n is constant, we can use the combined gas law: $\dfrac{P_1 V_1}{T_1} = \dfrac{P_2 V_2}{T_2}$.

Define the variables and select the variable to solve for.

The problem indicates that V_1 = 5.58 L, T_1 = 21 °C, P_1 = 25 mmHg, P_2 = 760 mmHg and T_2 = 0 °C; therefore, you need to solve for V_2, the final volume.

Algebraically isolate the unknown variable on one side of the equation.

Use algebra to manipulate the equation, so that V_2 is isolated:

$$V_2 = \frac{T_2 \times P_1 \times V_1}{P_2 \times T_1}$$

Substitute the known values into the equation and solve for the unknown variable.

REMEMBER, YOU MUST FIRST CONVERT TEMPERATURES TO KELVIN!

T_1 = 21 °C + 273 = 294 K

T_2 = 0 °C + 273 = 273 K

$$V_2 = \frac{273 \ \cancel{K} \times 25 \ \cancel{mmHg} \times 5.58 \ L}{760 \ \cancel{mmHg} \times 294 \ \cancel{K}} = 0.17 \ L \ \text{at STP}$$

Convert this volume into number of moles.

Express the conversion as a conversion factor:

$$\frac{1 \ mol}{22.4 \ L} \ \text{or} \ \frac{22.4 \ L}{1 \ mol}$$

Set up the calculation so that the supplied units cancel:

$$0.17 \ \cancel{L} \times \frac{1 \ mol}{22.4 \ \cancel{L}} = 7.6 \times 10^{-3} \ mol$$

Chapter 5

Solutions, Colloids, and Membranes

Chapter Summary

In this chapter, you have learned about how scientists and medical professionals classify and quantify the compounds and elements that make up a mixture. You have also learned how to distinguish mixtures and different types of solutions. Knowing how to interpret concentrations of solutions and understanding how to use concentrations play an important role in the health field, from preparing IV solutions to delivering correct drug dosages to patients. Understanding concentrations will provide insight into dialysis, osmosis, and other biochemical pathways that use membranes.

Section 5.1 Mixtures and Solutions

In a Nutshell: Solutions

There are two general classes of mixtures: heterogeneous and homogeneous. In a heterogeneous mixture, the components are unevenly distributed throughout the mixture. In a homogeneous mixture, the components are evenly distributed throughout the mixture. A homogeneous mixture is also known as a solution.

A solution consists of a solvent and one or more solutes. The solute is the substance that is present in the lesser amount and the solvent is the substance that is present in the greatest amount. A solution can contain more than one solute in a solvent. Polar solvents will dissolve polar solutes, and nonpolar solvents will dissolve nonpolar solutes, as characterized by the phrase "like dissolve like." When two liquids do no mix, they are said to be insoluble in one another, or immiscible. Most biological solutions consist of molecules or ions dissolved in water, known as aqueous solutions.

Worked Example #1

Identify the solute and solvent in each solution.

   a.   500 g of ethanol containing 2 g of iodine

   b.   50 mL of CO_2 and 245 mL of H_2O

   c.   3 g KCl and 235 g of water

In each case, identify which substance is present in the smallest amount and which substance is present in the greatest amount. The substance present in the greatest amount is the solvent, while the substance present in the smallest amount is the solute.

a. *Substance present in the smallest amount: iodine*

Substance present in the greatest amount: ethanol

Solute: iodine

Solvent: ethanol

b. *Substance present in the smallest amount: CO_2*

Substance present in the greatest amount: H_2O

Solute: CO_2

Solvent: H_2O

c. *Substance present in the smallest amount: KCl*

Substance present in the greatest amount: water

Solute: KCl

Solvent: water

Try It Yourself #1

Identify the solute and solvent in each solution.

a. 125 g of Cu and 7 g of Sn
b. 150 mL of ethanol and 25 mL of CO_2
c. 27 g O_2 and 235 g of water

a. *Substance present in the smallest amount:* _____

Substance present in the greatest amount: _____

Solute: _____

Solvent: _____

b. *Substance present in the smallest amount:* _____

Substance present in the greatest amount: _____

Solute: _____

    Solvent: _____

c. *Substance present in the smallest amount:* _____

    Substance present in the greatest amount: _____

    Solute: _____

    Solvent: _____

In a Nutshell: Molecules as Solutes

There are three important types of solutes present in biological solutions: molecules, ions, and gases. When a molecule dissolves in a solvent, the covalent bonds remain intact. When dissolved, the chemical structure of a molecule exists exactly as it did in its pure form, except that it surrounded by solvent molecules, rather than other solute molecules.

In a Nutshell: Ions as Solutes

The crystal lattice of most ionic compounds dissolves readily in water to form individual ions surrounded by many water molecules. A substance that produces ions when dissolved in water is known as an electrolyte. Solutions containing electrolytes conduct an electric current. The formula unit of an ionic substance allows you to determine which ions are produced and in what proportions when the ionic compound dissolves in water and forms an aqueous solution.

In a Nutshell: Gases as Solutes

Molecules in the gas phase can also dissolve in water. Blood contains the important dissolved gases oxygen, O_2, and carbon dioxide, CO_2.

Worked Example #2

For the following solutions, indicate whether the solute is a molecule or an electrolyte.

    a. $CaCl_2$

    b. sucrose

Identify the type of bond in type of bond in the solute; then determine the type of solute.

a. *CaCl₂ has an ionic bond. It produces ions when dissolved in water. Therefore, it is an electrolyte.*

$$CaCl_2 \ (s) \longrightarrow Ca^{2+} \ (aq) \ + \ 2\,Cl^- \ (aq)$$

(1 calcium ion)   *(2 chloride ions)*

b. *Sucrose has covalent bonds. Therefore, it is a molecule.*

Try It Yourself #2

For the following solutions, indicate whether the solute is a molecule or an electrolyte.

a. O_2

b. dextrose $(C_6H_{12}O_6)$

c. KCl

a. *Type of bond:* _____

Molecule or electrolyte: _____

b. *Type of bond:* _____

Molecule or electrolyte: _____

c. *Type of bond:* _____

Molecule or electrolyte: _____

Worked Example #3

How many ions of each type will be produced when $Ca(HCO_3)_2$ is dissolved in aqueous solution? Would you describe the solutes in $Ca(HCO_3)_2$ (aq) as molecules, electrolytes, or gases? Explain.

The formula unit for calcium hydrogen carbonate ($Ca(HCO_3)_2$) indicates that there is one calcium ion for every two polyatomic hydrogen carbonate anions (HCO_3^-) released:

$$Ca(HCO_3)_2 \ (s) \longrightarrow Ca^{2+} \ (aq) \ + \ 2\,HCO_3^- \ (aq)$$

(1 calcium ion)   *(2 hydrogen carbonate ions)*

Formula unit for calcium hydrogen carbonate: $Ca(HCO_3)_2$

Cation: Ca^{2+}

Number of cations produced: one

Anion: HCO$_3^-$

Number of anions produced: two

Ca(HCO$_3$)$_2$ produces ions when dissolved in water; therefore, it is an electrolyte.

Try It Yourself #3

How many ions of each type will be produced when Ca$_3$(PO$_4$)$_2$ is dissolved in aqueous solution?

Tools: Table 5-2

Formula unit for calcium phosphate: _____

Cation: _____

Number of cations produced: _____

Anion: _____

Number of anions produced: _____

Practice Problems for Mixtures and Solutions

Identify the solute and solvent for the following solutions. Is the solute a molecule, an electrolyte, or a gas? If the solute is an electrolyte, what are the ions produced and how many of each ion are produced?

    a.   A teaspoon of sugar and a teaspoon of cream dissolved in a cup of coffee

    b.   25 mL of water and 50 L of air

c. 1.23 g Na_3PO_4 and 235 g of water

d. 500 mL ethanol and 34 mL of CO_2

e. 175 mL water and 7 g $Mg(HCO_3)_2$

Section 5.2: Solution Concentrations

In a Nutshell: Measures of Solution Concentration

The concentration of a solution is a quantitative measure of how much solute is dissolved in a given quantity of solution. Concentration is always expressed as a ratio, where the amount of solute is given in the numerator and the total amount of solution (amount of solvent plus amount of solute) is given in the denominator. Concentrations are usually based on the mass of solute or the moles of solute. The units of concentration most commonly encountered in the medical field are mass/volume (m/v), % mass/volume (%m/v), moles/volume, and equivalents/volume.

In a Nutshell: mass/volume

When concentration is expressed as a ratio of mass to volume, the units of the solute are based on mass. The general form of the equation for determining a mass/volume concentration follows:

mass/volume

$$\frac{\text{mass}}{\text{volume}} = \frac{\text{mass of solute}}{\text{volume of solution}}$$

To solve solution concentration problems, we use a process called dimensional analysis. Recall that we used this process in Chapter 1.   Step 1: Express each conversion as two possible conversion factors. Step 2: Set up the calculation so that the supplied units cancel. As a final step in any calculation, you should double check that you have reported the correct number of significant figures in the final numerical answer.

Worked Example #4

Auralgan is used as a topical decongestant and analgesic. It comes in solution form and contains 54 mg of antipyrine per mL. What is the concentration of this solution in g/dL?

Express each conversion as two possible conversion factors.
In this problem, there are two conversions needed. The supplied unit is mg/mL, and the requested unit is g/dL. Therefore, we need to convert mg to g and mL to dL in order to express this concentration in g/dL. The two conversion factors are:

1st *conversion factor:* $\dfrac{1\,\text{mg}}{1\times10^{-3}\text{g}}$ or $\dfrac{1\times10^{-3}\text{g}}{1\,\text{mg}}$

and

2nd *conversion factor:* $\dfrac{1000\,\text{mL}}{10\,\text{dL}}$ or $\dfrac{10\,\text{dL}}{1000\,\text{mL}}$

Set up the calculation so that the supplied units cancel.
Begin with the supplied concentration. Choose the appropriate conversion factors for the numerator and denominator so that the supplied units (mg/mL) cancel and you are left with the answer in the requested units (g/dL). This calculation can be performed in one step as shown below:

$$\frac{54\,\text{mg}}{1\,\text{mL}} \times \frac{1\times10^{-3}\text{g}}{1\,\text{mg}} \times \frac{1000\,\text{mL}}{10\,\text{dL}} = 5.4\,\text{g/dL}$$

Check that the answer has the correct number of significant figures.

Try It Yourself #4

Amprenavir, used to treat HIV-1, comes as oral solution with a concentration of 15 mg of amprenavir per mL. What is this concentration in μg/L?

Express each conversion as two possible conversion factors.

Set up the calculation so that the supplied units cancel.

In a Nutshell: % mass/volume

When concentration is expressed as a ratio of % mass to volume, the units of the solute are based on mass. The % mass/volume is the ratio of mass of solute in grams to volume of solution in milliliters, reported as a percentage. The general form of the equation for determining a %mass/volume concentration is shown below:

$$\% \frac{mass}{volume} = \frac{g \text{ of solute}}{mL \text{ of solution}} \times 100$$

Remember that the units of % mass/volume must be in g/mL. Many solutions used in IV therapy have their concentrations reported in %m/v.

Worked Example #5

What is the % m/v of sucrose in a carbonated beverage that contains 39 g of sucrose in 355 mL of beverage?

The supplied units are g/mL. We will use the equation for % mass/volume.

$$\%\frac{\text{mass}}{\text{volume}} = \frac{\text{g of solute}}{\text{mL of solution}} \times 100$$

$$\%\frac{\text{mass}}{\text{volume}} = \frac{39 \text{ g sucrose}}{355 \text{ mL of beverage}} \times 100 = 11\% \text{ (m/v) sucrose}$$

Try It Yourself #5

What is the % m/v of I_2 in a solution prepared by dissolving 2.35 g I_2 in enough ethanol to make 155 mL of solution?

Tools: equation for % mass/volume

Worked Example #6

You have been asked to prepare 1.75 L of 0.90% NaCl (% m/v) solution for IV therapy. How many g of NaCl should you weigh out?

Express the conversions as conversion factors.
In this problem, the supplied units are liters and % solution. The requested unit is grams of solute. You will need to use the concentration as a conversion factor between the volume of solution (supplied unit) and the mass of solute (requested unit). You will also need to metric conversion between L and mL.

1st conversion factor: $\quad \dfrac{0.90 \text{ g}}{100 \text{ mL}} \quad$ or $\quad \dfrac{100 \text{ mL}}{0.90 \text{ g}}$

and

2nd conversion factor: $\quad \dfrac{1\ mL}{10^{-3}\ L} \quad$ or $\quad \dfrac{10^{-3}\ L}{1\ mL}$

Set up the calculation, so that the supplied units cancel.

Begin with the supplied concentration. Choose the appropriate conversion factors for the numerator and denominator, so that the supplied units cancel and you are left with the answer in the requested units. This calculation can be performed in one step as shown below:

$$1.75 \, \cancel{L} \times \frac{1 \, \cancel{mL}}{10^{-3} \, \cancel{L}} \times \frac{0.90 \, g}{100 \, \cancel{mL}} = 16 \, g$$

Try It Yourself #6

You have been asked to prepare 2.5 L of a 5% dextrose solution. How many g of dextrose should you weigh out?

Supplied units: _____

Express the conversions as conversion factors.

Set up the calculation, so that the supplied units cancel.

In a Nutshell: Molarity

Molarity, *M*, is defined as the number of moles of solute to 1 liter of solution: moles/liter.

$$\text{Molarity (M)} = \frac{\text{moles of solute}}{\text{L of solution}}$$

Worked Example #7

A solution having a volume of 5.8 L contains 34 mmol of CO_2 (carbon dioxide). What is the concentration of carbon dioxide in the solution in mol/L?

The supplied units are mmol and liters and the requested unit is mol/L.

Express the conversions as conversion factors:

$$\frac{1 \text{ mmol}}{10^{-3} \text{ mol}} \quad \text{or} \quad \frac{10^{-3} \text{ mol}}{1 \text{ mmol}}$$

Set up the calculation, so that the supplied units cancel:

$$M = \frac{34 \text{ mmol}}{5.8 \text{ L}} \times \frac{10^{-3} \text{ mol}}{1 \text{ mmol}} = 0.0059 \text{ mol/L}$$

Try It Yourself #7

A solution having a volume of 2.4 L contains 253 mmol of O_2. What is the concentration of oxygen in the solution in mol/L?

Supplied units: _____

Requested unit: _____

Express the conversions as conversion factors.

Set up the calculation, so that the supplied units cancel.

Worked Example #8

How many moles of potassium ions (K^+) are there in 3.2 L of 45 mM KCl?

The supplied units are liters and mmol/liters. The requested unit is moles.

Express the conversions as conversion factors.

1st conversion factor: $\quad \frac{45 \text{ mmol}}{L} \quad \text{or} \quad \frac{L}{45 \text{ mmol}}$

and

2nd conversion factor: $\dfrac{1\ \text{mmol}}{10^{-3}\ \text{mol}}$ or $\dfrac{10^{-3}\ \text{mol}}{1\ \text{mmol}}$

Set up the calculation, so that the supplied units cancel.

$$3.2\ \cancel{\text{L}} \times \dfrac{45\ \cancel{\text{mmol}}}{\cancel{\text{L}}} \times \dfrac{10^{-3}\ \text{mol}}{1\ \cancel{\text{mmol}}} = 0.14\ \text{moles}$$

Try It Yourself #8

How many moles of sodium ions (Na^+) are there in 7.2 L of 5.1 mM Na_3PO_4?

Supplied units: _____

Requested units: _____

Express the conversions as conversion factors.

Set up the calculation, so that the supplied units cancel.

In a Nutshell: equivalents/liter

Concentrations of ions are often expressed in equivalents /liter because this unit of concentration provides you with information about the number of charges per liter of solution. An equivalent, abbreviated eq, is the number of moles of charge. It is calculated by multiplying the number of moles of ion by the charge on the ion.

$$\text{Equivalents (eq)} = \text{moles of ion} \times \text{charge on ion}$$

To express the concentration of a solute using equivalents, write the number of equivalents of solute per liter of solution:

$$\frac{\text{equivalents}}{\text{L}} = \left(\frac{\text{eq}}{\text{L}}\right) = \frac{\text{equivalents of solute}}{\text{L of solution}}$$

Worked Example #9

A patient's blood test comes back with an iron level of 15 mmol/L. Convert this value to units of meq/L Fe^{3+}.

The supplied units are mmol/L. The requested units are meq/L. Use Table 5-4 to determine the number of equivalents per mole.

Express conversions as conversion factors:

$$\frac{3\,\text{meq}}{1\,\text{mmol}}Fe^{3+} \quad \text{or} \quad \frac{1\,\text{mmol}}{3\,\text{meq}}Fe^{3+}$$

Set up the calculation, so that the supplied units cancel:

$$\frac{15\,\cancel{\text{mmol}}}{1\,\text{L}} \times \frac{3\,\text{meq}}{1\,\cancel{\text{mmol}}}Fe^{3+} = 45\frac{\text{meq}}{\text{L}}Fe^{3+}$$

Try It Yourself #9

A patient's blood test comes back with a calcium level of 1.2 mmol/L. Convert this value to units of meq/L.

Tools: Table 5-4

Supplied units: _____

Requested units: _____

Express conversions as conversion factors.

Set up the calculation, so that the supplied units cancel.

Worked Example #10

 a. How many equivalents/L of citrate ion ($C_6H_4O_7{}^{3-}$) are there in a 0.54 M solution of $Na_3C_6H_4O_7$?

 b. How many eq/L of Na^+ are there in this solution?

Tools: Table 5-4

 a. *The supplied units are mol/L. The requested units are eq/L.*

  Express conversions as conversion factors:

$$\frac{3\ eq}{1\ mol}C_6H_4O_7^{3-} \quad or \quad \frac{1\ mol}{3\ eq}C_6H_4O_7^{3-}$$

  Set up the calculation, so that the supplied units cancel:

$$\frac{0.54\ \cancel{mol}}{1\ L} \times \frac{3\ eq}{1\ \cancel{mol}}C_6H_4O_7^{3-} = 1.6\ \frac{eq}{L}C_6H_4O_7^{3-}$$

 b. *Since the solution of $Na_3C_6H_4O_7$ is neutral, the eq/L of Na^+ ions in this solution must be equal to the eq/L of $C_6H_4O_7{}^{3-}$ ions.   Therefore, there are 1.6 eq/L Na^+ ions.*

Try It Yourself #10

 a. How many eq/L of calcium ions (Ca^{2+}) are there in a 1.8 M solution of $CaCl_2$?

 b. How many eq/L of chloride ions (Cl^-) are there in this solution?

Tools: Table 5-4

 Supplied units: _____

 Requested units: _____

 Express conversions as conversion factors.

 Set up the calculation, so that the supplied units cancel.

In a Nutshell: Dosage Calculations

Calculating dosages of drugs delivered as solutions is an important part of the healthcare worker's responsibility. Oral medications are commonly administered in liquid form and the total volume and drip rate of IV solutions must be calculated. At times, conversion between units of measurement may be necessary.

In oral suspensions, a drug in solid form is suspended as fine particles throughout a liquid. We will use dimensional analysis to calculate the proper dosage to give to a patient for oral suspensions. The concentration of the suspension can be used as a conversion factor between the mass of drug and volume of solution.

Worked Example #11

Gantrisin®, an antibiotic used to treat urinary tract infections, is prescribed for a child. The dosage is 2045 mg to be administered in a 24-hour period.   The suspension contains 500 mg in every 5 mL. How many mL of the suspension should be administered to the patient in the 24-hour period?

The supplied unit is 2045 mg of the drug. The requested unit is mL of suspension.

Express conversions as conversion factors.
You are given the concentration of the drug as m/v: 500 mg gantrisin per 5 mL of suspension. The concentration can be used as a conversion factor between the mass of drug and volume of solution:

$$\frac{500 \text{ mg}}{5 \text{ mL}} \quad \text{or} \quad \frac{5 \text{ mL}}{500 \text{ mg}}$$

Set up the calculation, so that the supplied units cancel.

$$2045 \text{ mg} \times \frac{5 \text{ mL}}{500 \text{ mg}} = 20 \text{ mL in 24 hrs}$$

Try It Yourself #11

Aristospan®, used to treat bursitis, is a suspension that contains 20 mg in every milliliter. A patient requires an injection of 15 mg of aristospan. How many mL should you administer to the patient?

Supplied units: _____

Requested units: _____

Express conversions as conversion factors.

Set up the calculations, so that the supplied units cancel.

In a Nutshell: Dosage of IV Solutions

Certain IV medications must be administered at a specific dosage per unit of time, known as flow rate. Flow rates are often given in mg per minute, µg per minute, or units per hour. Calculations can be made to determine flow rate or, when the flow rate is known, to assess the dosage being administered per unit time.

Worked Example #12

Levaquin® is a powerful antibiotic used to treat pneumonia. An order is given to administer 250 mg levaquin by IV over a period of one hour. The IV bag contains 5 mg levaquin in every 1 mL. What should the flow rate be in mL/min?

Express conversions as conversion factors.

There are two values that can be used as conversion factors: 250 mg per hour and 5 mg per 1 mL. The requested unit is the ratio mL/min. You will also need the conversion of hr to min.

1st *conversion factor:* $\dfrac{250 \text{ mg}}{1 \text{ hr}}$ or $\dfrac{1 \text{ hr}}{250 \text{ mg}}$

2nd *conversion factor:* $\dfrac{5 \text{ mg}}{\text{mL}}$ or $\dfrac{\text{mL}}{5 \text{ mg}}$

3rd *conversion factor:* $\dfrac{60 \text{ min}}{\text{hr}}$ or $\dfrac{\text{hr}}{60 \text{ min}}$

Set up the calculation, so that the supplied units cancel.

Remember that the answer should be in units of mL/min.

$$\frac{250 \text{ \sout{mg}}}{\text{\sout{hr}}} \times \frac{\text{\sout{hr}}}{60 \text{ min}} \times \frac{1 \text{ mL}}{5 \text{ \sout{mg}}} = 0.8 \text{ mL/min}$$

Try It Yourself #12

Retrovir IV is used to treat HIV patients. An order is given to infuse 64 mg of retrovir IV per hour. The concentration of the IV solution is 4 mg per mL. What should the flow rate be in mL/min?

Express conversions as conversion factors.

Set up the calculation so that the supplied units cancel.

Worked Example #13

Intravenous gammaglobulin is used to treat immunodeficiency orders. A patient requires 20 grams to be administered. The patient receives the drug at a flow rate of 50.9 mg per minute. How long (in hours) will the infusion last?

Express conversions as conversion factors.

You are supplied with the total amount of the drug, 20 grams. You can also use the flow rate as a conversion factor. You also need to convert min to hr and g to mg.

1st conversion factor: $\dfrac{50.9 \text{ mg}}{\text{min}}$ or $\dfrac{\text{min}}{50.9 \text{ mg}}$

2nd conversion factor: $\dfrac{60 \text{ min}}{\text{hr}}$ or $\dfrac{\text{hr}}{60 \text{min}}$

3rd conversion factor: $\dfrac{1 \text{ g}}{1000 \text{ mg}}$ or $\dfrac{1000 \text{ mg}}{1 \text{ g}}$

Set up the calculation, so that the units cancel.

$$20 \ \cancel{g} \times \frac{1000 \ \cancel{mg}}{1 \ \cancel{g}} \times \frac{\cancel{min}}{50.9 \ \cancel{mg}} \times \frac{1 \text{ hr}}{60 \ \cancel{min}} = 6.5 \text{ hrs}$$

Try It Yourself #13

Brevibloc is used for rapid control of the ventricular rate during surgery. An order is given to administer brevibloc at a rate of 3.2 mg per minute. The concentration of the solution is 10 mg brevibloc per mL. This drug is usually administered for a 4-minute time period. How many mL were delivered in 4 minutes?

Supplied units: _____
Requested units: _____

Express conversions as conversion factors.

Set up the calculation, so that the units cancel.

Practice Problems for Solution Concentrations

1. Every 5 mL of an oral suspension of amoxicillin contains 200 mg of amoxicillin. What is the concentration of the oral suspension in g/dL?

2. You are asked to prepare 2.75 L of 3.3% dextrose (% m/v) solution for IV therapy. How much dextrose do you need to weigh out?

3. A patient's blood test shows that there is 0.9 mmol/L of magnesium in her blood. What is this concentration in mol/L? How many meq/L of Mg^{2+} are there in her blood?

4. Indocin IV is used to close the ductus arteriosus, an arterial shunt, in premature infants. An order is given to administer 0.238 mg of indocin by IV over a period of 30 minutes. The concentration of the indocin solution is 0.05 mg/mL. What should the flow rate be in mL/min?

Extension Topic 5-1: Converting between Mass and Mole-Based Concentration Units

Often it is convenient to know the number of solute molecules or ions in a solution, expressed in moles rather than the mass of the solute molecules or ions. In order to convert the mass of a solute into the moles of solute, you will need to know the molar mass of the solute. To find the number of moles of an ion, first you must find the number of moles of the compound that contains the ion and then convert that number to moles of ion.

Worked Example #14

Calculate the molarity (mol/L) of Cl^- in half-normal saline solution, which is 0.45% (m/v) NaCl.

Rewrite the % m/v concentration as a fraction, such that you have a specific mass divided by a specific volume.

$$0.45\% \text{ (m/v) NaCl} = \frac{0.45 \text{ g NaCl}}{100 \text{ mL solution}}$$

Convert mass into moles.

Convert the numerator, 0.45 g NaCl, into moles NaCl. Use the molar mass of NaCl, 58.44 g/mol, as the conversion factor. Note, we are not changing the denominator (the 100 mL solution) yet.

$$\frac{0.45 \text{ g NaCl}}{100 \text{ mL solution}} \times \frac{1 \text{ mol NaCl}}{58.44 \text{ g NaCl}} = \frac{7.7 \times 10^{-3} \text{ mol NaCl}}{100 \text{ mL solution}}$$

Convert moles of the compound to moles of the requested ion.

Use the subscripts in the chemical formula to determine the ratio of moles of ion to moles of compound. The chemical formula for NaCl shows that there is 1 mol of Cl^- ions for every 1 mol of NaCl, $\dfrac{1 \text{ mol } Cl^-}{1 \text{ mol NaCl}}$.

$$\frac{7.7 \times 10^{-3} \text{ mol NaCl}}{100 \text{ mL solution}} \times \frac{1 \text{ mol } Cl^-}{1 \text{ mol NaCl}} = \frac{7.7 \times 10^{-3} \text{ mol } Cl^-}{100 \text{ mL solution}}$$

Convert the supplied volume units into the requested volume units.

In this problem you convert the denominator, 100 mL solution, into liters of solution. Use the conversion factor 1 mL = 10^{-3} liter.

$$\frac{7.7\times10^{-3}\ \text{mol Cl}^-}{100\ \cancel{\text{mL}}} \times \frac{1\ \cancel{\text{mL}}}{10^{-3}\ \text{L}} = \frac{0.077\ \text{mol Cl}^-}{\text{L}}$$

Try It Yourself #14

Calculate the molarity (mol/L) of Na^+ ions in a 45 mg/mL solution of Na_3PO_4.

Rewrite the concentration as a fraction with the volume in the denominator.

Convert mass into moles in the numerator. (Use the molar mass of the compound.)

Convert moles of compound to moles of ion. (Use the chemical formula to determine the ratio of ions to compound.)

Convert the supplied volume units into the requested volume units.

Worked Example #15

What is the concentration (in g/mL) of dextrose in a 35 mM dextrose solution? The molar mass of dextrose is 180.2 g/mol.

Rewrite the molarity as a fraction, such that you have a specific number of moles divided a specific volume:

$$35 \text{ mM dextrose} = \frac{35 \text{ mmol dextrose}}{1 \text{ L solution}}$$

Convert moles to mass.

Convert the numerator, mmol of dextrose, to g of dextrose. Use the molar mass of dextrose (180.2g/mol) as a conversion factor. You will also need the conversion factor, 1 mmol = 10^{-3} mol.

$$\frac{35 \text{ mmol dextrose}}{1 \text{ L solution}} \times \frac{10^{-3} \text{ mol dextrose}}{1 \text{ mmol dextrose}} \times \frac{180.2 \text{ g dextrose}}{1 \text{ mol dextrose}} = \frac{6.3 \text{ g dextrose}}{1 \text{ L solution}}$$

Convert supplied volume units to requested volume units.

In this problem, you convert the denominator, 1 liter of solution, into mL of solution. Use the conversion factor 1 mL = 10^{-3} liter.

$$\frac{6.3 \text{ g dextrose}}{1 \text{ L solution}} \times \frac{10^{-3} \text{ L}}{1 \text{ mL}} = \frac{6.3 \times 10^{-3} \text{ g dextrose}}{1 \text{ mL}}$$

Try It Yourself #15

What is the % (m/v) $NaHCO_3$ in a 4.2 mM sodium bicarbonate solution? The molar mass of sodium carbonate is 84.0 g/mol.

Rewrite the concentration as a fraction with the volume in the denominator.

Convert moles into mass in the numerator. (Use the molar mass of the compound.)

Convert the supplied volume units into the requested volume units.

Convert mass/volume (g/mL) to a percentage by multiplying by 100.

Practice Problems for Converting Between Mass and Mole-Based Concentration Units

1. Calculate the molarity of a Tylenol suspension that contains 160 mg of Tylenol in every 4 mL. The molar mass of Tylenol is 151.16 g/mol.

2. What is the % (m/v) iodine in a 0.12 M tincture of iodine solution? The molar mass of iodine is 253.81 g/mol.

Section 5.3: Colloids and Suspensions

In a Nutshell: Colloids

Colloids and suspensions are two other types of mixtures. A colloid is a homogeneous mixture. It contains particles that are much larger than the solute particles of a solution. The size of the colloidal particles is between 1 nm and 1 μm. The component of the colloid that is present in the greatest amount is known as the medium. In a colloid, the particles and the continuous medium can be either a gas, a liquid, or a solid; however, both cannot be gases.

In a Nutshell: Suspensions

A suspension is a heterogeneous mixture. The particles in a suspension are not uniformly distributed throughout the medium and eventually settle. The particles in a suspension are larger than 1 μm and can be filtered out. The component of the suspension that is present in the greatest amount is the dispersion medium. In a suspension, the particles are usually solid or liquid, while the dispersion medium is a solid, liquid, or a gas.

Worked Example #16

Do the following samples represent solutions, colloids, or suspensions?

    a. a glass of soda
    b. cement
    c. mud

Solutions

    a. *A glass of soda is a solution. There is a gas (carbon dioxide) dissolved in a liquid in a homogenous mixture.*
    b. *Cement is colloid. The particles (i.e., rock, sand) are large and evenly distributed throughout the medium.*
    c. *Mud is a suspension. The dirt particles will settle after some time.*

Try It Yourself #16

Do the following samples represent solutions, colloids, or suspensions?

    a. whipped cream
    b. milk

c. sugar and cream in coffee

d. paint

Solutions

    a. Whipped cream is_____

    b. Milk is_____

    c. Sugar and cream in coffee is_____

    d. Paint is_____

Practice Problems for Colloids and Suspensions

Do the following samples represent solutions, colloids, or suspensions?

    1. flour in water

    2. half-normal saline

    3. mayonnaise

    4. dust particles in air

    5. barium enema

    6. cheese

    7. champagne

Section 5.4: Membranes, Osmosis, and Dialysis

In a Nutshell: Semipermeable Membranes

A membrane is a structure that acts as a barrier between two environments. A semipermeable membrane is a membrane that allows passage of certain small molecules and ions across the membrane, while preventing the passage of larger molecules and ions. Simple diffusion is the simplest way for a molecule to pass through a membrane. Simple diffusion is the spontaneous movement of a molecule or ion from a region of higher concentration to a region of lower concentration.

In a Nutshell: Osmosis and Dialysis

Osmosis and dialysis describe the movement of solvent and solute across a semipermeable membrane. Osmosis refers to the flow of solvent though a semipermeable membrane. Dialysis refers to the flow of small solutes through a semipermeable membrane. The physical properties and chemical composition of the membrane determine whether osmosis, dialysis, or a combination of the two occurs.

Osmosis is a form of simple diffusion. The solvent (usually water) flows from a region of lower solute concentration (higher water concentration) to a region of higher solute (lower water concentration), across a semipermeable membrane. The flow of water through a membrane is governed by the total number of solute particles, not their mass, size, or identity.

The terms, hypertonic, hypotonic, and isotonic, are frequently used to describe the relative concentrations of the two solutions on either side of the semipermeable membrane. Hypertonic describes the solution with higher solute concentration; hypotonic describes the solution with the lower solute concentration; and isotonic describes solutions with equal solute concentrations. Osmosis can be stopped by placing external pressure on the hypertonic solution. The minimum amount of pressure that must be applied to the hypertonic solution to stop water from flowing across the membrane from the hypotonic solution is known as the osmotic pressure.

Dialysis is simple diffusion of small solutes across a semipermeable membrane from the area of higher solute concentration to the area of lower solute concentration. Dialysis can be used to separate solutes from colloidal particles, so the identity of the particles in solution does matter.

Worked Example #17

Consider the following solutions separated by a semipermeable membrane. Indicate the direction of water flow between solution A and solution B, or state if no flow occurs. Which solution is hypertonic?

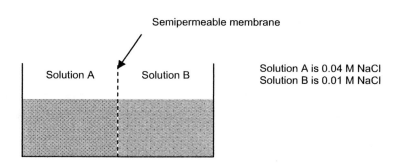

Semipermeable membrane

Solution A   Solution B

Solution A is 0.04 M NaCl
Solution B is 0.01 M NaCl

Since the question asks about water flowing across the membrane, the process involved is osmosis. Solution A has a higher solute concentration and lower water concentration. The water will flow from solution A to solution B. Solution A is hypertonic, solution B is hypotonic.

Try It Yourself #17

Consider the following solutions separated by a semipermeable membrane. Indicate the direction of water flow between solution A and solution B, or state if no flow occurs. Which solution is hypotonic?

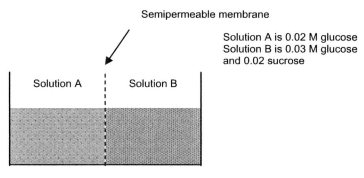

Semipermeable membrane

Solution A   Solution B

Solution A is 0.02 M glucose
Solution B is 0.03 M glucose
and 0.02 sucrose

What is flowing across the membrane? _____

What process is involved? _____

Which solution has the higher concentration? _____

Which solution has the lower concentration? _____

Which solution is hypertonic? _____

Which solution is hypotonic? _____

Which way will the water flow? _____

Worked Example #18

Biochemists often use dialysis to purify a protein by separating it from a salt. One compartment contains a solution of salt (small molecule) and protein (large molecule). The other compartment contains pure water. The two compartments are separated by a semipermeable membrane. Describe how you would separate the salt from the protein, based on the principles of dialysis.

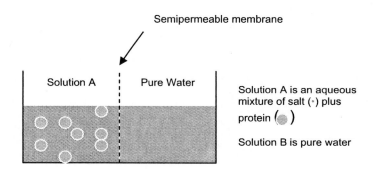

Solution

Since salt is a small molecule, it will diffuse through the membrane from solution A to the pure water side, B, from higher concentration to lower concentration, until the concentrations as equal. The protein will be unable to diffuse through the membrane because of its large size. By periodically replacing solution B with water, the salt can be separated from the protein. In the end, solution A will contain only protein. The solutions of water will contain salt.

Try It Yourself #18

When a patient undergoes kidney dialysis, small molecules such as urea and creatine are separated from the blood. One compartment contains the patient's blood and the other compartment contains the dialysate, a solution similar in composition to blood. The two compartments are separated by a semipermeable membrane. Describe how urea and creatine are separated from blood by dialysis.

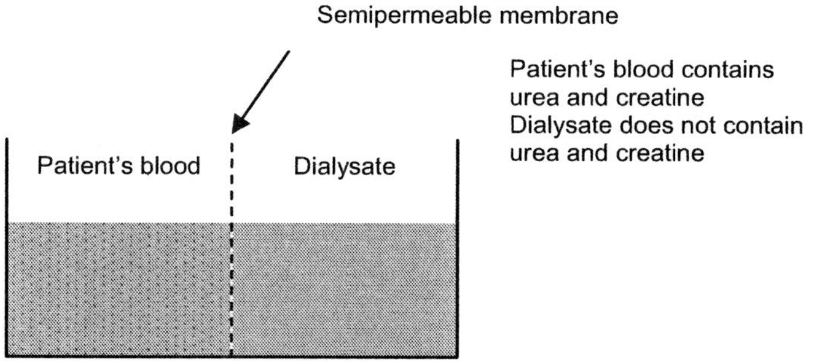

Semipermeable membrane

Patient's blood   Dialysate

Patient's blood contains urea and creatine
Dialysate does not contain urea and creatine

Side with lower concentration of urea and creatine: _____

Side with higher concentration of urea and creatine: _____

Which way will urea and creatine flow? _____

Practice Problems for Membranes, Osmosis, and Dialysis

1. Two solutions are separated by a semipermeable membrane. Solution A contains 3.6 mmol glucose and solution B contains 7.2 mmol glucose. Which solution is hypertonic? In which direction will the water flow between solution A and solution B?

2. Explain how a solution of glucose (a small molecule) and glycogen (a large molecule) can be separated using dialysis.

Chapter 5 Quiz

1. An IV solution contains 0.45% (% m/v) NaCl.

   a. What is the solute in this solution?

   b. What is the solvent?

   c. Is the solute a molecule, an electrolyte, or a gas?

   d. What is the concentration of the solution in mg/mL?

2. Singulair® is used to treat allergies in infants. The drug is prepared as follows; 4 mg of Singulair® is dissolved in 5 mL of baby formula. What is the concentration of this solution in g/dL?

3. Minocin® is used to treat an infection caused by a microorganism. An order is given to administer 200 mg. The suspension contains 50 mg in every 5 mL. How mL of the suspension should be administered to the patient?

4. Remifentanil hydrochloride is used as an analgesic with anesthesia during surgery. An order is given to administer 15.9 μg/min. The IV solution contains 20 μg/mL. What is the flow rate in mL/hr?

5. A patient's blood test shows a Mg^{2+} ion concentration of 1.7 meq/L. What is the concentration of Mg^{2+} in mol/L?

6. A patient's blood test shows that 3 mL contains 0.0126 mmol cholesterol. What is the concentration of cholesterol in mol/L?

7. What is the % m/v of solution prepared by dissolving 0.192 g NaCl in 30 mL water?

8. Identify the following as a mixture, solution, colloid, or suspension:
   a. milk

   b. concrete

   c. brass

   d. whipped cream

   e. mud

f.   mist

g.   fog

h.   a glass of champagne

i.   tea

9.   Two solutions are separated by a semipermeable membrane. Solution A contains 0.06
     M glucose and solution B contains 0.02 M glucose and 0.04 M sucrose. In what
     direction will water flow between solution A and solution B?

10. The urea from a solution containing urea and platelet cells can be separated by dialysis.
     Explain.

Chapter 5
Answers to Additional Exercises

5.41   The two components of a solution are the solute and the solvent. The solvent is present in the greater amount.

5.43   a. solute: tin and sometimes phosphorus; solvent: copper b. solute: carbon dioxide, sugar, flavorings; solvent: water. c. solute: carbon dioxide, ethanol; solvent: water d. solute: isopropyl alcohol; solvent: water.

5.45   Concentration is a measure of how much solute is dissolved in a given quantity of solution.

5.47   The supplied units are 125 μg/dL. The requested units are mg/L. The conversion factors needed are

$$\frac{1\times10^{-6}\ g}{1\ \mu g}\ \text{and}\ \frac{1\ mg}{1\times10^{-3}g}\ \text{and}\ \frac{10\ dL}{1\ L}.$$

$$\frac{125\ \cancel{\mu g}}{1\ \cancel{dL}}\times\frac{1\times10^{-6}\ \cancel{g}}{1\ \cancel{\mu g}}\times\frac{1\ mg}{1\times10^{-3}\ \cancel{g}}\times\frac{10\ \cancel{dL}}{1\ L}=1.25\ mg/L$$

Iron is an electrolyte.

5.49   The supplied units are 14 g/dL. The requested units are mg/mL. The conversion factors needed are

$$\frac{1\ mg}{1\times10^{-3}g}\ \text{and}\ \frac{10\ dL}{1\ L}\ \text{and}\ \frac{1\times10^{-3}L}{1\ mL}.$$

$$\frac{14\ \cancel{g}}{\cancel{dL}}\times\frac{1\ mg}{1\times10^{-3}\ \cancel{g}}\times\frac{10\ \cancel{dL}}{1\ \cancel{L}}\times\frac{1\times10^{-3}\ \cancel{L}}{1\ mL}=140\ mg/mL$$

5.51   The supplied units are 500 mL and 0.3% solution. The requested unit is grams of solute. The conversion factor is $\dfrac{0.3\ g}{100\ mL}$. $500\ \cancel{mL}\times\dfrac{0.3\ g}{100\ \cancel{mL}}=1.5\ g$

5.53   The supplied units are mmol/L. The requested units are mol/L. The conversion factor is $\dfrac{1\times10^{-3}\ mol}{1\ mmol}$.

$$\frac{12\ \cancel{mmol}}{L}\times\frac{1\times10^{-3}\ mol}{1\ \cancel{mmol}}=1.2\times10^{-2}\ mol/L$$

5.55   The supplied units are pmol/L. The requested units are mol/L. The conversion

factor is $\dfrac{1\text{ mol}}{1\times10^{12}\text{ pmol}}$.

$$\frac{120\ \cancel{\text{pmol}}}{\text{L}}\times\frac{1\text{mol}}{1\times10^{12}\ \cancel{\text{pmol}}}=1.2\times10^{-10}\text{ mol}/\text{L}$$

5.57   The supplied units are the number of ions and liters. The requested unit is mmol/L.

The conversion factors are $\dfrac{1\text{ mol}}{6.02\times10^{23}\text{ ions}}$ and $\dfrac{1\text{ mmol}}{1\times10^{-3}\text{mol}}$ and $\dfrac{1\text{ mol }K_2CO_3}{2\text{ mol }K^+}$

$$\frac{9.65\times10^{25}\ \cancel{K^+\ \text{ions}}}{3\text{ L}}\times\frac{1\ \cancel{\text{mol }K^+}}{6.02\times10^{23}\ \cancel{K^+\ \text{ions}}}\times\frac{1\ \cancel{\text{mol }K_2CO_3}}{2\ \cancel{\text{mol }K^+}}\times\frac{1\text{ mmol}}{1\times10^{-3}\ \cancel{\text{mol }K_2CO_3}}=2.67\times10^{4}\text{ mmol/L}$$

5.59   The supplied unit is mmol/L. The requested unit is meq/L. Table 5-4 shows that we

should use $\dfrac{1\text{ meq}}{1\text{ mmol}}$ as the conversion factor.

$$\frac{4.0\ \cancel{\text{mmol}}}{\text{L}}\times\frac{1\text{ meq}}{1\ \cancel{\text{mmol}}}=4.0\frac{\text{meq}}{\text{L}}$$

5.61   The IU takes into account the biological effect of the solute.

5.63   The supplied units are units of Betapen-VK. The requested units are mL.   The

conversion factors are $\dfrac{125\text{ mg}}{200,000\text{ units}}$ and $\dfrac{5\text{ mL}}{125\text{ mg}}$.

$$300,000\ \cancel{\text{units}}\times\frac{125\ \cancel{\text{mg}}}{200,000\ \cancel{\text{units}}}\times\frac{5\text{ mL}}{125\ \cancel{\text{mg}}}=7.5\text{ mL}$$

5.65   The supplied unit is mg/hr. The requested unit is mL/min. The conversion factors

are $\dfrac{2\text{ mL}}{125\text{ mg}}$ and $\dfrac{1\text{ hr}}{60\text{ min}}$.

$$\frac{250\ \cancel{\text{mg}}}{6\ \cancel{\text{hr}}}\times\frac{2\text{ mL}}{125\ \cancel{\text{mg}}}\times\frac{1\ \cancel{\text{hr}}}{60\text{ min}}=0.011\text{ mL/min}$$

5.67   The supplied unit is mL/hr. The requested unit is mg/hr. The conversion factor is

$\dfrac{50\text{ mg}}{250\text{ mL}}$.

$$\frac{60\ \cancel{\text{mL}}}{\text{hr}}\times\frac{50\text{ mg}}{250\ \cancel{\text{mL}}}=12\text{ mg/hr}$$

5.69   The supplied units are lb and mg erythromycin. The requested unit is mg

erythromycin/kg body weight. The conversion factors are $\dfrac{1\ kg}{2.2\ lb}$ and $\dfrac{24\ hr}{1\ day}$.

$$\frac{300\ mg\ erythromycin}{75\ \cancel{lb} \times 6\ \cancel{hr}} \times \frac{2.2\ \cancel{lb}}{1\ kg} \times \frac{24\ \cancel{hr}}{1\ day} = 35\ \frac{mg\ erythromycin}{kg \times day}.$$ Yes, it is within

the recommended range.

5.71   The supplied unit is g/mL. The requested unit is mol/L. The conversion factor are

the molar mass of glucose, 180.2 g/mol, and $\dfrac{1000\ mL}{1\ L}$.

$$\frac{3.3\ \cancel{g}}{100\ \cancel{mL}} \times \frac{1\ mol}{180.2\ \cancel{g}} \times \frac{1000\ \cancel{mL}}{1\ L} = 0.18\ mol/L$$

5.73   The supplied unit is mg/dL. The requested unit is mol/L. The conversion factors are

the molar mass of Na_3PO_4, 163.94 g/mol and $\dfrac{1 \times 10^{-3}\ g}{1\ mg}$ and $\dfrac{10\ dL}{1\ L}$ and

$\dfrac{3\ mol\ Na^+}{1\ mol\ Na_3PO_4}$.

$$\frac{0.6\ \cancel{mg\ Na_3PO_4}}{1\ \cancel{dL}} \times \frac{1 \times 10^{-3}\ \cancel{g}}{1\ \cancel{mg}} \times \frac{10\ \cancel{dL}}{1\ L} \times \frac{1\ \cancel{mol\ Na_3PO_4}}{163.94\ \cancel{g}} \times \frac{3\ mol\ Na^+}{1\ \cancel{mol\ Na_3PO_4}} = 1 \times 10^{-4}\ mol/L$$

5.75   In a suspension, the particles are analogous to the solute of the solution. The
         dispersion medium is analogous to the solvent of a solution.

5.77   a. solution b. colloid c. solution and colloid d. solution e. solution f. colloid

5.79   The solutes in whole blood are less than 1 nm in size and include ions such as Ca^{2+},
         molecules such as glucose, and gases such as oxygen. The colloidal particles in
         whole blood range in size from 1 nm to 100 nm and include proteins and starch. The
         suspended particles in whole blood are greater than 100 nm in size and include red
         blood cells, white blood cells, and platelets.

5.81   A semipermeable membrane allows the passage of small molecules and ions
         across the membrane while preventing the passage of larger molecules and ions.

5.83   Simple diffusion is the spontaneous movement of a molecule or ion from a region of
         higher concentration to a region of lower concentration.

5.85   a. The solution with the higher concentration is solution A. The water will flow from
             solution B to solution A.

b. The solution with higher concentration is solution B. The question asks about water flow (i.e., osmosis), so the identity of the solutes does not matter, just the total concentration of the solutes. The water will flow from solution A to solution B.

c. The solutions are isotonic. No net flow of water will occur between solutions.

5.87   In osmosis, water flows from a hypotonic solution to a hypertonic solution.

5.89   No, I would not expect dialysis to occur. With isotonic solutions, there is no difference in concentration between the two solutions and so net crossing of the membrane from either side will occur.

5.91   Creatine will diffuse from solution A to pure water because that is the direction of lower solute concentration. Solution B can be periodically replaced with pure water to encourage further diffusion of creatine from solution A. Eventually, almost all of the creatine will be separated from solution A. The globulin remains in solution A because it is too large a molecule to cross the semipermeable membrane.

5.93   In order to compare the two solutions, we first need to convert the concentrations to molarities. For solution A, the supplied unit is g/mL. The requested unit is mol/L.

The conversion factors are the molar mass of NaCl, 58.44247 g/mol, and $\dfrac{1\ mL}{1\times10^{-3}L}$.

$$\frac{0.45\ \cancel{g}}{100\ \cancel{mL}}\times\frac{1\ mol}{58.44247\ \cancel{g}}\times\frac{1\ \cancel{mL}}{1\times10^{-3}L}=0.078\ mol/L\ NaCl$$

For solution B, the supplied unit is meq/L. The requested unit is mol/L. Using Table 5-4, we see that there are 2 meq/mmol for Mg^{2+}. The conversion factors are 2 meq/mmol and $\dfrac{1\ mmol}{1\times10^{-3}mol}$.

$$\frac{0.95\ \cancel{meq}}{L}\times\frac{1\ \cancel{mmol}}{2\ \cancel{meq}}\times\frac{1\times10^{-3}mol}{1\ \cancel{mmol}}=4.8\times10^{-4}\ mol/L\ MgCl_2$$

Solution A contains 0.078 M NaCl, and an electrolyte which yields two ions when it dissolves. Thus, the effective concentration of ions is 0.078 M × 2, or 0.16 M. Solution B contains 4.8×10^{-4} M Na_2S, and an electrolyte which yields three ions when it dissolves, and its effective ion concentration is 4.8×10^{-4} M × 3, or 0.0014 M. Solution B has the lower concentration of ions, so water will flow from compartment B to compartment A.

5.95   The walls of the arterial capillaries present in the kidneys are the membranes that selectively allow waste products to diffuse from the blood stream into the kidney.

Urea, ions, glucose, and amino acids are small enough to pass through these membranes.

5.97   In hemodialysis, ions and other solutes, such as urea and creatine, flow out of the blood into the dialysate. The concentrations of ions, urea, and creatine are lower in the dialysate than in the blood. These solutes will diffuse from a region of higher concentration to a region of lower concentration.

5.99   The two types of kidney dialysis are hemodialysis and peritoneal dialysis. Hemodialysis needs to be performed at a medical center and removes the waste products and excess water from the patient's blood using an artificial kidney. Peritoneal dialysis can be performed at home and uses the patient's abdominal cavity to remove the waste products.

5.101 The supplied unit is meq/mL. The requested unit is mol/L. Table 5-4 shows that there is 1 meq/mmol for K^+. The conversion factors needed are $\dfrac{1\,mL}{1\times10^{-3}L}$ and

$\dfrac{1\,mmol}{1\times10^{-3}\,mol}$.

$$\frac{10\ \cancel{meq}}{500\ \cancel{mL}}\times\frac{1\ \cancel{mL}}{1\times10^{-3}L}\times\frac{1\ \cancel{mmol}}{1\ \cancel{meq}}\times\frac{1\times10^{-3}\,mol}{1\ \cancel{mmol}}=0.02\ mol/L$$

5.103 Solution A contains 0.25 KCl, and an electrolyte which yields two ions when it dissolves. Thus, the effective concentration of ions is 0.25 M × 2, or 0.50 M. Solution B contains 0.25 M Na_2S, and an electrolyte which yields three ions when it dissolves. Thus, its effective concentration of ions is 0.25 M × 3, or 0.75 M. Solution A has the lower concentration of ions, so water will flow from compartment A to compartment B.

Chapter 6

Hydrocarbons and Structure

Chapter Summary

In this chapter, you started your focus on the fundamentals of organic chemistry. You have learned about the four types of hydrocarbons: alkanes, alkenes, alkynes, and aromatic hydrocarbons. You were introduced to the basic rules for naming these hydrocarbons. You also studied the structural characteristics of these compounds, including the conformations that alkanes can adopt, and learned about structural isomers and geometric isomers. Most biological compounds are organic compounds; therefore, understanding organic chemistry will help you understand the chemistry of the human body.

Section 6.1 Hydrocarbons

In a Nutshell: Types of Hydrocarbons

Hydrocarbons are molecules that contain exclusively hydrogen and carbon. Pure hydrocarbons are divided into four categories: alkanes, alkenes, alkynes, and aromatic hydrocarbons. Alkanes contain only single bonds, alkenes contain one or more carbon-carbon double bond, and alkynes contain one or more carbon-carbon triple bond. Aromatic hydrocarbons have a unique ring structure containing several multiple bonds. Alkanes are classified as saturated hydrocarbons because they contain the maximum number of hydrogen atoms for a given number of carbon atoms. The general formula for a saturated hydrocarbon is C_nH_{2n+2}, where *n* is equal to the number of carbon atoms in the formula. Alkenes, alkynes, and aromatic compounds are classified as unsaturated hydrocarbons because they contain less than the maximum number of hydrogen atoms per carbon atom.

Worked Example #1

What is the chemical formula for a saturated hydrocarbon with 10 carbon atoms?

The formula for a saturated hydrocarbon is C_nH_{2n+2}. Substituting the number 10 for n *gives the number of hydrogen atoms: (2 × 10) + 2 = 22. Therefore, the formula for a saturated hydrocarbon with 10 carbon atoms is $C_{10}H_{22}$.*

Try It Yourself #1

What is the chemical formula for a saturated hydrocarbon with 12 carbon atoms?

Number of carbon atoms: _____

Number of hydrogen atoms: _____

Chemical formula: _____

Worked Example #2

From the chemical formula, determine which one of the hydrocarbons listed below is a saturated hydrocarbon.

    a. C_6H_6

    b. C_6H_{14}

    c. C_6H_{10}

For six carbons, a saturated hydrocarbon should contain 14 hydrogen atoms. When 6 is substituted for n in the formula $C_nH_{2n+2,}$ we see that there are 14 hydrogen atoms. Therefore (b), C_6H_{14}, is a saturated hydrocarbon.

Try It Yourself #2

From the chemical formula, determine which one of the hydrocarbons listed below is a saturated hydrocarbon.

    a. C_8H_{18}

    b. C_8H_{14}

    c. C_8H_{16}

For a saturated hydrocarbon with eight carbon atoms:

Number of hydrogen atoms: _____

Chemical formula: _____

Worked Example #3

For each of the structural formulas shown below, identify whether it is an alkane, alkene, or alkyne.

a. 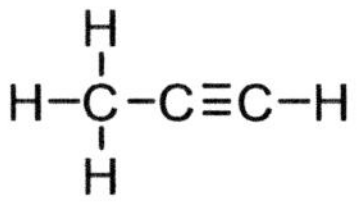
```
    H
    |
H−C−C≡C−H
    |
    H
```

b.
```
  H H H H H H
  | | | | | |
H−C−C−C−C−C−C−H
  | | | | | |
  H H H H H H
```

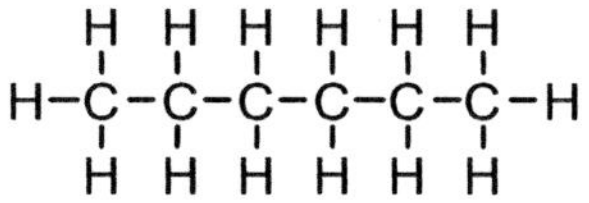

c.
```
       H  H H
        \ | |
     H   C−C−H
      \ /    |
       C=C   H
      /   \
     H     H
```

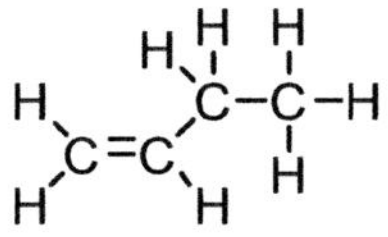

a. *The compound is an alkyne because it contains a carbon-carbon triple bond.*

b. *The compound is an alkane because it contains only carbon-carbon single bonds.*

c. *The compound is an alkene because it contains a carbon-carbon double bond.*

Try It Yourself #3

For each of the structural formulas shown below, identify whether it is an alkane, alkene, or alkyne.

a. 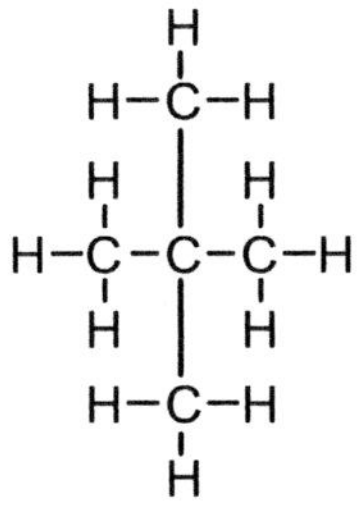
```
      H
      |
    H−C−H
    H | H
    | | |
  H−C−C−C−H
    | | |
    H | H
    H−C−H
      |
      H
```

b. 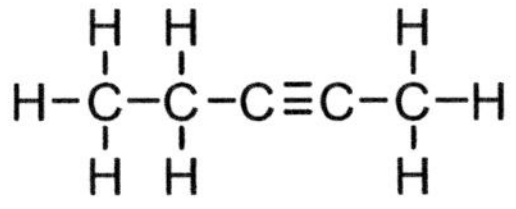
```
  H H       H
  | |       |
H−C−C−C≡C−C−H
  | |       |
  H H       H
```

c.

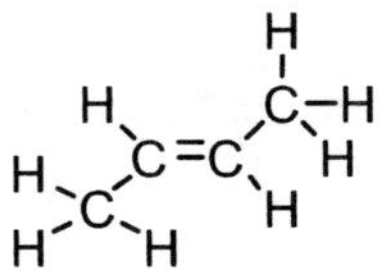

a. Type of carbon-carbon bonds present: _____

   Type of hydrocarbon: _____

b. Type of carbon-carbon bonds present: _____

   Type of hydrocarbon: _____

c. Type of carbon-carbon bonds present: _____

   Type of hydrocarbon: _____

In a Nutshell: Physical Properties of Hydrocarbons

All hydrocarbons are nonpolar molecules. Nonpolar molecules interact through dispersion forces, the weakest of the intermolecular forces of attraction. Consequently, hydrocarbons have low boiling points. Hydrocarbons are insoluble in water because they are nonpolar. A hydrophobic—water fearing—compound is insoluble in water and soluble in nonpolar substances. A hydrophilic—water loving—compound is soluble in water and insoluble in nonpolar substances.

Worked Example #4

Based on your everyday experience, which of the following household substances is(are) hydrophobic?

   a. sunscreen

   b. olive oil

   c. vinegar

Sunscreen and olive oil are hydrophobic. They do not dissolve in water. Vinegar dissolves in water.

Try It Yourself #4

Based on your everyday experience, which of the following household substances is(are) hydrophobic?

    a. baking powder

    b. margarine

    c. lipstick

    a. *Does the substance dissolve in water?_____*

        It is _____.

    b. *Does the substance dissolve in water?_____*

        It is _____.

    c. *Does the substance dissolve in water?_____*

        It is _____.

Practice Problems for Hydrocarbons

    1. Which of the following structures represent a saturated hydrocarbon?

        a. C_3H_6

        b. C_4H_{10}

        c. $C_{12}H_{24}$

        d. $C_{11}H_{22}$

        e. $C_{13}H_{28}$

2. Based on your everyday experience, which of the following household substances are hydrophobic?

    a. motor oil

    b. balsamic vinegar

    c. candle wax

    d. rubbing alcohol

3. For each of the structural formulas shown below, identify whether it is an alkane, alkene, or alkyne.

    a.

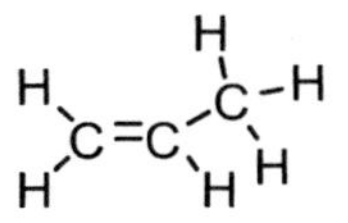

    $$H_2C=CH-CH_3$$

    b.

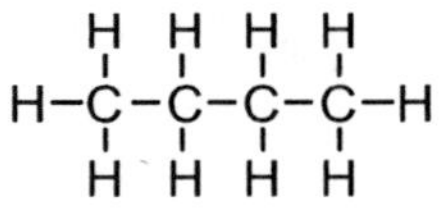

    $$H-CH_2-CH_2-CH_2-CH_3... (H-C-C-C-C-H)$$

    c.

    $$H-CH_2-C\equiv C-CH_2-H$$

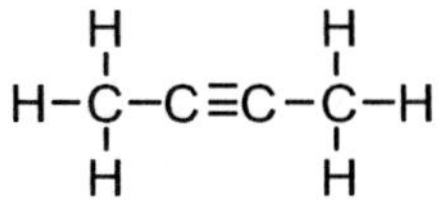

Section 6.2 Saturated Hydrocarbons: The Alkanes

In a Nutshell: Alkane Conformations

Alkanes are hydrocarbons that contain only carbon-carbon single bonds as well as carbon-hydrogen bonds; they contain no multiple bonds. A molecule has free rotation about a carbon-carbon single bond (unless the bond is part of a ring or other structural constraint). Consequently, a molecule can exist in many different rotational forms, known as conformations. Anytime a structure can be converted into another structure by merely rotating about one or more carbon-carbon single bonds, it indicates that the two structures are the same compound, just different conformations.

Worked Example #5

Do the following pairs of molecules represent different conformations of the same molecule or different molecules? If they are different molecules, explain what makes them different.

a.

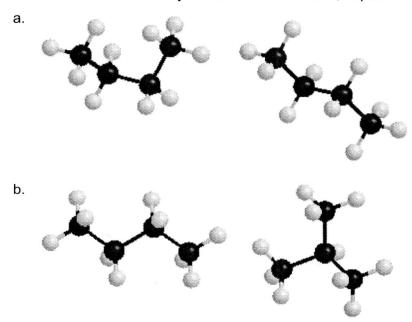

b.

Solutions

a. *The two molecules are the same molecule; they are conformations of the same molecule. The carbon-carbon bond in the middle of the molecule has been rotated.*

b. *These two molecules are different. They have a different connectivity of the atoms. The first model has the four carbon atoms connected in a continuous chain. The second model has a central carbon atom bonded to three other carbon atoms.*

Try It Yourself #5

Do the following pairs of molecules represent different conformations of the same molecule or different molecules? If they are different molecules, explain what makes them different.

a.

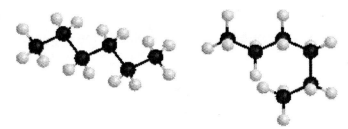

b.

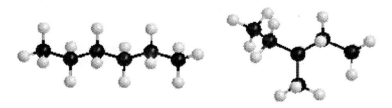

a. *The number of carbon and hydrogen atoms in the molecules is the (same or different)* _____.

The connectivity of the carbon atoms is the (same or different)

_____.

They are the (same or different) molecule(s) _____.

b. *The number of carbon and hydrogen atoms in the molecules is the (same or different)* _____.

The connectivity of the carbon atoms is the (same or different)

_____.

They are the (same or different) molecule(s) _____.

In a Nutshell: Structural Isomers

Molecules with the same chemical formula but a different connectivity of atoms are known as structural isomers. Structural isomers are also known as constitutional isomers; they are different chemical compounds with different chemical properties and different chemical names. The structure of a linear chain of carbon atoms is known as the straight-chain isomer because of its end-to-end arrangement of carbon atoms. A branched chained isomer has branch points along the hydrocarbon chain. The number of structural isomers that exist for any chemical formula increases as the number of carbon atoms in the chemical formula increases.

Worked Example #6

Which of the following pairs are not structural isomers?

a.

H H H H
| | | |
H-C-C-C-C-H   and

H
|
H-C-H
|
H H | H
| | | |
H-C-C-C-C-H
| | | |
H H H H

With left structure also:
H-C-H
|
H

b.

H H H H
| | | |
H-C-C-C-C-H   and

H H H H H
| | | | |
H-C-C-C-C-C-H
| | | | |
H H H H H

With left structure also:
H-C-H
|
H

c.

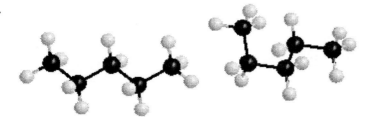

a.   The two molecules are not structural isomers. They both have a four-carbon chain with a CH$_3$ group bonded to the second carbon from the end of the chain. Note: it doesn't matter which end you start counting.

b.   The two molecules are structural isomers. They have the same number of carbon atoms, but the connectivity of the carbon atoms is different.

c.   The two molecules are not structural isomers. They have the same number of carbon atoms, but the atoms have been rotated around a bond.

Try It Yourself #6

Which of the following pairs are not structural isomers?

a.

H H H H
| | | |
H-C-C-C-C-H
| | | |
H H H H

H
|
H-C-H
|
H | H
| | |
H-C-C-C-H
| | |
H H H

b.

c.

a.   The molecules _____ structural isomers.

b.   The molecules _____ structural isomers.

c.   The molecules _____ structural isomers.

Worked Example #7

Write the structural formula for two of the nine structural isomers for C_7H_{16}.

Start with the straight chain isomer.

For the next isomer, move a CH_3 group from the end carbon to the next carbon over.

To find the next isomer, keep moving the CH_3 to the next carbon on the backbone.

Try It Yourself #7

Write the structural formula for two more of the nine structural isomers for C_7H_{16} (not the ones listed in Worked Example #7).

Structure of first isomer:

Structure of second isomer:

In a Nutshell: Cycloalkanes

A cycloalkane is a molecular ring created when a carbon-carbon bond is formed between the two ends of an alkane chain. Cycloalkanes are written as polygons with three or more sides. A carbon atom is present at each corner of the polygon with two hydrogen atoms attached. The most common ring sizes encountered in nature are five-and six-membered rings. Three- and four-membered rings are less common because the rings are strained.

Worked Example #8

How many carbon atoms are there in the following cycloalkanes?

a.

b.

   a. *The polygon has seven corners; therefore, there are seven carbon atoms in this cycloalkane.*

b.   *The polygon has eight corners; therefore, there are eight carbon atoms in this cycloalkane.*

Try It Yourself #8

How many carbon atoms are there in the following cycloalkanes?

a.

b.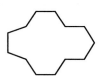

a.   *Number of corners in the polygon:* _____
     Number of carbon atoms in cycloalkane: _____
b.   *Number of corners in the polygon:* _____
     Number of carbon atoms in cycloalkane: _____

Practice Problems for Saturated Hydrocarbons: The Alkanes

1.   How many carbon atoms are there in the following cycloalkanes?

a.

b.

c.

2. Do the following pairs of molecules represent different conformations of the same molecule or different molecules? If they are different molecules, explain what makes them different.

a.

```
                                    H
                                    |
                                 H-C-H
                                    |
                                 H-C-H
    H  H  H  H  H         H  H     |
    |  |  |  |  |         |  |     |
 H-C-C-C-C-C-H        H-C-C-C-H
    |  |  |  |  |         |  |  |
    H  H  H  H  H         H  H  H
```

b.

```
                                    H
                                    |
                                 H-C-H
                                    |
                                 H-C-H
    H  H  H  H  H         H  H     |     H
    |  |  |  |  |         |  |     |     |
 H-C-C-C-C-C-H        H-C-C-C-C-H
    |  |  |  |  |         |  |  |  |
    H  H  H  H  H         H  H  H  H
```

3. Which of the following pairs are structural isomers?

a.

```
    H  H  H  H              H  H  H  H
    |  |  |  |              |  |  |  |
 H-C-C-C-C-H        H-C-C-C-C-H
    |  |  |  |              |  |  |  |
    H  H  H  H              H     H  H
                              |
                           H-C-H
                              |
                              H
```

b.

```
    H  H  H  H  H              H  H  H  H  H  H
    |  |  |  |  |              |  |  |  |  |  |
 H-C-C-C-C-C-H        H-C-C-C-C-C-C-H
    |  |  |  |  |              |  |  |  |  |  |
    H  |  H  H  H              H  H  H  H  H  H
       |
    H-C-H
       |
       H
```

c.

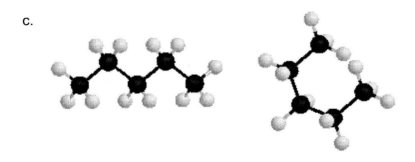

Section 6.3 Writing Structures

In a Nutshell: Condensed Structural Formulas

Structural formulas depict the connectivity of atoms in an organic compound by showing all covalent bonds and atoms. Two simpler ways of writing organic structures have been developed that are in common use: condensed structural formulas and skeletal line structures.

In a condensed structural formula, each carbon atom and its attached hydrogen atom or atoms is written as a group: CH, CH_2, or CH_3. Each group is then written sequentially according to its order in the chain. Bonds are omitted, except in the case of branch points.

In writing the condensed formula for a straight-chain alkane, the first and last carbons are written as a CH_3 group, while the other carbon atoms are written as a CH_2 group. In writing the condensed formula for long, straight-chain alkanes, the repeating CH_2 groups are often indicated by writing a single CH_2 enclosed in parentheses followed by a subscript that indicates the number of repeating CH_2 groups.

In writing the condensed formula for a branched-chain alkane, the same guidelines are followed as for writing straight-chain alkenes; any branch point along the chain is indicated by inserting a bond (pointing either up or down) at the carbon atom where the branch occurs.

Worked Example #9

For each condensed structural formula below, determine whether it represents a straight-chain alkane or a branched-chain alkane, then write the corresponding structural formula.

   a. $CH_3CH_2CH_2CH_2CH_2CH_2CH_2CH_3$

   b.
$$\begin{array}{c} CH_3 \\ | \\ CH_3CHCH_2CH_2CH_3 \end{array}$$

   c.
$$\begin{array}{c} CH_3 \\ | \\ CH_3CH_2CHCH_2CHCH_2CH_3 \\ | \\ CH_2CH_3 \end{array}$$

a. *Straight-chain alkane because there are no bonds indicating branch points along the chain.*

```
    H  H  H  H  H  H  H  H
    |  |  |  |  |  |  |  |
H — C— C— C— C— C— C— C— C— H
    |  |  |  |  |  |  |  |
    H  H  H  H  H  H  H  H
```

b. *Branched-chain alkane. There is one branching group off the second carbon in the chain, indicated by the bond shown projecting from the carbon atom.*

```
          H
          |
        H—C—H
    H   |  H  H  H
    |   |  |  |  |
H — C — C— C— C— C— H
    |   |  |  |  |
    H   H  H  H  H
```

c. *Branched-chain alkane. There are two branching groups off the third fifth carbons in the chain, indicated by the bonds shown projecting from the carbon atom.*

```
              H
              |
            H—C—H
    H   H   |   H  H  H  H
    |   |   |   |  |  |  |
H — C — C — C — C— C— C— C— H
    |   |   |   |  |  |  |
    H   H   H   H  |  H  H
                   H
                   |
              H — C — C— H
                   |  |
                   H  H
```

Try It Yourself #9

For each condensed structural formula below, determine whether it represents a straight-chain alkane or a branched-chain alkane, then write the corresponding structural formula.

a. CH_2CH_3
     $CH_3CHCH_2CHCH_3$
              CH_3

b. $CH_3(CH_2)_5CH_3$

c. CH_3
     CH_3CH
          CH_3

a. *Are there branching points in the chain? _____*

It is a _____ *chain alkane.*

Structural formula:

b. *Are there branching points in the chain?* _____

It is a _____ *chain alkane.*

Structural formula:

c. *Are there branching points in the chain?* _____

It is a _____ *chain alkane.*

Structural formula:

In a Nutshell: Skeletal Line Structures

Skeletal line structures are a convenient shorthand for writing complex chemical structures and have a clean and uncluttered appearance. There are general rules for writing skeletal line structures:

1. All carbon-carbon single bonds are shown as a single line, ——.

2. Double bonds are shown as two parallel lines, $=$, and triple bonds are shown as three parallel lines, $\equiv$.

3. The chemical symbol for carbon, C, is omitted. The presence of a carbon atom is implied whenever two lines join, *as well as at the end of a line.* A continuous chain of carbon atoms is represented as a zigzag arrangement of lines.

4. Atoms other than carbon and hydrogen (heteroatoms, such as oxygen, nitrogen, sulfur, and phosphorus) must be written-in in order to distinguish them from carbon atoms.

5. Carbon-hydrogen bonds, as well as the H atoms bonded to carbon, are omitted from skeletal line structures all together.

- Hydrogen atoms attached to a heteroatom, however, must be written in.

- Because a carbon atom always has four bonds, you can determine the number of H atoms bonded to a particular carbon atom in a skeletal line structure by counting the number of bonds and subtracting the value from four:

   Number of C-H bonds on a C atom = 4 − (number of lines to C atom)

The unique connectivity of atoms determines the identity of a given structural isomer, just as it did with structural formulas. The guidelines for drawing structures are somewhat flexible, so you will often find that a combination of the skeletal line structure and condensed formula is used.

Worked Example #10

Write the skeletal line structure that corresponds to each of the condensed structural formulas shown below:

   a. $CH_3(CH_2)_5CH_3$

   b.
   $$CH_3CHCH_2CHCH_3$$
   with substituents CH_2CH_3 and CH_3

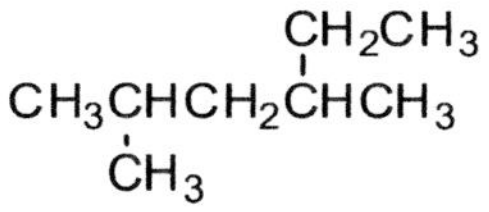

Solutions

   a.    Draw a line for every carbon-carbon bond.

   b. 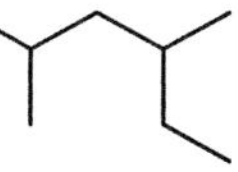   Draw a line for every carbon-carbon bond.

Try It Yourself #10

Write the skeletal line structure that corresponds to each of the condensed structural formulas shown below:

   a.
   $$CH_3CHCHCH_2CH_3$$
   with substituents CH_2CH_3 and CH_2CH_3

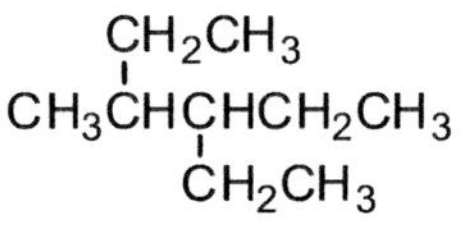

   b. $CH_3(CH_2)_9CH_3$

   a. *Skeletal line structure:*

b. Skeletal line structure:

Worked Example #11

Write the structural formula and the condensed structural formula for the skeletal line structures shown below.

a.

b.

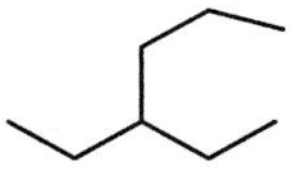

Solutions

a. *The end of each line and the intersection of two lines represent a carbon atom. The condensed structural formula is $CH_3CH_2CH_2CH_3$. The structural formula is:*

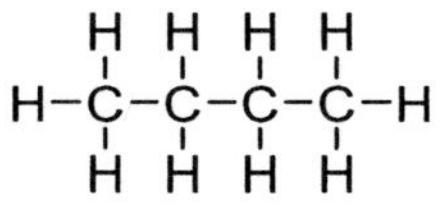

b. *The end of each line and the intersection of two lines represent a carbon atom. The condensed structural formula is* $CH_3CH_2\overset{\overset{\displaystyle CH_2CH_2CH_3}{|}}{C}HCH_2CH_3$ *. The structural formula is:*

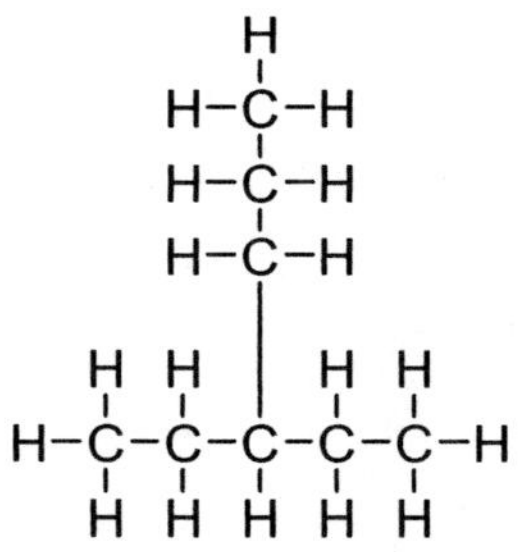

Try It Yourself #11

Write the structural formula and the condensed structural formula for the skeletal line structures shown below.

a.

b.

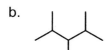

a. *Structural formula:*

      Condensed structural formula:

b. *Structural formula:*

      Condensed structural formula:

Practice Problems for Writing Structures

1. Write the structural formula and the condensed structural formula for the skeletal line structures below.

    a.

b.

c. ⤢

2. Write the skeletal line structure that corresponds to each of the condensed structural formulas below.

a.

$$CH_3CH_2\overset{\overset{\displaystyle CH_2CH_3}{|}}{C}HCH_2\overset{\overset{\displaystyle \cdot}{|}}{C}HCH_3$$
$$\qquad\qquad\qquad CH_3$$

b. $CH_3(CH_2)_{10}CH_3$

c.

$$CH_3\overset{\overset{\displaystyle CH_3}{|}}{C}H\overset{\overset{\displaystyle \cdot}{|}}{C}HCH_2CH_2CH_2CH_3$$
$$\qquad\qquad CH_3$$

3. In Questions 1 and 2, which molecules are branched-chain alkanes and which molecules are straight-chain alkanes?

Section 6.4 Unsaturated Hydrocarbons: Alkenes and Alkynes

In a Nutshell: Alkenes

An alkene is a hydrocarbon that contains one or more carbon-carbon double bonds. An alkene containing two carbon-carbon double bonds is referred to as a diene. An alkene containing several carbon-carbon double bonds is called a polyene. The bond angles around each of the carbon atoms that form a double bond are 120°. In the skeletal line structure an alkene is represented by two parallel lines.

Worked Example #12

Identify the skeletal line structure below as a simple alkene, diene, or polyene.

a.

b.

c.

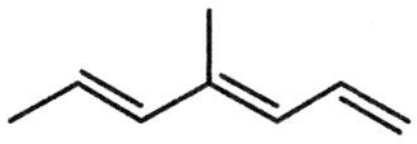

a. *There are two double bonds in the structure, therefore it is a diene.*

b. *There is one double bond in the structure, therefore it is a simple alkene.*

c. *There are more than two double bonds in the structure, therefore it is a polyene.*

Try It Yourself #12

Identify the skeletal line structure below as a simple alkene, diene, or polyene.

a.

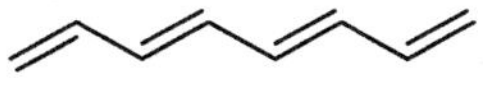

b.

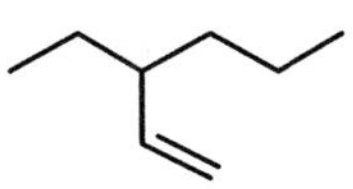

c.

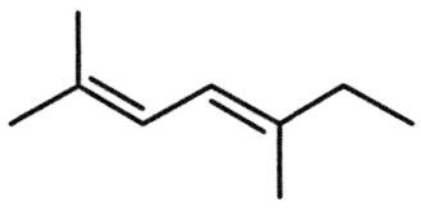

a. *Number of double bonds:* _____

   Type of alkene: _____

b. *Number of double bonds:* _____

   Type of alkene: _____

c. *Number of double bonds:* _____

   Type of alkene: _____

In a Nutshell: Geometric Isomers

Molecules are not free to rotate about carbon-carbon double bonds; therefore, geometric isomers are possible. Geometric isomers are compounds with the same chemical formula and same connectivity of atoms, but with a different three-dimensional orientation as the result of the double bond. When two groups are on the same side of the double bond, the geometric isomer is called the cis isomer. When two groups are on opposite sides of the double bond, the geometric isomer is called the trans isomer. The cis/trans designation must be included when naming geometric isomers and is only used when there are two carbon-containing groups on the double bond. To draw the geometric isomer of an alkene, switch the two groups attached to either double-bond carbon atom.

Worked Example #13

Draw the geometric isomer for the following compounds. Indicate if the isomer shown is cis or trans.

a.

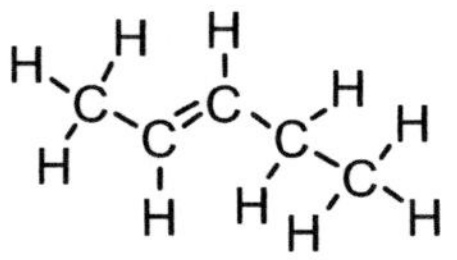

b.

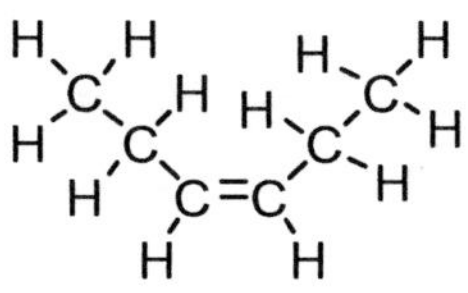

a. To draw the geometric isomer, exchange the two groups attached to one carbon of the double bond.

The isomer shown in the question is the trans isomer; the two groups (CH₃ and CH₂CH₃) are on opposite sides of the double bond.

b. To draw the geometric isomer, exchange the two groups attached to one carbon of the double bond.

The isomer shown in the question is the cis isomer, the two groups (CH₃CH₂ and CH₃CH₂) are on the same side of the double bond.

Try It Yourself #13

Draw the geometric isomer for the following compounds. Indicate if the isomer shown is cis or trans.

a.

b.

a. Are the two groups on the same side of the double bond? _____

    The geometric isomer shown is: _____

    The other geometric isomer: _____

    The structure of the other geometric isomer:

b. Are the two groups on the same side of the double bond? _____

The geometric isomer shown is: _____

The other geometric isomer: _____

The structure of the other geometric isomer:

In a Nutshell: Alkynes

An alkyne contains one or more carbon-carbon triple bond. Alkynes have a linear geometry around the carbon atoms that form the carbon-carbon triple bond.   Alkynes are not common in nature.

Practice Problems for Unsaturated Hydrocarbons: Alkenes and Alkynes

1. Identify the structure below as a simple alkene, a diene, or a polyene.

a.

b.

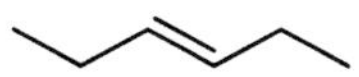

c.

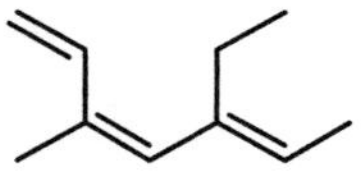

2. In the following molecule, identify the alkene and the alkyne.

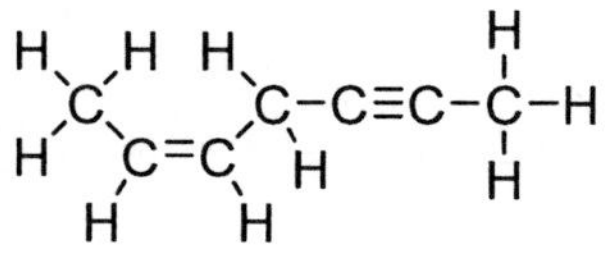

3. Write the geometric isomer for the compounds shown. Indicate whether the isomer that you have drawn is the cis or trans isomer.

a.

b.

Section 6.5 Naming Hydrocarbons

In a Nutshell: Naming Straight-Chain Hydrocarbons and Cycloalkanes

Every organic compound has a unique name based on the IUPAC system. An IUPAC name is composed of three parts: a prefix, a root, and a suffix. In the IUPAC naming system, the names of straight-chain alkanes do not have a prefix. To assign an IUPAC name to a straight-chain alkane, use the following rules: 1) assign the root, 2) assign the suffix, 3) assign a locator number to the root if a multiple bond is present. To assign the root, count the number of carbons in the chain. The suffix indicates the type of hydrocarbon: alkane, alkene, or alkyne. The suffix for an alkane or cycloalkane is "ane;" for an alkene, "ene;" and for an alkyne, "yne." A locator number is used to indicate the location of a multiple bond in an alkene or alkyne. Number the carbon atoms in the chain starting form the end closest to the multiple bond. The number corresponding to the first carbon atom of the multiple bond is the locator number. Geometric isomers may also need a cis or trans designation. A cycloalkane is assigned the root corresponding to the number of carbon atoms in the ring with the additional prefix "cyclo" inserted before the root name.

Worked Example #14

Write the IUPAC name for the following compounds.

a.

b.

c.

Solutions

a. *Nonane. There are nine carbon atoms in a straight chain. The root is nonane.*
   Since the compound is an alkane, the suffix remains "ane."

b. *Cyclohexane. There are six carbon atoms in a ring. The root is hexane. Since the*
   six carbon atoms are in a ring, the prefix "cyclo" is inserted before the root.

c. *2-pentyne. There are five carbon atoms in the chain, so the root is pentane. The*
   suffix is changed to "yne" to yield pentyne because there is a triple bond in the
   molecule. The chain is numbered from the left because that end is closer to the
   triple bond. The first carbon atom of the triple bond is carbon 2, so the locator
   number is 2.

Try It Yourself #14

Write the IUPAC name for the following compounds:

a.

b.

c.

Solutions

a. *Number of carbon atoms in the ring: _____*

   Root name: _____

   Prefix: _____

   IUPAC name: _____

b. *Number of carbon atoms in the chain: _____*

   Root name: _____

   Suffix ending for the carbon-carbon multiple bond: _____

   Locator number for the double bond: _____

   Cis or trans designation for the geometric isomer: _____

   IUPAC name: _____

c. *Number of carbon atoms in the chain: _____*

   Root name: _____

Suffix ending: _____

IUPAC name: _____

Worked Example #15

Draw the skeletal line structure of the following compound, *trans*-3-hexene.

Solution

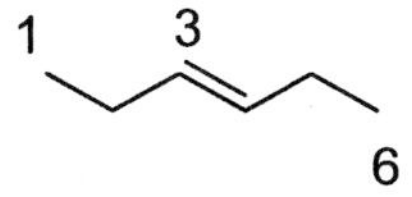

Write a zigzag structure showing six points and place a double bond between the third and fourth carbons. To indicate the trans isomer, place the two large groups on the opposite sides of the double bond.

Try It Yourself #15

Draw the skeletal line structure of the *cis*-3-octene.

Skeletal line structure:

In a Nutshell: Naming Branched-Chain Hydrocarbons

In branched-chain hydrocarbons one or more hydrogen atoms along the main carbon chain are "substituted" by one or more smaller carbon atom chains. Any branch replacing a hydrogen atom is called a "substituent." In the IUPAC naming system, the substituents are assigned a name that appears as a prefix in front of the root. To assign an IUPAC name to a branched-chain alkane, use the following rules: 1) assign the root, 2) assign the suffix, 3 – 5) assign the prefix (which includes: naming the substituent, assigning a locator name to the substituent, and assembling the prefix), 6) assemble the IUPAC name.

The root name is determined by finding the longest continuous carbon chain, containing the double or triple bond if one exists. The suffix is changed to reflect whether the molecule is an alkane, alkene, or alkyne. A locator number is placed immediately before the root name to indicate the location of the first carbon of a multiple bond. The substituents are assigned a name and a locator number that appears as a prefix in front of the root. The substituents are named according to the number of carbon atoms in the substituent chain. The ending on the substituent is changed from "ane" to "yl." The substituent is then assigned a locator number based on the numbering of the main chain. Once the substituent has been named, the prefix name is assembled by listing the substituents in alphabetical order. If a substituent is repeated more than once along the chain, the prefixes "di," "tri," or "tetra" are used. The last step is to assemble the IUPAC name by putting together the prefix, root, and suffix in that order. Numbers are separated from letters with a dash and numbers are separated from other numbers with a comma.

In a Nutshell: Finding the Main Chain

The main chain is not always the one that lies horizontally on the page. For alkenes and alkynes, the main chain must contain both carbon atoms of the multiple bond, even if there is a longer chain.

Worked Example #16

Provide the IUPAC name for the structure shown below:

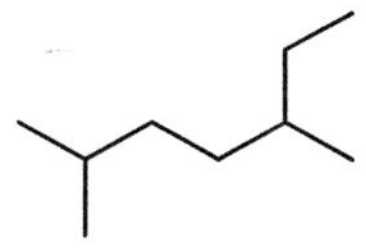

Assign the root.

The longest continuous chain of carbon atoms contains seven carbon atoms, so according to Table 6-4 the root name is heptane.

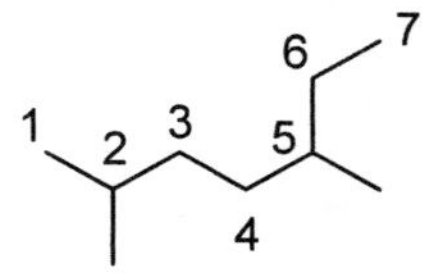

Assign the suffix.

The suffix remains "ane" because the compound is an alkane.

Name each substituent.

There are two substituents, one on carbon atom number two and one on carbon atom number five of the main chain. Both substituents contain only one carbon atom; its root name is methane, which is changed to methyl.

Assign a locator number to the substituents.

The locator number 2 is assigned to one of the substituents because it is located on the second carbon of the main chain. The locator number 5 is assigned to one of the substituents because it is located on the fifth carbon of the main chain.

Assemble the prefix.

Place the locator number in front of the substituent names. Since the two substituents are the same (both methyl), the additional prefix of "di" is used. The prefix is 2,5-dimethyl.
Assemble the IUPAC name.

Write the prefix followed by the root and suffix. The name of the compound is 2,5-dimethylheptane.

Try It Yourself #16
Write the IUPAC name for the following branched chain alkane.

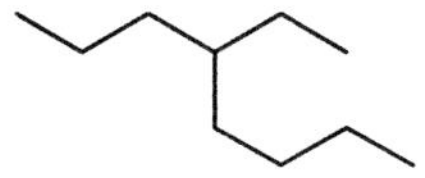

Assign the root.
Number of carbon atoms in the main chain: _____

Root name: _____

Assign the suffix.
Suffix: _____

Name each substituent.
Number of carbon atoms in substituent: _____

Root name for substituent: _____

Substituent name: _____

Assign a locator number to the substituent.

Locator number: _____

Assemble the prefix.

Prefix: _____

Assemble the IUPAC name.

IUPAC name: _____

Worked Example #17

Write the skeletal line structure of 2-methyl-3-ethyl-2-pentene.

Solution

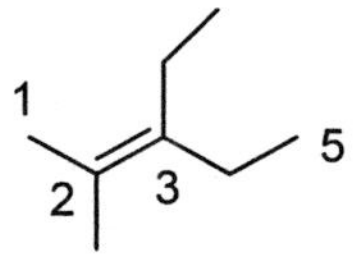

The root name "pentene" indicates a five-carbon chain. The "ene" indicates that there is a double bond, and the locator number "2" indicates that the double bond is at carbon 2. A methyl group is placed on carbon 2 and an ethyl group is placed on carbon 3.

Try It Yourself #17

Write the skeletal line structure of 3-methyl-4-ethyloctane.

Solution

Number of carbon atoms in the main chain: _____

Substituents on carbon number(s): _____

Type of substituent: _____

Skeletal line structure:

Practice Problems for Naming Hydrocarbons

1. Write the IUPAC name of the following compounds.

   a.

   b.

   c.

   d.

2. Write the structural formula and skeletal line structure for the following compounds.

   a. *trans*-3-hexene

   b. cyclohexane

    c. 2-methylpentane

    d. 2,3,4,6-tetramethylheptane

Section 6.6 Aromatic Hydrocarbons

In a Nutshell: Benzene

The simplest aromatic hydrocarbon is benzene. Each carbon atom in benzene has a trigonal planar geometry and every C-C-C bond angle is 120°, giving the overall molecule a flat two-dimensional shape. There are six electrons in the benzene ring that are delocalized over all six carbon atoms of the ring. In order to depict delocalized electrons, resonance structures are drawn. The actual structure of benzene is understood to be a hybrid of its resonance structures. The delocalization of electrons minimizes electron-electron repulsions so that the molecule is more stable. The more stable the molecule, the less likely it is to undergo chemical reactions. The extra stability of the aromatic compounds places them in a category separate from alkenes.

In a Nutshell: Naming Substituted Benzenes

Benzene is the IUPAC name for C_6H_6 and is the root name for compounds containing substituents attached to benzene. To assign an IUPAC name to a substituted benzene, the following rules are used: 1) assign the root and 2) assign the prefix. No suffix is needed to assign an IUPAC name. The prefix provides the identity of the substituent on the benzene ring. Substituents are named using the same naming system used to name the substituents on hydrocarbon chains. When one substituent is present, no locator number is needed. When one or two substituents are attached to the benzene ring, locator numbers must be included in the prefix to indicate the relative positions of the substituents on the ring. The locators are numbered in a way that gives the lower set of numbers to the substituents.

Worked Example #18

Provide the IUPAC name for the following compound.

CH₃

CH₂CH₃

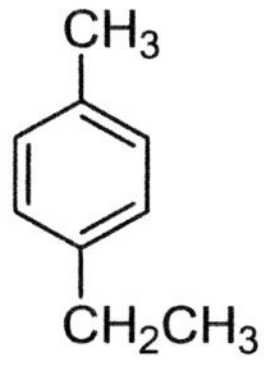

Assign the root.

The root is benzene.

Assign the prefix.

Locator numbers are needed because there are substituents on the benzene ring. The numbers 1 and 4 are used.

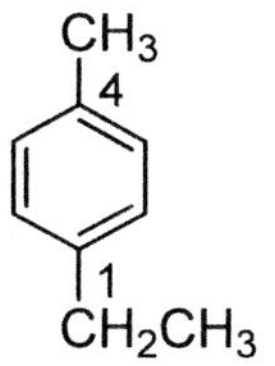

CH₃
4

1
CH₂CH₃

The methyl group is on carbon 4 and the ethyl group is on carbon 1.

The IUPAC name is 1-ethyl-4-methylbenzene.

Try It Yourself #18

Provide the IUPAC name for the following compound.

H₃C   CH₃

            CH₃

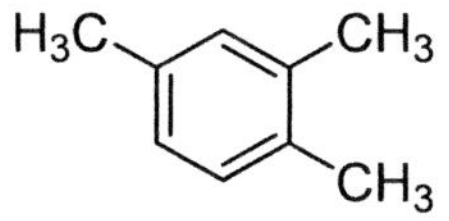

Assign the root.

Root name: _____

Name each substituent.

Number of carbon atoms in substituent: _____

Root name for substituent: _____

Substituent name: _____

Assign a locator number to the substituent. _____

Locator number: _____

Assemble the prefix.

Prefix: _____

Assemble the IUPAC name.

IUPAC name: _____

Worked Example #19

Provide a structure for 1-methyl-3-propylbenzene.

Solution

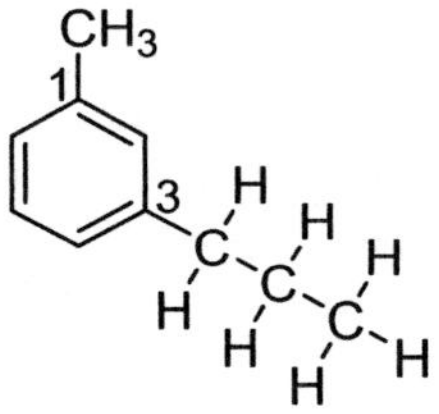

A methyl group is attached to carbon 1 and a propyl (three carbon atoms) group is attached to carbon 3.

Try It Yourself #19

Provide a structure for 1,2,4-triethylbenzene.

Structure:

Practice Problems for Aromatic Hydrocarbons

1. Write the IUPAC name of the following compounds.

   a.

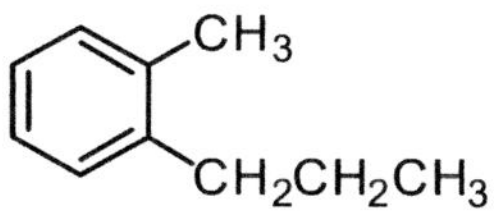

      CH₃
      CH₂CH₂CH₃

   b.

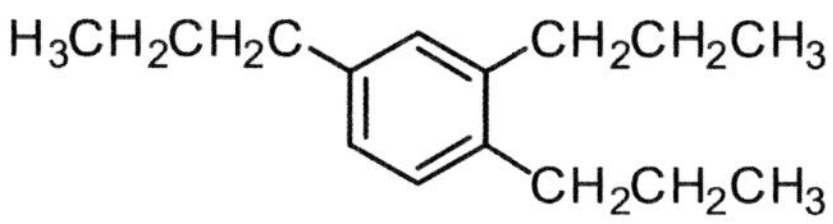

      H₃CH₂CH₂C.   CH₂CH₂CH₃
      CH₂CH₂CH₃

   c.

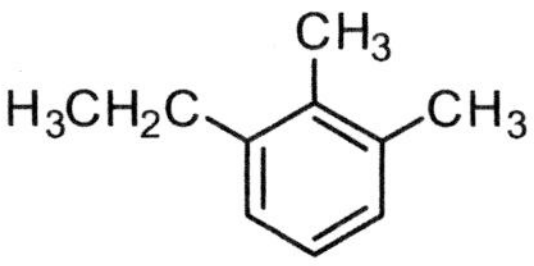

      CH₃
      H₃CH₂C.   CH₃

2. Provide a structure for the following compounds:

   a. 5-ethyl-1,3-dimethylbenzene

   b. 1-butyl-3-methylbenzene

   c, 1,2-diethylbenzene

Chapter 6 Quiz

1. Which of the following compounds represent a saturated hydrocarbon:

    a. $C_{12}H_{26}$

    b. $C_{10}H_{20}$

    c. $C_{15}H_{32}$

    d. C_6H_6

    e. C_9H_{14}

2. Are hydrocarbons hydrophilic or hydrophobic? Explain.

3. What is the C-C-C bond angle in the following compounds?

    a. propane

    b. propene

    c. propyne

4. Identify the structure below as a simple alkene, a diene, a polyene, an alkyne, or an aromatic hydrocarbon:

a.

b.

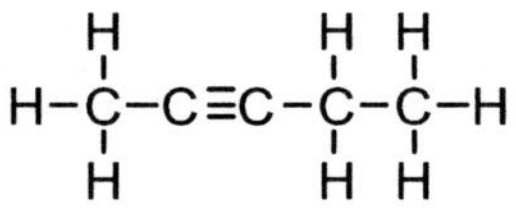

$$H-\underset{\underset{H}{|}}{\overset{\overset{H}{|}}{C}}-C\equiv C-\underset{\underset{H}{|}}{\overset{\overset{H}{|}}{C}}-\underset{\underset{H}{|}}{\overset{\overset{H}{|}}{C}}-H$$

c.

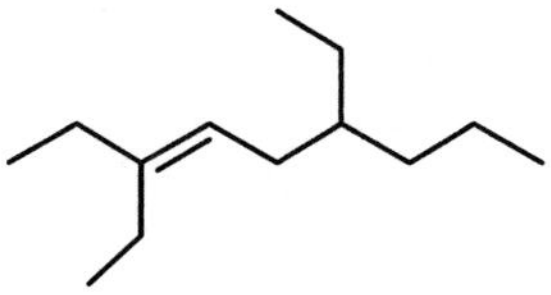

d.

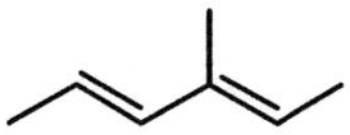

e.

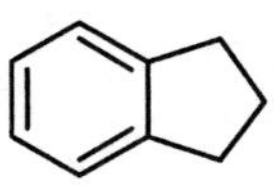

5. Identify the following pairs of molecules as structural isomers, conformational isomers or geometric isomers.

a.

b.

c.

6. Provide a skeletal line structure for the following compounds:

a. 2-ethyl-3-methyl-1-propylbenzene

b.   *cis*-5-ethyl-2,2-dimethyl-3-nonene

c.   *trans*-4-methyl-2-pentene

d.   2,5-dimethyl-3-pentyne

e.   2,4-dimethyl-3-ethylhexane

f.   cyclodecane

7. Write the IUPAC name for the following compounds:

a.

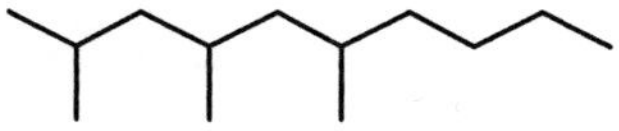

b.

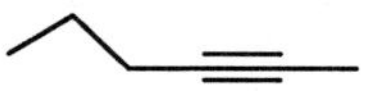

c.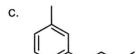

d.

e.

8. Draw the structural formula, the condensed structural formula, and the skeletal line formula for C_3H_8.

9. Draw the structural formula, the condensed structural formula, and the skeletal line structure for 3-ethyl-4-methyloctane.

10. Write the IUPAC name for the following compounds:

a. $CH_3(CH_2)_7CH_3$

b. 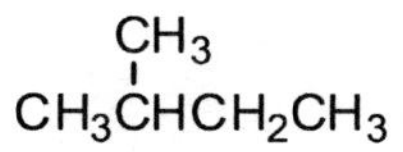 CH_3
   $\overset{|}{CH_3CHCH_2CH_3}$

c. 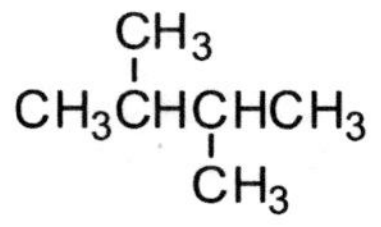 CH_3
   $\overset{|}{CH_3CHCHCH_3}$
       $\overset{|}{CH_3}$

Chapter 6
Answers to Additional Exercises

6.45   Organic compounds contain carbon atoms. Inorganic compounds do not necessarily contain carbon.

6.47   The four types of hydrocarbons are alkanes, alkenes, alkynes, and aromatic hydrocarbons.

6.49   a. False, organic compounds can be synthesized in a laboratory. b. True, pharmaceuticals can be isolated from plants and animals. c. True, pharmaceuticals can be prepared in the chemical laboratory. d. True, Vaseline is a hydrocarbon.

6.51   a. hydrophobic

b. insoluble in water

c. soluble in other hydrocarbons

6.53   Water has a higher boiling point than methane. Water molecules form hydrogen bonds with other water molecules. Hydrogen bonds are stronger than the dispersion forces holding the methane molecules together, so heat is needed to separate water molecules and form a gas.

6.55   Alkanes are hydrocarbons that contain only carbon-carbon single bonds and carbon-hydrogen bonds. They contain no multiple bonds. They are classified as saturated hydrocarbons because they contain the maximum number of hydrogen atoms for a given number of carbon atoms.

6.57   Every atom in an alkane must have a tetrahedral geometry; therefore, the overall shape of the molecule takes on a zig-zag appearance when the chain has three or more carbon atoms. Yes, it can have other shapes due to the different conformations that arise from C-C bond rotation.

6.59   The geometry of a carbon atom in an alkane is tetrahedral.

6.61   The geometry of a carbon atom in a triple bond of an alkyne is linear.

6.63   a. Different compounds because they have different formulas. The two compounds are C_3H_8 and C_2H_6. b. Different compounds because they are structural isomers. One compound has four carbons in the main chain and the other compound has three carbons in the main chain. c. Different compounds because they have different formulas. The two compounds are C_4H_{10} and C_5H_{12}. d. Identical compounds, but different conformations.

6.65   The structure on the left has five carbons in a straight chain. The structure on the right has three carbons in the main chain and two CH_3 groups branching off the middle carbon. All structural isomers have the same chemical formula.

6.67   a. CH_3
               |
       $CH_3CHCHCH_2CH_2CH_2CH_3$
               |
               CH_3

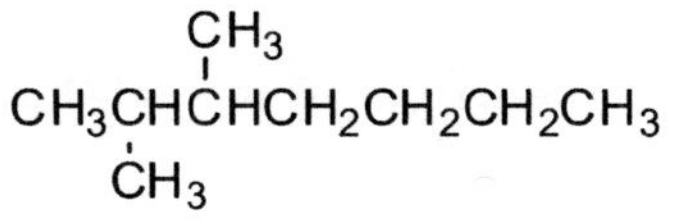

       b. $CH_3CHCH_2CH_2CH_2CH_2CH_3$
                 |
                 CH_2CH_3

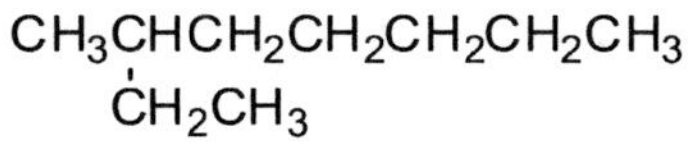

       c. CH_3
               |
       $CH_3CCH_2CH_3$
               |
               CH_3

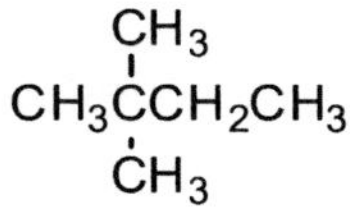

       d. $CH_3CH_2CH_2CH_3$

       e. $CH_3CHCH_2CH_3$
                 |
                 CH_3

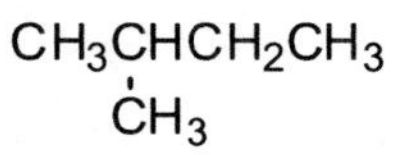

       f. $CH_3CH_2CH_2CH_2CH_2CH_2CH_3$

6.69   a.

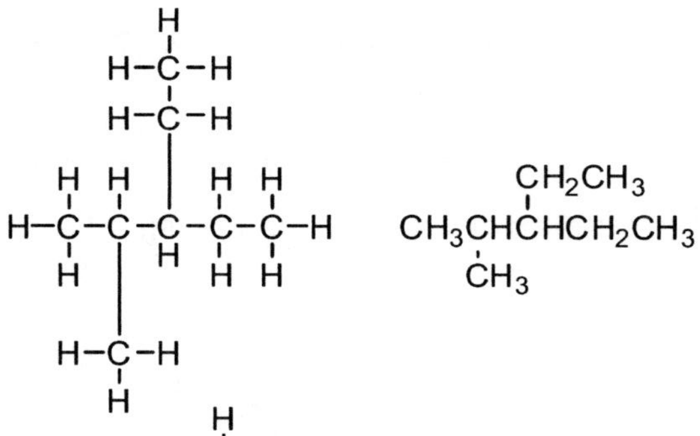

       CH_2CH_3
       |
       $CH_3CHCHCH_2CH_3$
       |
       CH_3

       b.

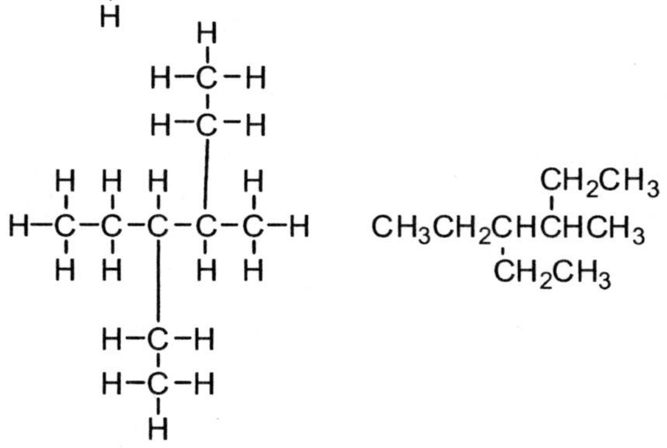

       CH_2CH_3
       |
       $CH_3CH_2CHCHCH_3$
       |
       CH_2CH_3

       c.

       CH_3
       |
       $CH_3CCHCH_2CH_3$
       | |
       CH_3
       CH_3

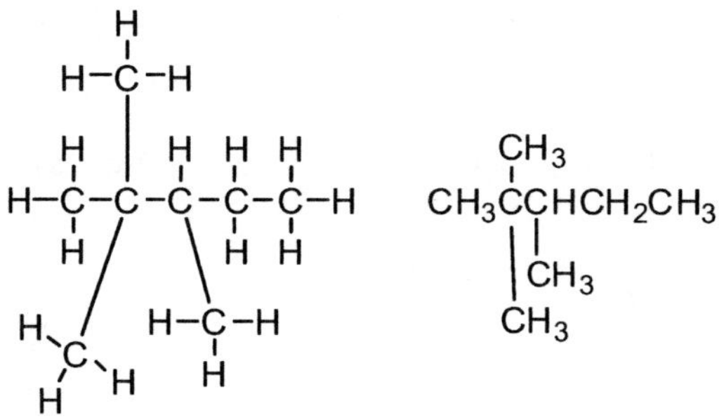

d.

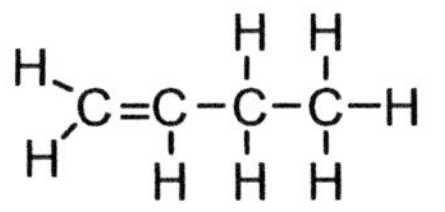

CH₂CHCH₂CH₃ → $CH_2CHCH_2CH_3$

6.71   a.

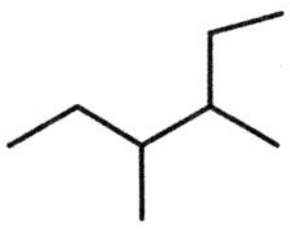

b.

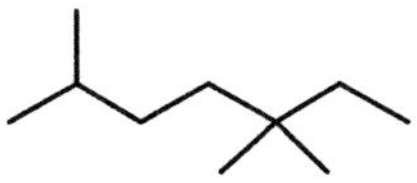

c.

6.73   Three carbons in a ring  ▷. Five carbons in a ring ⬠. The ring with five carbons is more commonly found in nature. The three-carbon ring is strained; the bond angles are forced to be 60° when they should be 109.5°.

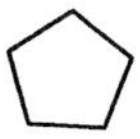

6.75

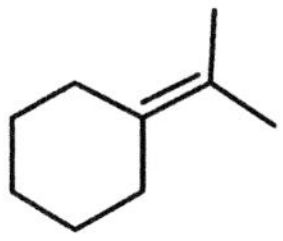

simple alkene    diene 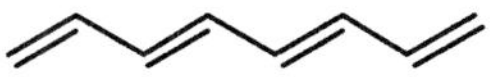   polyene 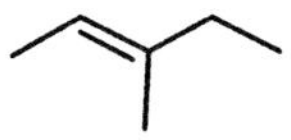   simple alkene

6.77

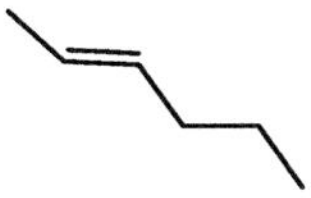

6.79   Yes, the rotational freedom around a carbon-carbon double bond is different from that around a carbon-carbon single bond. There is no rotation about a carbon-carbon double bond, while there is rotation about a carbon-carbon single bond.

6.81   The C-C-C bond angle is 180° because the carbon atoms are part of a carbon-carbon triple bond.

6.83   The three parts of an IUPAC name are the prefix, the root, and the suffix.

6.85   a. Heptane. There are seven carbon atoms in the chain.

        b. Butane. There are four carbon atoms in the chain.

c. Propane There are three carbon atoms in the chain.

6.87

hexane   2-methylpentane   3-methylpentane

2,2-dimethylbutane   2,3-dimethylbutane

If two compounds end up having the same name, then they are the same compound, even though they are different conformations.

6.89   a. 3-methylhexane. There are six carbon atoms in the main chain and a methyl group located at carbon 3.

b. 4-propylheptane. There are seven carbon atoms in the main chain and a propyl group located at carbon 4.

c. 2,3,3-trimethylhexane. There are six carbon atoms in the main chain and one methyl group located at carbon 2 and two methyl groups located at carbon 3.

6.91   a. Cyclooctane. There are eight carbon atoms in the ring.

b. Cycloheptane. There are seven carbon atoms in the ring.

c. Cyclopentane. There are five carbon atoms in the ring.

6.93   a. *trans*-2-pentene. There are five carbon atoms in the main chain. The double bond is a trans double bond starting at carbon 2.

b. 1-hexene. There are six carbon atoms in the main chain with a double bond starting at carbon 1.

c. 2,3-dimethyl-2-pentene. There are five carbon atoms in the main chain. The double bond starts at carbon 2. There are two methyl groups on carbons 2 and 3.

d. 3-ethyl-2-hexene. There are six carbons in the main chain with the double bond starting at carbon 2. There is an ethyl group on carbon 3.

e. Cyclopentene. There are five carbon atoms and a double bond in the ring.

f. Cycloheptene. There are seven carbon atoms and a double bond in the ring.

6.95   a.

The root name hexene indicates a six-carbon chain. The suffix "ene" indicates a double bond which starts at carbon 3 and is cis.

b. The root name pentene indicates a five-carbon chain. The suffix "ene" indicates a double bond that starts at carbon 2 and is trans.

c.

The root name pentene indicates a five-carbon chain. The suffix "ene" indicates a double bond that starts at carbon 2 and is cis. There is a methyl group on carbon 4.

d.

The root name heptene indicates a seven-carbon chain. The suffix "ene" indicates a double bond that starts at carbon 3 and is trans. There is an ethyl group on carbon 5.

6.97   a. 3-ethyl-4-octyne. There are eight carbon atoms in the main chain with a triple bond starting at carbon 4. There is an ethyl group on carbon 3.

b. 1-heptyne. There are seven carbon atoms in the main chain with the triple bond starting at carbon 1.

c. 3-hexyne. There are six carbon atoms in the main chain with the triple bond starting at carbon 3.

d. 2-methyl-5-decyne. There are ten carbon atoms in the main chain with the triple bond starting at carbon 5. There is a methyl group on carbon 2.

e. 2,5,6-trimethyl-3-heptyne. There are seven carbon atoms in the main chain with the triple bond starting at carbon 3. There are three methyl groups located on carbons 2, 5, and 6.

6.99   Benzene is classified as an unsaturated hydrocarbon because each carbon does not have the maximum number of hydrogen atoms attached.

6.101 Bezene and cyclohexane both have six carbon atoms in a ring. Benzene has six hydrogen atoms and three carbon-carbon double bonds and is more stable due to the delocalization of the electrons. The carbon atoms in benzene are trigonal planar. Cyclohexane has 12 hydrogen atoms and only carbon-carbon single bonds. In cyclohexane the carbon atoms are tetrahedral.

6.103

6.105 a. propylbenzene

b. 1,3-dimethylbenzene

c. 1-ethyl-2-methylbenzene

6.107 a.

PABA

The "benz" part of the root name appears in the IUPAC name for PABA.

b. The substitution on the aromatic ring is 1,4-.

6.109 a. Vitamins E and K_1 contain aromatic rings. Vitamins A and D_3 can be classified as polyenes.

b. There are 13 carbons in the long hydrocarbon chain. It is a branched hydrocarbon.

c. The two heteroatoms in vitamin E are oxygen atoms.

d. These vitamins are hydrophobic. They contain mostly carbon and hydrogen atoms and thus are similar to fat molecules.

e. Since these vitamins are hydrophobic, they are not soluble in water (the main component of urine). Therefore, they would not be readily excreted in urine.

f. Vitamin A plays a role in the chemistry of vision.

g. There are six substituents on the aromatic ring in vitamin E.

h. Vitamin D_3 and cholesterol are very similar in structure. Vitamin D_3 is missing one carbon-carbon bond in the second six-carbon ring and has three more carbon-carbon double bonds than cholesterol. Yes, it seems plausible that the body can synthesize cholesterol out of vitamin D_3.

6.111 a. Yes. Some electrons should be delocalized.

b. No. Naphthalene should not be chemically reactive.

c. Yes. Naphthalene and anthracene should be flat molecules. All of the carbon atoms are trigonal planar.

d. The bond angles are 120°.

e.

6.113 A carcinogen is a molecule or compound that causes cancer. Benzene is a carcinogen.

6.115

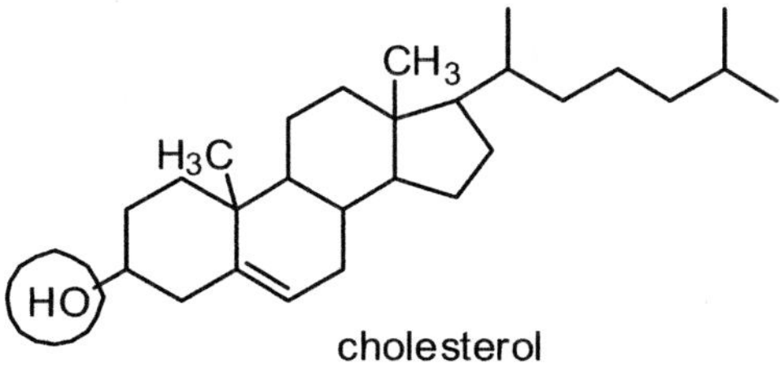

cholesterol

6.117 Cholesterol is used to build cell membranes and make bile, vitamin D, and many important steroid hormones.

6.119 Plaque contains cholesterol and fats and can build up along the artery walls, restricting blood flow.

6.121 The carotid artery is most likely blocked if a person has a stroke.

6.123 The liver synthesizes cholesterol.

Chapter 7

Organic Functional Groups

Chapter Summary

In this chapter, you continued your focus on the fundamentals of organic chemistry. You have been introduced to all the major functional groups in organic chemistry and biochemistry. You have learned about organic compounds that contain oxygen, nitrogen, and phosphorus atoms. You have learned to identify these functional groups within natural products and pharmaceuticals. Knowing the structure of functional groups is key to understanding the way they behave in a chemical reaction.

Section 7.1 C—O-Containing Functional Groups: Alcohols and Ethers

Certain combinations of atoms and bonds are seen frequently; these characteristic arrangements of atoms and bonds are known as functional groups because they function in a characteristic way in a chemical reaction. There are two functional groups derived from water: alcohols and ethers. An alcohol is an R-OH group, and an ether is an R-OR group, where R is one or more carbon atoms. The carbon atom attached to the alcohol group must be a saturated carbon. The carbon atoms attached to the oxygen atom in an ether can be a saturated carbon or an aromatic carbon. Alcohols can be classified as primary (1°), secondary (2°), or tertiary (3°) depending on whether they have one, two, or three carbon atoms, respectively, on the carbon atom bearing the OH group. An OH group is sometimes called a hydroxyl group. When writing the skeletal line structure of an alcohol or ether, the oxygen atoms must always be written in, in order to distinguish the oxygen atom from a carbon atom. In addition, hydrogen atoms bonded to a heteroatom must always be written in; only those hydrogen atoms attached to carbon atoms may be omitted.

In a Nutshell: Naming Alcohols and Ethers

Follow these guidelines to name an alcohol: 1) assign the root, 2) assign the suffix, 3) assign a locator number to the root indicating the location of the alcohol, 4) assign a prefix. To assign the root, follow the IUPAC rules established for hydrocarbons to name the root. For the main chain, select the longest continuous chain that contains the alcohol functional group. To assign the suffix, change the suffix of the IUPAC name from "ane" to "anol." To assign a locator number, begin numbering the main chain from the end that is closest to the OH group. Place the locator number followed by a dash in front of the root name. Assign a prefix if the main chain contains substituents.

Common names are used to name simple ethers. The common name for an ether is constructed by naming each R group as though it were a substituent, with an "yl" ending, then ending the name with "ether." If both R groups are identical, the prefix "di" is added before the name of the R groups.

Worked Example #1

Identify the following compounds as alcohols or ethers. If the compound is an alcohol, indicate whether it is a primary, secondary, or tertiary alcohol.

a.
$$\begin{array}{cccc} & H & H & & H \\ & | & | & & | \\ H- & C- & C- & O- & C-H \\ & | & | & & | \\ & H & H & & H \end{array}$$

b.
$$\begin{array}{ccc} H & H & H \\ | & | & | \\ H-C-C-C-O-H \\ | & | & | \\ H & H & H \end{array}$$

c.
$$\begin{array}{cccc} & & H & & \\ & & | & & \\ & H & O & H & H \\ & | & | & | & | \\ H- & C- & C- & C- & C-H \\ & | & & | & | \\ & H & & H & H \\ & & | & & \\ & & H-C-H & & \\ & & | & & \\ & & H & & \end{array}$$

a. *Ether. There is a methyl group and an ethyl group attached to the oxygen.*

b. *Primary alcohol. There is an OH group present indicating the alcohol group. There is only one carbon atom attached to the carbon bearing the OH group, so it is a primary alcohol.*

c. *Tertiary alcohol. There is an OH group present indicating the alcohol group. There are three carbon atoms attached to the carbon bearing the OH group, so it is a tertiary alcohol.*

Try It Yourself #1

Identify the following compounds as alcohols or ethers. If the compound is an alcohol, indicate whether it is a primary, secondary, or tertiary alcohol.

a.
```
            H
            |
     H  H  O  H  H
     |  |  |  |  |
  H--C--C--C--C--C--H
     |  |  |  |  |
     H  H  H  H  H
```

b.
```
     H  H  H      H  H
     |  |  |      |  |
  H--C--C--C--O--C--C--H
     |  |  |      |  |
     H  H  H      H  H
```

c.
```
                     H
                     |
     H  H  H  H  H   O
     |  |  |  |  |   |
  H--C--C--C--C--C---C--H
     |  |  |  |  |  |
     H  |  H  H  H  H
        |
      H--C--H
        |
      H--C--H
        |
        H
```

a. Atoms bonded to oxygen: _____

    Alcohol or ether: _____

    If it is an alcohol, number of carbon atoms attached to the carbon atom bearing the

    OH group: _____

    Type of alcohol: _____

b. Atoms bonded to oxygen: _____

    Alcohol or ether: _____

    If it is an alcohol, number of carbon atoms attached to the carbon atom bearing the

    OH group: _____

    Type of alcohol: _____

c. Atoms bonded to oxygen: _____

    Alcohol or ether: _____

    If it is an alcohol, number of carbon atoms attached to the carbon atom bearing the

    OH group: _____

Type of alcohol: _____

Worked Example #2

Assign an IUPAC name to the alcohols and a common name to the ethers in the Worked Example #1.

a. *The two substituents attached to the oxygen atom are methyl and ethyl. The common name for this ether is ethyl methyl ether.*

$$H-\overset{3}{\underset{H}{\overset{H}{C}}}-\overset{2}{\underset{H}{\overset{H}{C}}}-\overset{1}{\underset{H}{\overset{H}{C}}}-O-H$$

b. *1-propanol.* *The main chain that contains the alcohol group has three carbon atoms. The root name is propane. Since the molecule is an alcohol, the suffix is changed to "anol." The alcohol group is located on carbon 1.*

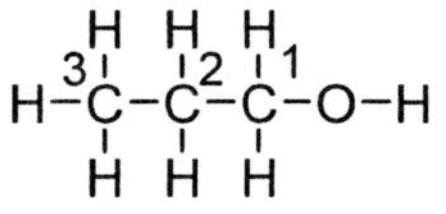

$$\begin{array}{c} H \\ | \\ H-\overset{1}{\underset{H}{\overset{H}{C}}}-\overset{2}{\underset{H}{\overset{O}{C}}}-\overset{3}{\underset{H}{\overset{H}{C}}}-\overset{4}{\underset{H}{\overset{H}{C}}}-H \\ \qquad | \\ \qquad H-\overset{}{\underset{H}{C}}-H \\ \qquad\qquad | \\ \qquad\qquad H \end{array}$$

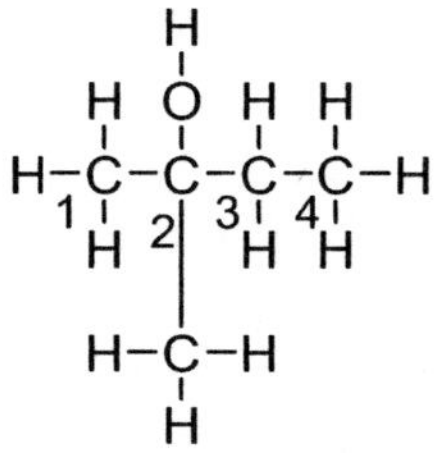

c. *2-methyl-2-butanol.* *The main chain that contains the alcohol group has four carbon atoms. The root name is butane. Since the molecule is an alcohol, the suffix is changed to "anol." The alcohol is located on carbon 2. There is a methyl group also located on carbon 2.*

Try It Yourself #2

Assign an IUPAC name to the alcohols and a common name to the ethers in the Try It Yourself #1.

a. *Number of carbon atoms in the main chain:* _____

Root name: _____

Suffix ending for an alcohol: _____

Locator number for the alcohol: _____

IUPAC name: _____

b. Substituents attached to the oxygen atom: _____

Common name: _____

c. Number of carbon atoms in the main chain: _____

Root name: _____

Suffix ending for an alcohol: _____

Locator number for the alcohol: _____

Substituent name: _____

Locator number for the substituent: _____

IUPAC name: _____

In a Nutshell: Phenols

When an O—H group is attached to a carbon atom that is part of an aromatic ring, the functional group is no longer considered an alcohol, but a phenol. The phenol functional group includes the entire benzene ring together with the OH group. To name a phenol, follow the rules for naming substituted benzenes, but the root name is changed from benzene to phenol. Begin numbering the carbon atoms at the carbon atom bonded to the OH group and then proceed clockwise or counterclockwise, in whichever direction gives the lower set of numbers.

Worked Example #3

Assign an IUPAC name to the following molecules:

a.

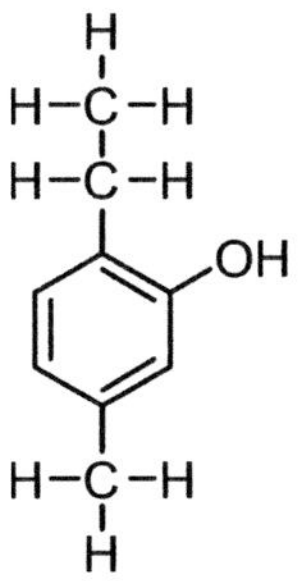

b.

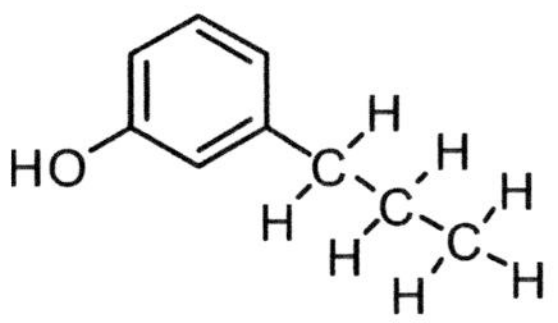

a. *2-ethyl-5-methyphenol*

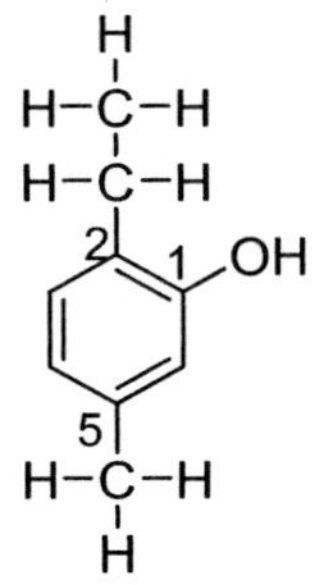

b. *3-propylphenol.*

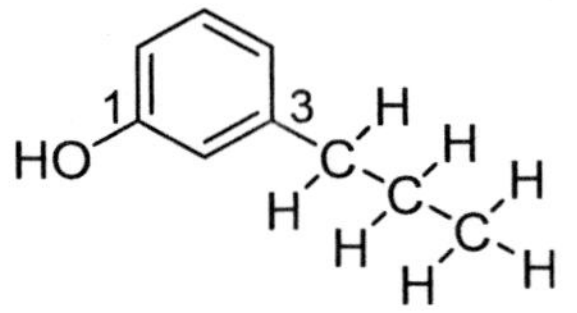

Try It Yourself #3

Assign an IUPAC name to the following molecules:

a.

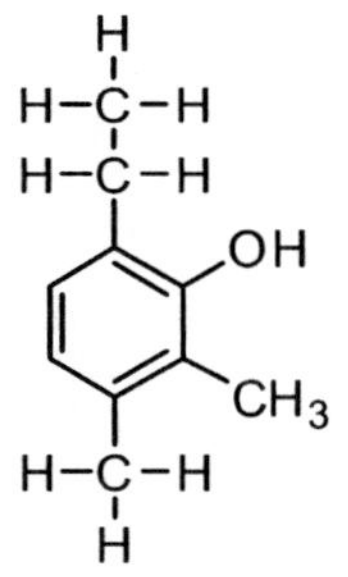

b.

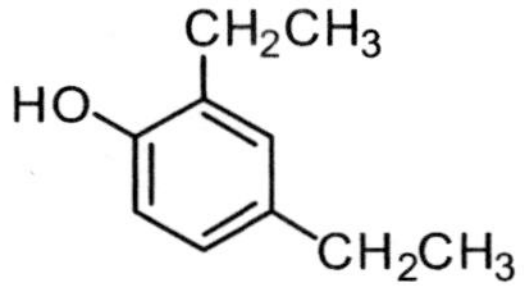

a. *Subsituents on phenol:* _____

   Locator number for substituents: _____

   IUPAC name: _____

b. *Subsituents on phenol:* _____

   Locator number for substituents: _____

IUPAC name: _____

In a Nutshell: Hydrogen Bonding in Alcohols and Ethers

Ethers cannot hydrogen bond with other ether molecules because they lack an OH bond and thus they have relatively low boiling points. Ethers with two or three carbon atoms are soluble in water because water can form hydrogen bonds with the oxygen atom. Alcohols and phenols are capable of hydrogen bonding with one another. The solubility of an alcohol in water depends on the relative proportion of hydrocarbon structure to the number of OH groups in the overall molecular structure.   In general, the more OH groups there are in a molecule, the more water soluble it is and the more extensive the hydrocarbon portion is, the less water soluble it is.

Practice Problems for C—O-Containing Functional Groups: Alcohols and Ethers

1. Indicate whether the following molecules are alcohols or ethers. If it is an alcohol, indicate whether it is a primary, secondary, or tertiary alcohol or a phenol.

    a.

    $$H-\underset{\underset{H}{|}}{\overset{\overset{H}{|}}{C}}-\underset{\underset{H}{|}}{\overset{\overset{H}{|}}{C}}-O-\underset{\underset{H}{|}}{\overset{\overset{H}{|}}{C}}-\underset{\underset{H}{|}}{\overset{\overset{H}{|}}{C}}-\underset{\underset{H}{|}}{\overset{\overset{H}{|}}{C}}-\underset{\underset{H}{|}}{\overset{\overset{H}{|}}{C}}-H$$

    b.

    $$H-\underset{\underset{H}{|}}{\overset{\overset{H}{|}}{C}}-\underset{\underset{H}{|}}{\overset{\overset{H}{|}}{C}}-\underset{\underset{H}{|}}{\overset{\overset{H}{|}}{C}}-\underset{\underset{H}{|}}{\overset{\overset{H}{|}}{C}}-\underset{\underset{H}{|}}{\overset{\overset{O-H}{|}}{C}}-\underset{\underset{H}{|}}{\overset{\overset{H}{|}}{C}}-\underset{\underset{H}{|}}{\overset{\overset{H}{|}}{C}}-H$$

c.

d.

e.

2. Assign an IUPAC name to the molecules shown in Question 1.

3. Do you expect any of the molecules in Question 1 to be soluble in water? If so, which ones? Explain.

Section 7.2 C=O-Containing Functional Groups

Several functional groups contain a carbon-oxygen double bond, commonly known as a carbonyl group. The carbon atom of a carbonyl group is always attached to two groups or atoms. One group is usually an "R" group. The other group is a hydrogen atom, another R group, or a heteroatom. The identity of the carbonyl-containing functional group depends on what this atom or group is. The main carbonyl-containing functional groups are aldehydes, ketones, carboxylic acids, esters, thioesters, and amides. The geometry around the carbonyl carbon is trigonal planar. The carbonyl group is polar because oxygen is more electronegative than carbon.

In a Nutshell: Aldehydes

An aldehyde is a carbonyl group attached to a hydrogen atom and an R group. Formaldehyde is the sole exception because it has two H atoms attached to the carbonyl group. In condensed structural notation an aldehyde is written as RCHO. When writing the skeletal line structure for an aldehyde, the carbon atom of the aldehyde is not written in, but the hydrogen atom attached to the carbonyl group is shown. To name an aldehyde following the IUPAC naming system, the suffix is changed from "ane" to "anal." No locator name is needed for straight-chain aldehydes because the aldehyde functional group appears at the end of the chain. In branched-chain aldehydes, numbering begins with the carbonyl carbon.

In a Nutshell: Ketones

A ketone is a carbonyl group bonded to two R groups that are both carbon atoms. When writing condensed structural formulas, a ketone is written as RCOR. To name a ketone using IUPAC rules, the suffix is changed from "ane" to "anone," and a locator number is inserted in front of the root to indicate the location of the carbonyl carbon along the main chain. Number the main chain starting from the end closest to the carbonyl group.

Worked Example #4

Identify the following compounds as aldehydes or ketones. Write the IUPAC name for them.

a.

b.

a. *This molecule is an aldehyde. It has a hydrogen atom attached to the carbonyl*

group. The IUPAC name is 2-ethylpentanal.   *There are five carbon*
atoms in the chain with the aldehyde. The root is pentane. The suffix changes from
"ane" to "anal." The ethyl group is attached to carbon 2 and is added as a prefix.

b. *This molecule is a ketone. It has two carbon atoms attached to the carbonyl group.*
The IUPAC name is 3-hexanone. There are six carbon atoms in the chain. The root

is hexane. The suffix changes from "ane" to "anone."   *The locator*
number is 3 since the carbonyl group is located at carbon 3.

Try It Yourself #4

Identify the following compounds as aldehydes or ketones. Write the IUPAC name for them.

a.

b.

a. *Atoms attached to carbonyl carbon:* _____

Aldehyde or ketone: _____

Number of carbon atoms in ring: _____

Root name: _____

Suffix ending: _____

IUPAC name: _____

b. *Atoms attached to carbonyl carbon:* _____

Aldehyde or ketone: _____

Number of carbon atoms in chain: _____

Root name: _____

Suffix ending: _____

IUPAC name: _____

In a Nutshell: Carboxylic Acids

A carboxylic acid is a carbonyl group attached directly to an O—H group. The condensed structural formula for a carboxylic acid is written as either an RCO_2H or $RCOOH$. There are three other functional groups that contain a carbonyl carbon attached to a heteroatom: esters, thioesters, and amides. Because esters, thioesters, and amides are derived from carboxylic acids through chemical reactions these three functional groups are often called carboxylic acid derivatives.

In a carboxylic acid, the carbonyl group and the O-H together act as a single unit and thus constitute one functional group. Carboxylic acids and alcohols are classified as two distinct functional groups. To name a carboxylic acid using the IUPAC naming system, the suffix is changed from "ane" to "anoic acid." No locator number is needed, though numbering of the carbon atoms in the chain begins with the carbonyl carbon.

The carboxylic acid functional group readily loses a proton to form a carboxylate ion, RCO_2^-. Whether a carboxylic acid exists in its neutral or ionic form depends on its environment. In the body carboxylic acids exist in their ionic form. In its ionic form, RCO_2^-, the suffix in the name of the carboxylic acid changes from "ic acid" to "ate."

Carboxylic acids are capable of hydrogen bonding because they contain the strong O—H bond dipole and there is a partial negative charge on both oxygen atoms of the carboxylic acid. Carboxylic acids are soluble in water provided that the hydrocarbon portion of the R group is not large enough to outweigh the polar characteristics of the carboxylic acid functional group. The ionic form of a carboxylic acid is even more water soluble than the neutral form.

238   Chapter 7

Long hydrocarbon chains containing a carboxylic acid are known as fatty acids. Some compounds contain two or more carboxylic acid functional groups, such as citric acid.

Worked Example #5

Indicate whether the following compounds are the neutral form or the ionic form of the carboxylic acid. Write the IUPAC name for the carboxylic acid.

a.

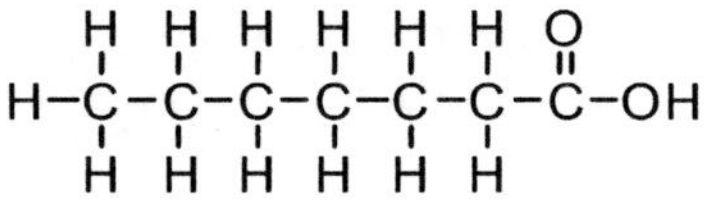

b.

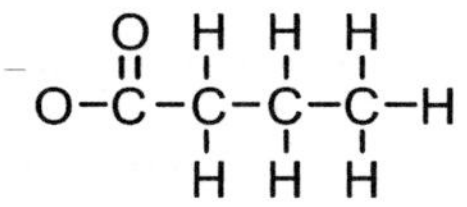

   a.   *This molecule is a neutral carboxylic acid. The IUPAC name is heptanoic acid. There are seven carbon atoms in the chain, so the root name is heptane. The suffix is changed from "ane" to "anoic acid."*

   b.   *This molecule is the ionic form of the carboxylic acid. The IUPAC name for the carboxylic acid is butanoic acid. There are four carbon atoms in the chain, so the root name is butane. The suffix is changed from "ane" to "anoic acid."*

Try It Yourself #5

Indicate whether the following compounds are the neutral form or the ionic form of the carboxylic acid. Write the IUPAC name for the carboxylic acid.

a.

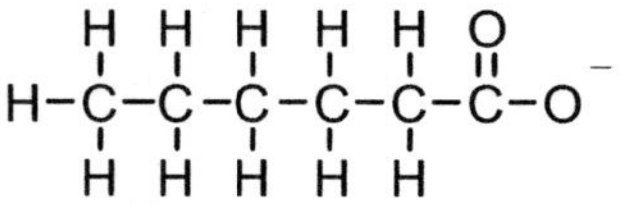

b.

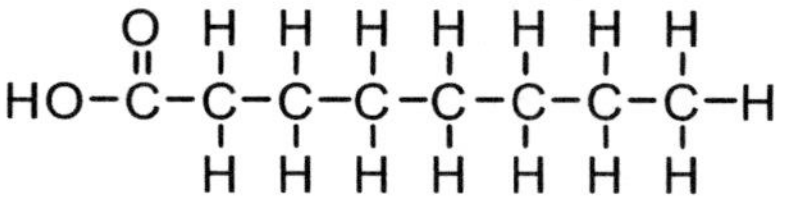

a. *Ionic form or neutral form:* _____

    Number of carbon atoms in chain: _____

    Root name: _____

    Suffix ending: _____

    IUPAC name: _____

b. *Ionic form or neutral form:* _____

    Number of carbon atoms in chain: _____

    Root name: _____

    Suffix ending: _____

    IUPAC name: _____

In a Nutshell: Esters and Thioesters

An ester has an O—R group and an R group attached to the carbonyl carbon. A thioester has an S—R group and an R group attached to the carbonyl carbon. Esters and thioesters contain two R groups, one attached to the carbonyl carbon and the other attached to the oxygen or sulfur atom. These two R groups may be the same (R = R') or they may be different (R ≠ R'). An ester is written as RCO_2R' or RCOOR' in condensed structural notation; a thioester is written as RCOSR'. The ester functional group is present in a variety of biomolecules, including triglycerides. The thioester functional group is less common, but some are involved in metabolism, such as acetyl coenzyme A.

Worked Example #6

Americaine® is an anesthetic lubricant. Circle and label the ester functional group. Circle and label the aromatic ring functional group.

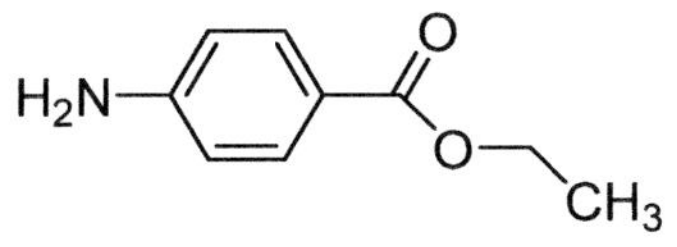

Solution

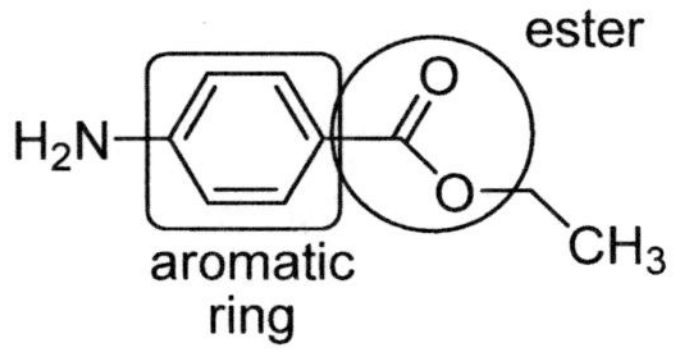

Try It Yourself #6

The structure of aspirin (acetylsalicylic acid) is shown below. Circle and label the ester functional group.   What other functional groups are present in the molecule?

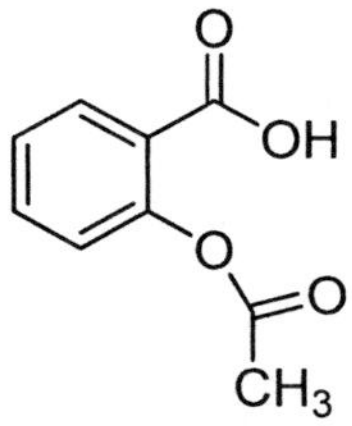

Extension Topic 7-1: Assigning the IUPAC Name for an Ester

To assign an IUPAC name for an ester, use the following rules: 1) assign the root, 2) assign the suffix, 3) assign the prefix. To assign the root, count the number of carbon atoms in the main chain containing the carbonyl carbon. Assign the root. For branched-chain esters, number the chain starting with the carbonyl carbon. To assign the suffix, change "ane" to "anoate." To assign the prefix, name the R group attached to the oxygen atom as you would any substituent with a "yl" ending, and insert the prefix in front of the root name.

Worked Example #7

Write the IUPAC name for the following ester.

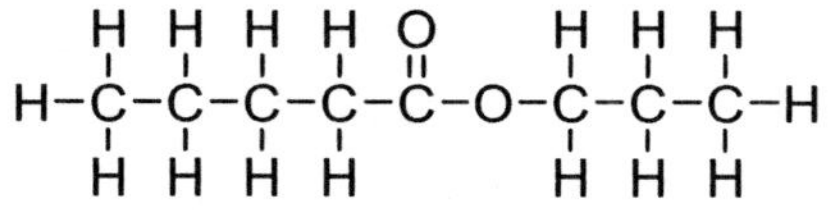

Propyl pentanoate. There are five carbon atoms in the chain containing the carbonyl carbon. The root name is pentane. The suffix changes from "ane" to "anoate." There are three carbon atoms in the chain attached to the oxygen atom. The substituent name is propyl.

Try It Yourself #7

Write the IUPAC name for the following ester.

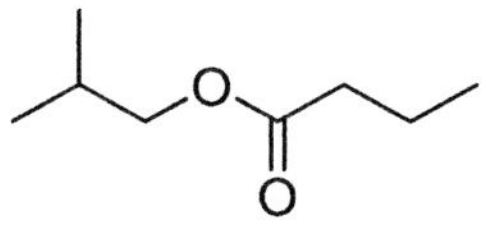

Number of carbon atoms in the main chain containing the carbonyl carbon: _____

Root name: _____

Suffix change: _____

Number of carbon atoms in the chain attached to the oxygen atom: _____

Substituent name for the chain: _____

IUPAC name for the ester: _____

Worked Example #8

Write the skeletal line structure for the 1-methylethyl butanoate.

Solution

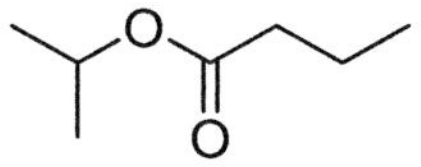

The root name is butane. There are four carbon atoms in the chain containing the carbonyl carbon. The substituent attached to the oxygen atom is an ethyl group with a methyl group on carbon 1.

Try It Yourself #8

Write the skeletal line structure for cyclopentyl propanoate.

Solution

Number of carbon atoms in the main chain: _____

Structure of the substituent attached to the oxygen atom: _____

Structure of the ester:

In a Nutshell: Amides

An amide functional group contains a nitrogen atom attached to carbonyl carbon. The nitrogen atom will be bonded to two other atoms in addition to the carbonyl carbon. These atoms may be hydrogen atoms or carbon atoms (R groups). The condensed structural formula for an amide appears as $CONH_2$, CONHR, CON(R)R', depending on the number of R groups attached to the nitrogen atom.

To name simple amides, count the number of carbon atoms in the carbon chain containing the carbonyl carbon and assign the root. Change the suffix from "ane" to "anamide." If the nitrogen atom contains one or two R groups, they are named as substituents, using the "yl" ending, which is inserted as a prefix following the letter N- to indicate that the substituent is located on the nitrogen atom rather than on the main chain.

The amide functional group is very common in nature. Molecules containing amides include proteins, endorphins, and penicillin antibiotics.

Worked Example #9

The structure for Tylenol®, acetaminophen, follows. Circle and label the amide functional group. How many R groups are attached to the nitrogen atom? What other functional groups are present in acetaminophen?

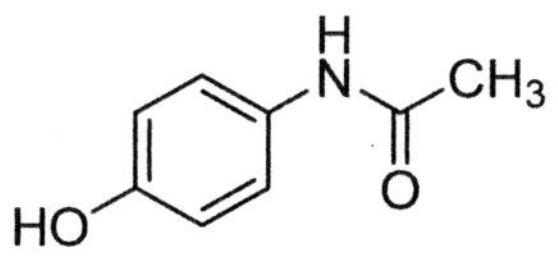

Solution

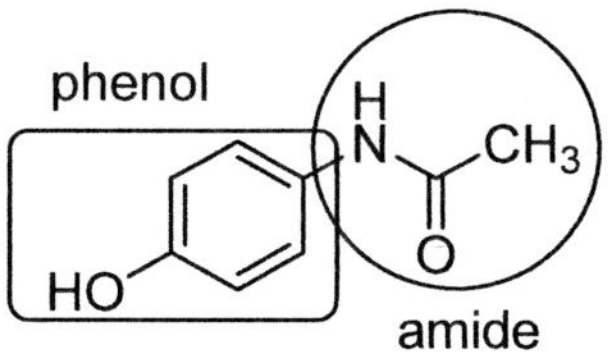

There is one R group attached to the nitrogen atom; it is the aromatic ring.

Try It Yourself #9

Keflex® is an antibiotic. Circle and label the amide functional groups. What other functional groups are present in Kelfex?

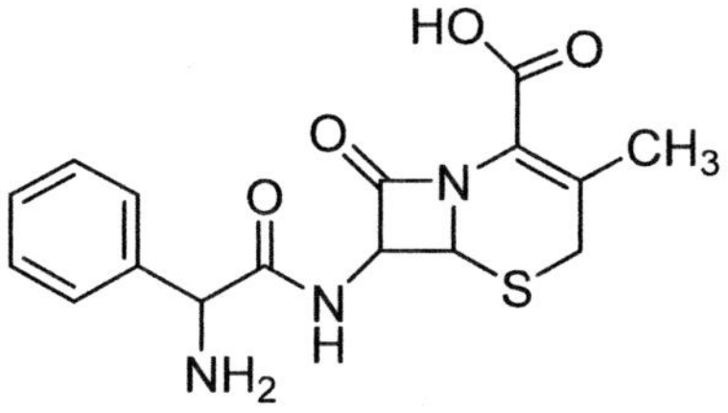

Practice Problems for C=O-Containing Functional Groups

1. Identify the following carbonyl-containing functional groups:

a.

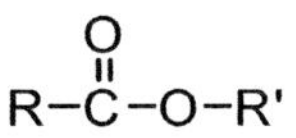

b.

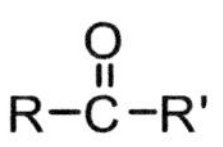

c.

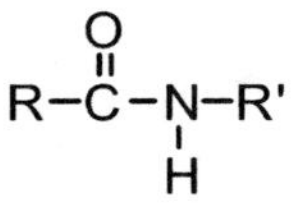

$$R-\overset{\overset{\displaystyle O}{\|}}{C}-\underset{\underset{\displaystyle H}{|}}{N}-R'$$

d.

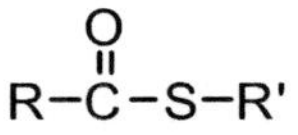

$$R-\overset{\overset{\displaystyle O}{\|}}{C}-S-R'$$

e.

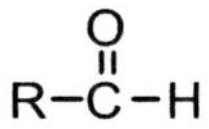

$$R-\overset{\overset{\displaystyle O}{\|}}{C}-H$$

f.

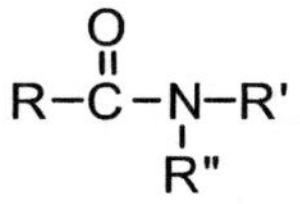

$$R-\overset{\overset{\displaystyle O}{\|}}{C}-\underset{\underset{\displaystyle R''}{|}}{N}-R'$$

g.

$$R-\overset{\overset{\displaystyle O}{\|}}{C}-O-H$$

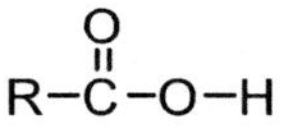

2. Depo Provera (medroxyprogesterone acetate) is used as a contraceptive. Its structure is shown below. Circle and label the carbonyl-containing functional groups.

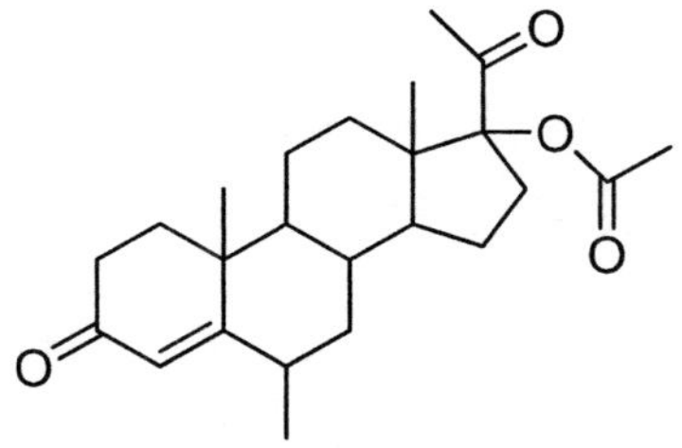

3. Write the IUPAC name for the following molecules:

a.

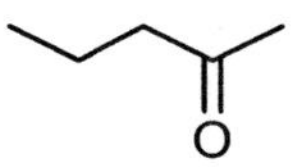

b.

c.

Section 7.3 C—N-Containing Functional Groups: Amines

Amines are functional groups containing a nitrogen atom with three single bonds, each attached to either a carbon atom or a hydrogen atom. The carbon atoms on the amine can be saturated or aromatic. The amines are designated 1°, 2°, and 3° if there are one, two, or three R groups, respectively, attached to the nitrogen atom. The condensed structural formula will appear as $-NH_2$ for a primary amine, $-NHR$ for a secondary amine, and $-NR_2$ for a tertiary amine. The nitrogen atom is trigonal pyramidal and has bond angles of 109.5°.

In a Nutshell: Naming Amines

The common names for simple amines are assembled in a manner similar to ethers; the R groups are named as though they are substituents, with a "yl" ending, and are followed by the term "amine."

In a Nutshell: Alkaloids and Complex Amines

Amines are a common functional group in drug molecules such as antidepressants and opioid analgesics. The amine functional group is also present in many natural brain chemicals, such as dopamine and adrenaline. They are also present in compounds that are derived from plants; these amines derived from plants are known as alkaloids.

In a Nutshell: Hydrogen Bonding in Amines

Amines containing one or two N-H bonds are capable of hydrogen bonding. Amines with a small hydrocarbon component are soluble in water. If the hydrocarbon portion of the amine is large, the molecule will be insoluble in water.

In a Nutshell: Ionic and Neutral Forms of an Amine

The nonbonding electrons on the amine nitrogen atom readily form a bond to a proton to make an N—H bond, giving the nitrogen atom fourth bond. This process converts the neutral amine into a polyatomic ion analogous to the ammonium ion. Whether the amine exists in its neutral or ionic form depends on its environment. In the cell, amines are usually in their ionic form. The ionic form of the amine is soluble in water because there is a full positive charge on the polyatomic ion.

Worked Example #10

Wellbutrin® is an antidepressant.

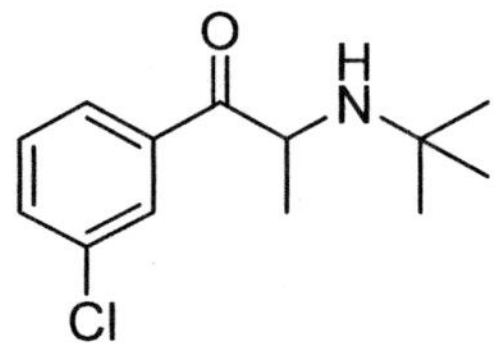

a. Circle and label the amine functional group in Wellbutrin.
b. Is the amine a primary, secondary, or tertiary amine?
c. What other functional groups are present in this molecule?
d. Draw the ionic form of Wellbutrin.

Solutions

a.

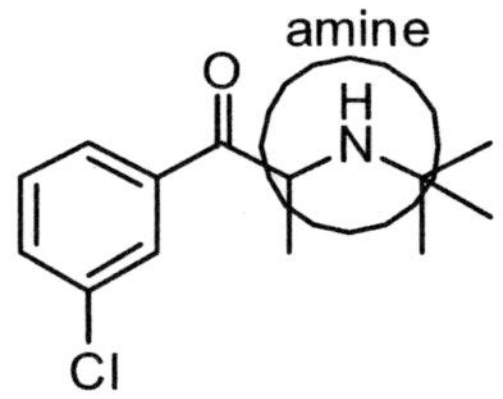

b. *The amine is a secondary amine. There are two carbon groups attached to the nitrogen atom.*

c. *There are a ketone and an aromatic ring in the structure of Wellbutrin.*

d.

Try It Yourself #10

Paxil ® (paroxetine hydrochloride) is an orally administered psychotropic drug,

a. Circle and label the amine functional group in Paxil.

b. Is the amine a primary, secondary, or tertiary amine?

c. Draw the ionic form of Paxil.

Solutions

a.

    b. *Number of carbon groups attached to nitrogen:* _____

       Amine is: _____

    c. *Ionic form of Paxil:*

Extension Topic 7-2: Assigning the IUPAC Name for an Amine

To assign an IUPAC name for an amine, use the following rules: 1) assign the root, 2) assign the suffix, 3) assign a locator number to the root indicating the location of the amine, 4) assign the prefix. To assign the root, identify the R group attached to the nitrogen that has the longest chain and assign the root. Number the chain from the end closest to the nitrogen atom. To assign the suffix, change "ane" to "anamine." A locator number is used to indicate the carbon atom on the main chain that contains the amine. Place the locator number followed by a dash in front of the root name. To assign the prefix, the other R groups attached to the nitrogen atom are named as you would any substituent; however, instead of a locator number, the prefix "N-" is used. If there are two R groups they are written in alphabetical order.

Worked Example #11

Write the structure of N,N-ethylmethyl-2-butanamine.

Solution

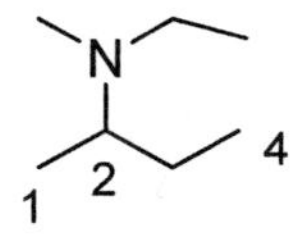

The root name is butane. There are four carbons in the chain containing the amine. The amine is located on carbon 2. There are an ethyl group and a methyl group attached to the nitrogen.

Try It Yourself #11

Write the structure of N-ethyl-3-methyl-2-pentanamine.

Solution

Number of carbons in the chain containing the amine: _____

Position of the amine: _____

Substituents on the nitrogen atom: _____

Structure of N-ethyl-3-methyl-2-pentanamine:

Worked Example #12

Write the IUPAC name for the following amine.

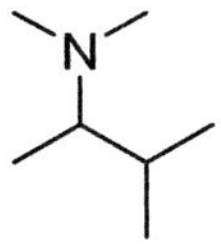

Solution

N,N-dimethyl-3-methyl-2-butanamine. There are four carbon atoms in the main chain.

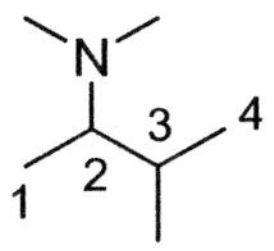

The root name is butane. The suffix changes from "ane" to "anamine." There is a methyl group attached to carbon 3 of the main chain, so methyl is added as a prefix to the amine name. There are two methyl groups attached to the nitrogen atom; they are indicated by the N,N-dimethyl.

Try It Yourself #12

Write the IUPAC name for the following amine.

Solution

Number of carbon atoms in the main chain containing the amine: _____

Root name: _____

Suffix change: _____

Locator number of the amine: _____

Substituents attached to the nitrogen atom: _____

IUPAC name: _____

Practice Problems for C—N-Containing Functional Groups: Amines

1. Are the amines shown the ionic form or the neutral form?

    a.

    b.   H—C—C—C—C—NH₂

    c.

d.

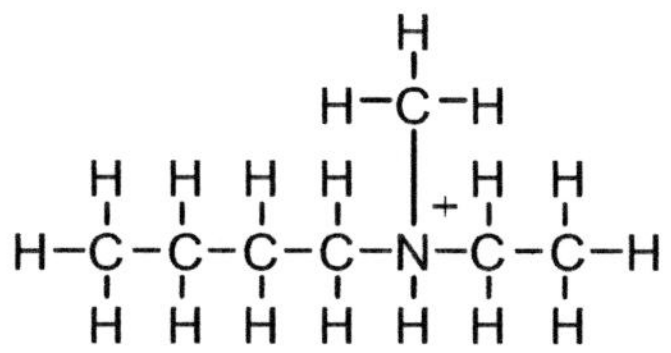

$$\begin{array}{c}
\quad\quad\quad\quad H \\
\quad\quad\quad H-C-H \\
\quad H\ H\ H\ H\quad\ |\ H\ H \\
H-C-C-C-C-N^{+}-C-C-H \\
\quad H\ H\ H\ H\ H\ H\ H
\end{array}$$

2. Label the amines in Question 1 as primary, secondary, or tertiary amines.

3. Avelox is a broad spectrum antibiotic. Circle and label the amines in the structure of avelox.

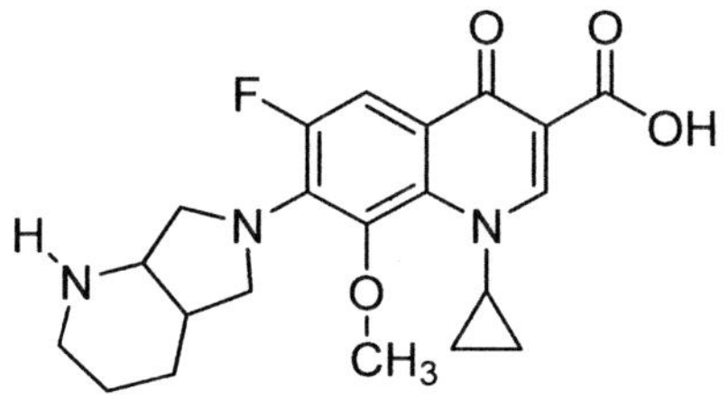

4. Write the IUPAC name for the following amines:

a.

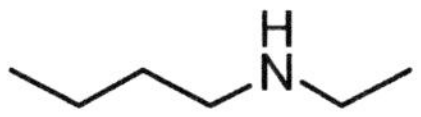

b.

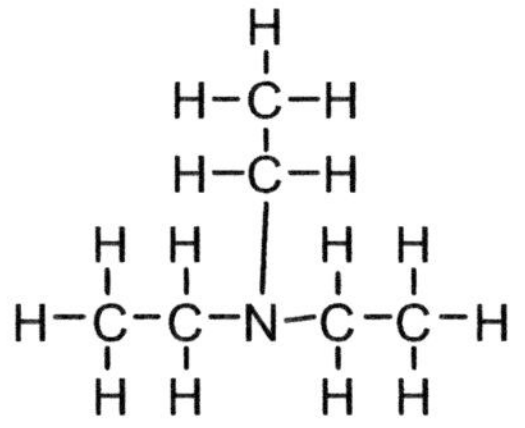

$$\begin{array}{c}
\quad\quad\quad H \\
\quad\quad H-C-H \\
\quad\quad H-C-H \\
\quad H\ H\quad\ |\ H\ H \\
H-C-C-N-C-C-H \\
\quad H\ H\quad\ H\ H
\end{array}$$

Section 7.4 P=O Containing Functional Groups

The last functional groups to consider are derived from phosphoric acid (H_3PO_4). These functional groups have one phosphorous-oxygen double bond and three phosphorus-oxygen single bonds. The functional groups derived from phosphoric acid are found in important biomolecules, such as DNA, RNA, and ATP.

Phosphoric acid can lose one, two, or all three hydrogen atoms to form three different polyatomic ions. The number of hydrogen atoms lost depends on the aqueous environment. The most abundant form in the cell is the ion that has lost two hydrogen atoms. This form, HPO_4^{2-}, is called monohydrogen phosphate or inorganic phosphate. When one or more of the hydrogen atoms is replaced by one or more carbon atoms (an R group), the result is an organic molecule called a phosphate ester. A unique characteristic of phosphate esters is that the phosphate group can be connected to additional phosphate groups, P_i's, via a bond called the phosphoanhydride bond. The phosphoanhydride bond forms between one of the oxygen atoms of the phosphate group and the phosphorus atom of another phosphorus group. The attachment of two phosphate groups to a carbon atom creates a diphosphate ester, while the attachment of three phosphate groups creates a triphosphate ester.

Worked Example #13

Identify the structures below as mono, di, or triphosphate esters. Place a box around each phosphate group.

a.

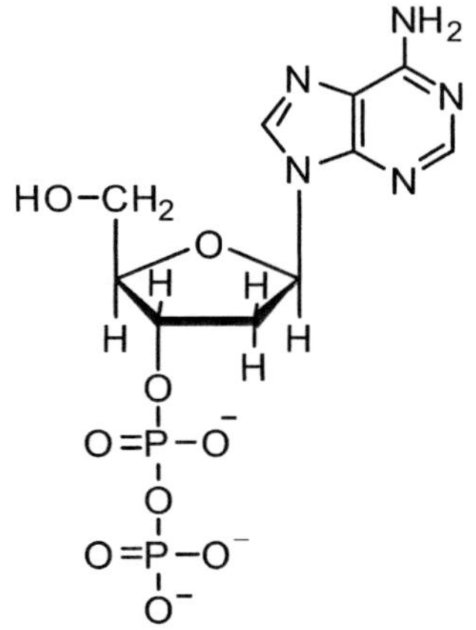

b.

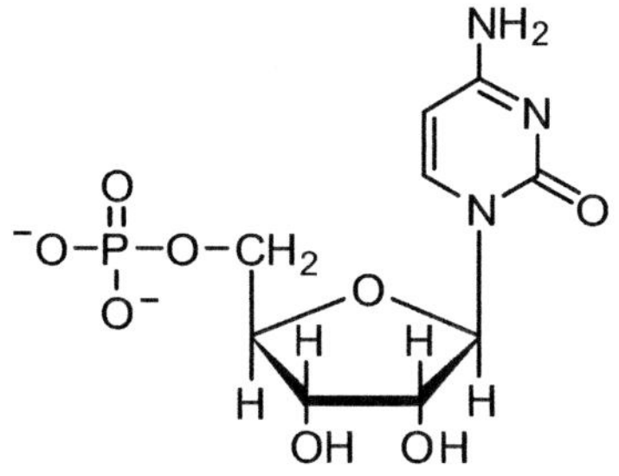

Solution

a. *This molecule is a diphosphate ester. There are two phosphate groups.*

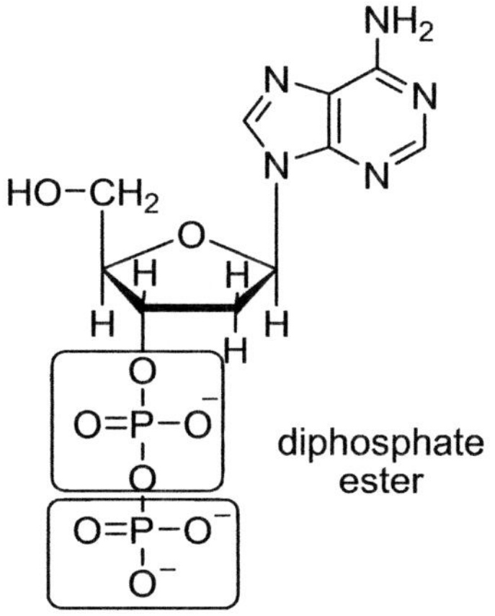

diphosphate
ester

b. *This molecule is a monophosphate ester. There is one phosphate group.*

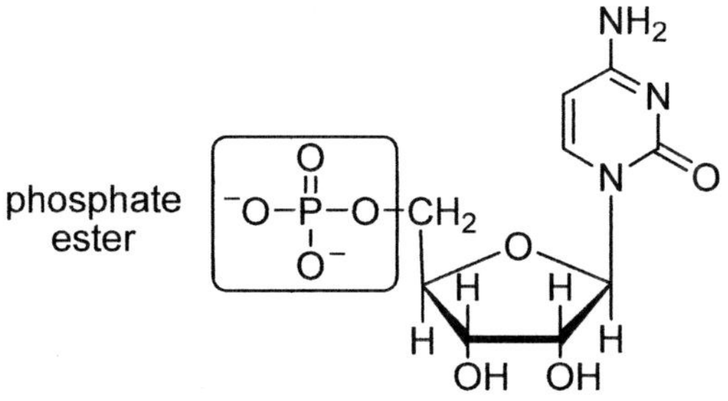

phosphate
ester

Try It Yourself #13

Identify the structures that follow as mono, di, or triphosphate esters. Place a box around each phosphate group.

a.

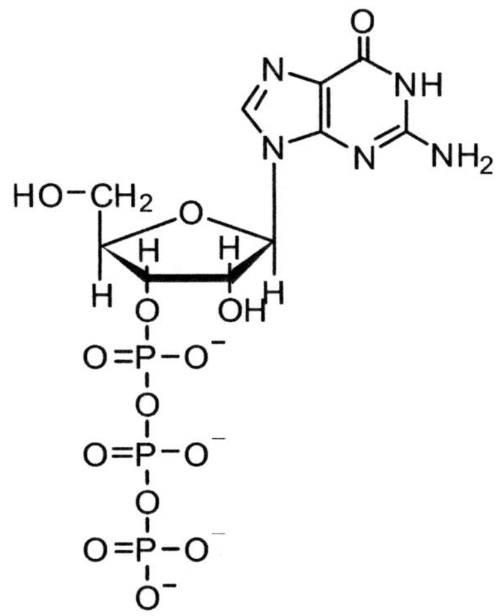

HO−CH₂

O=P−O⁻

O=P−O⁻

O=P−O⁻

b.

⁻O−P−O−P−O−CH₂

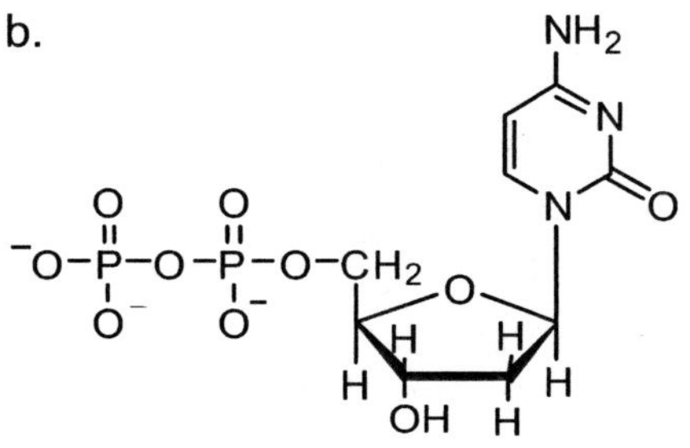

Solutions

a. *Number of phosphate groups:* _____

   Type of phosphate: _____

   Draw a box around each phosphate group.

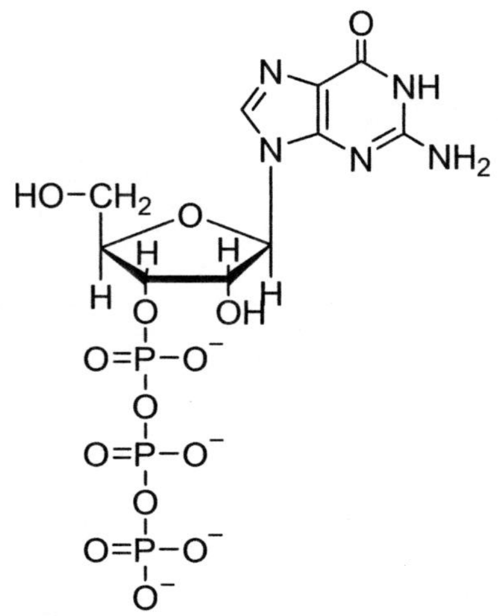

HO−CH₂

O=P−O⁻

O=P−O⁻

O=P−O⁻

b. *Number of phosphate groups:* _____

Type of phosphate: _____

Draw a box around each phosphate group.

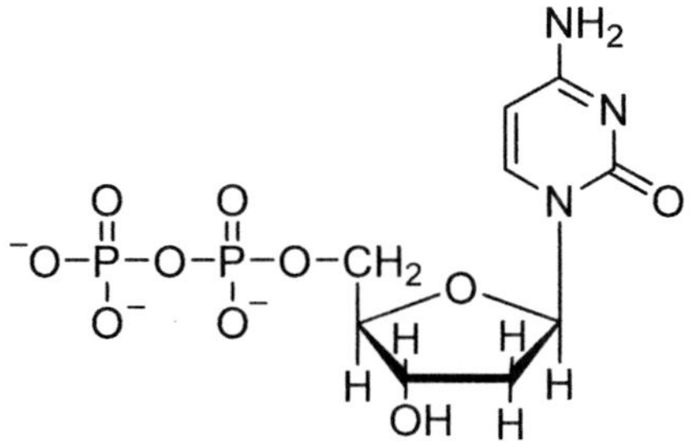

Practice Problems for P=O Containing Functional Groups

1. Identify the structures below as mono, di, or triphosphate esters. Place a box around each phosphate group.

a.

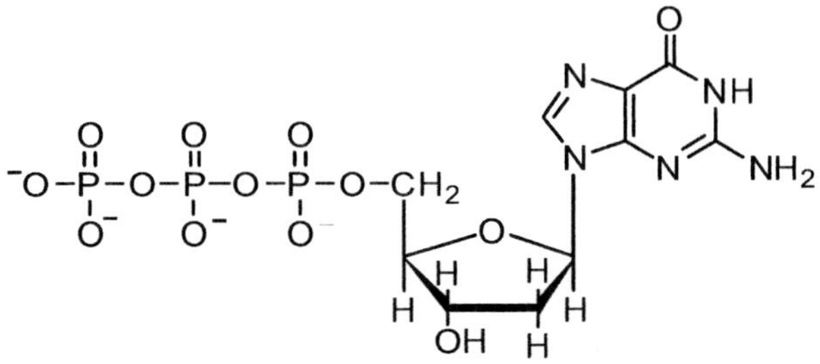

b.

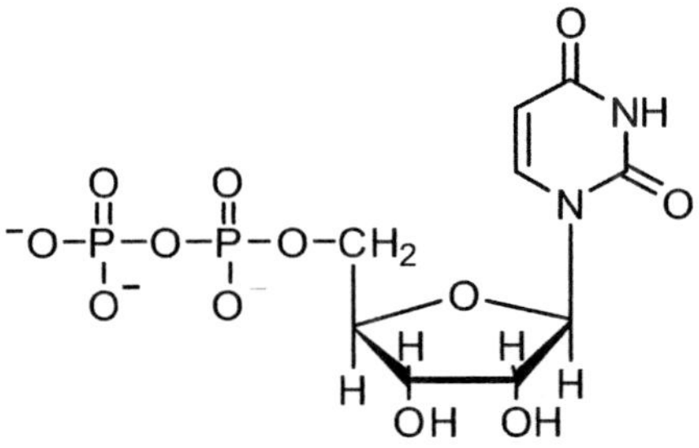

2. What is the total charge on the molecules shown in Question 1?

Chapter 7 Quiz

1. Doxycycline is a broad spectrum antibiotic. Circle and label the functional groups in the structure shown below.

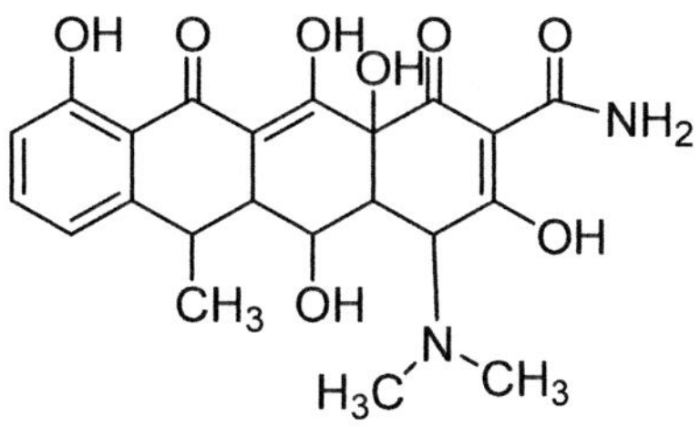

2. Inderal is a beta-adrenergic receptor blocking agent. It is used in the management of hypertension. Circle and label the functional groups in the structure shown below.

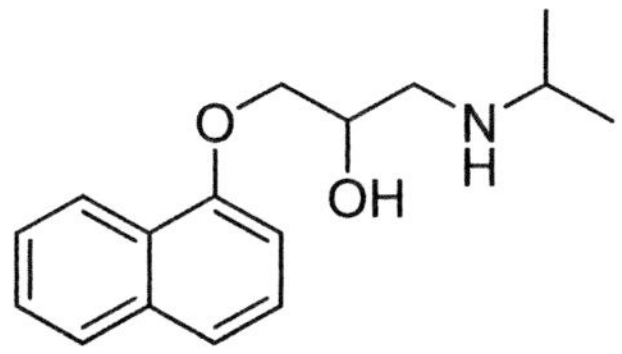

Inderal

3. Ritalin is a mild central nervous system stimulant. It is used in the treatment of attention deficit disorders. Circle and label the functional groups in the structure shown below. Is the amine in its neutral or ionic form?

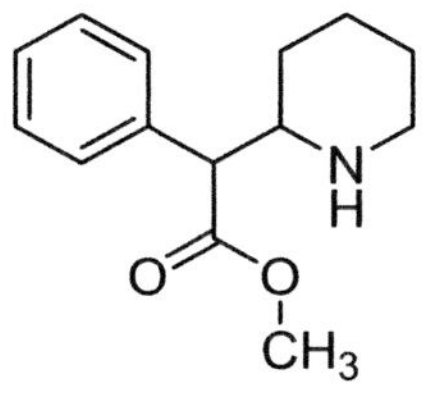

Ritalin

4. Glyceraldehyde 3-phosphate is an intermediate of glycolysis, a metabolic pathway. Circle and label the functional groups in the structure shown below.

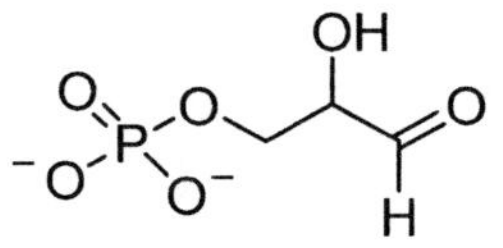

Glyceraldehyde 3-phosphate

5. Lotensin® (Benzapril hydrochloride) is used in the treatment of hypertension. It is an ACE (angiotension-converting enzyme) inhibitor.

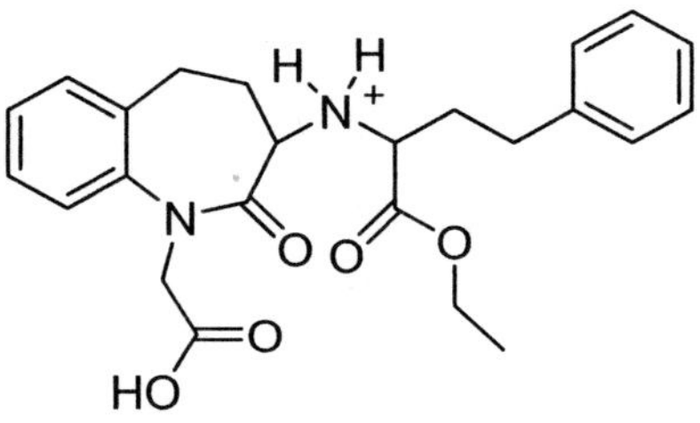

Lotensin

   a. Circle and label the following functional groups in Lotensin: 2 aromatic rings, an ester, an amide, an amine, and a carboxylic acid.
   b. Is the amine in its ionic or neutral form?

   c. Is the carboxylic acid in its neutral or ionic form?

6. Aldactone® (Spironolactone) is a diuretic used to treat edema in patients with congestive heart failure, hypertension, and hypokalemia (low blood potassium). Circle and label the functional groups in Aldactone.

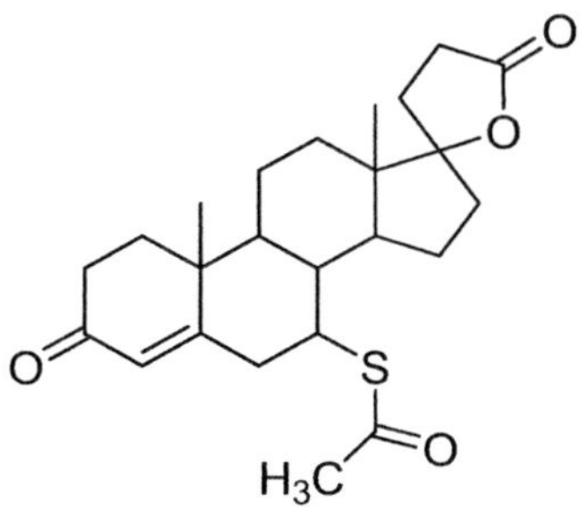

Aldactone

7. Assign a common name to the ethers and an IUPAC name for all other molecules shown below:

a.

b. $CH_3CH_2CH_2CH_2CH_2CH_2CH_2CHO$

c.

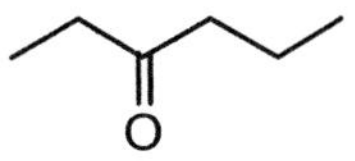

d. $CH_3CH_2CH_2CH_2CH_2CH_2CH_2CH_2OH$

e.

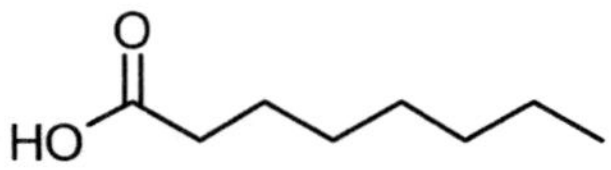

8. Write a skeletal line structure for the following molecules:

   a. heptanal

   b. 3-heptanone

   c. pentanoic acid

   d. dimethyl ether

   e. 2-methyl-3-pentanol

9. Classify the following amines and alcohols as primary, secondary, or tertiary amines or alcohols.

   a.

   b.

   c.

   d.

10. Are the neutral forms or the ionic forms of the following amines and carboxylic acids shown?

   a.

   b.

c.

d.

Chapter 7
Answers to Additional Exercises

7.31 a. 1-hexanol. There are six carbon atoms in the main chain. The root name is hexane. The suffix changes from "ane" to "anol." The OH group is located on carbon 1. Primary alcohol. The carbon atom bonded to the OH group is only attached to one other carbon atom. $CH_3(CH_2)_4CH_2OH$

b. 2-pentanol. There are five carbon atoms in the main chain. The root name is pentane. The suffix changes from "ane" to "anol." The OH group is located on carbon 2. Secondary alcohol. The carbon atom bonded to the OH group is

attached to two other carbon atoms. $CH_3CH_2CH_2\overset{\underset{|}{OH}}{C}HCH_3$

c. Cyclopentanol. There are five carbon atoms in the ring. The root name is cyclopentane. The suffix changes from "ane" to "anol." No locator number is needed. Secondary alcohol. The carbon atom bonded to the OH group is bonded to two other carbon atoms.

d. Cyclobutanol. There are four carbon atoms in the ring. The root name is cyclobutane. The suffix changes from "ane" to "anol." No locator number is needed. Secondary alcohol. The carbon atom bonded to the OH group is bonded to two other carbon atoms.

e. 2-methyl-3-pentanol. There are five carbon atoms in the longest carbon chain containing the OH group. The root name is pentane. The suffix changes from "ane" to "anol." The OH group is located on carbon 3. There is a methyl group on carbon 2; therefore, 2-methyl is added as a prefix. It is a secondary alcohol. The carbon atom bonded to the OH group is bonded to two other carbon atoms.

7.33

Estradiol

Betamethasone

7.35   The alcohol, 2-methyl-1-butanol, will be more soluble in water. The OH group of the alcohol can hydrogen bond with water. The hydrocarbon cannot hydrogen bond with water.

7.37   An ether has two carbon atoms (R groups) attached to an oxygen atom. An alcohol has one carbon atom (R group) and a hydrogen atom attached to an oxygen atom.

7.39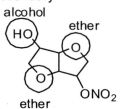

7.41   a. Propanal. There are three carbon atoms on the main chain. The root name is propane. The suffix is changed from "ane" to "anal."

   b. Pentanal. There are five carbon atoms on the main chain. The root name is pentane. The suffix is changed from "ane" to "anal."

   c. 2-hexanone. There are six carbon atoms on the main chain. The root name is hexane. The suffix is changed from "ane" to "anone." The ketone is located on carbon 2.

   d. 3-pentanone. There are five carbons in the main chain. The root name is pentane. The suffix is changed from "ane" to "anone."

   e. Cyclohexanone. There are six carbons in the ring. The root name is cyclohexane. The suffix is changed from "ane" to "anone."

7.43   a. and b.

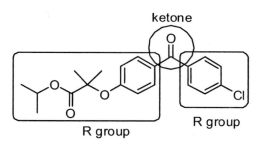

7.45   a.   A carboxylic acid has an R group and an OH group attached to a carbonyl carbon; an ester has an R group and an OR group attached to a carbonyl carbon.

   b.   A carboxylic acid has an R group and an OH group attached to a carbonyl carbon; a thioester has an R group and an SR group attached to a carbonyl carbon.

   c.   An ester has an oxygen atom attached to a carbonyl group and an R group; a thioester has a sulfur atom attached to a carbonyl group and an R group.

   d.   A carboxylic acid has an R group and an OH group attached to a carbonyl carbon; an amide has an R group and a nitrogen atom attached to the carbonyl carbon. The nitrogen atom in an amide may be attached to one or more hydrogen atoms or one or more R groups.

   e.   An amide has a nitrogen atom attached to a carbonyl carbon and may be attached to one or more hydrogen atoms or one or more R groups; an amine does not have a carbonyl carbon. An amine only has one or more hydrogen atoms or one or more R groups attached to a nitrogen atom.

   f.   A carboxylic acid has an R group and an OH group attached to a carbonyl carbon; an alcohol has an OH group, but does not contain a carbonyl group.

7.47   a.

   b.

   c.   Yes, tartaric acid should be soluble in water. The alcohol groups and the carboxylic acid groups can hydrogen bond with water molecules.

7.49   a.

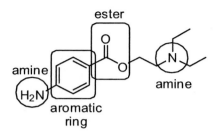

b.   Both amine groups are in their neutral form.

c.   The amine on the left is attached to one R group. The amine on the right is attached to three R groups.

7.51   a.

$$O-C-C-C-C-S-CoA$$

thioester

b.

$$O-C-C-C-C-S-CoA$$

carboxylate
ion

This functional group is in its ionic form.

7.53

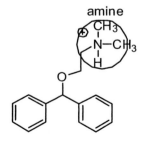

amide

There are two methyl groups attached to the nitrogen atom.

7.55   a.

amine

b.   The amine in Benadryl is in its ionic form.

c.

ether

aromatic ring   aromatic ring

7.57

monohydrogen phosphate

It is often called inorganic phosphate and abbreviated P_i.

7.59   A person who has schizophrenia has an excess amount of dopamine in the brain; a person who has Parkinson's disease has a decreased amount of dopamine in the brain.

7.61   neurotransmitters; amine functional group.

7.63   Dopamine cannot pass through the blood-brain barrier, to reach the brain where it is needed.

7.65   Removal of a carboxylic acid functional group converts L-dopa into dopamine.

7.67   Parkinson's disease involves loss of dopamine-producing neurons.

7.69   a-d.

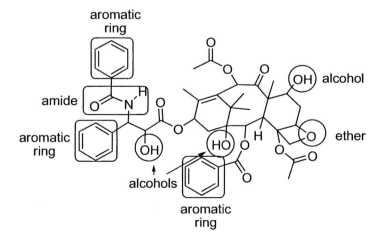

e-g.

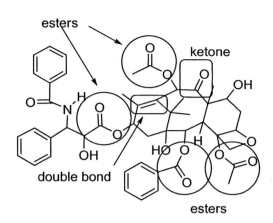

Chapter 8

Chemical Reaction Basics

Chapter Summary

In this chapter, you studied what happens to compounds when they undergo chemical reactions. To understand the relationship between energy and chemical reactions as well as reaction rates, you need to learn the basics of a chemical reaction. You learned how matter and energy are interrelated in a chemical reaction and that mass and energy are conserved in a chemical reaction. You learned about reaction kinetics and the role of different variables that affect the rate of a reaction. Understanding these reaction basics will give you insight into the structure and function of biomolecules.

Section 8.1 Writing and Balancing a Chemical Equation

In a Nutshell: Kinetic-Molecular View of a Chemical Reaction

In a chemical reaction, bonds break and new bonds are formed, yielding new chemical substances. Intimately connected with chemical reactions is the transfer of energy. Usually, when two atoms or molecules collide, they bounce of one another unchanged. Occasionally, however, a collision occurs with sufficient energy and in the proper orientation that a bond is broken or a new bond is made or both, creating a new chemical substance. This change in chemical structure occurs in a process called a chemical reaction. The reacting substances are known as reactants and the new substance(s) formed are known as products. In a chemical reaction, the same number and type of atoms appear in the product(s) as in the reactant(s), except that they are arranged differently. The conservation of atoms in a chemical reaction is a fundamental law of nature known as the law of conservation of mass: matter can be neither created nor destroyed.

In a Nutshell: Writing a Chemical Equation

A chemical reaction is represented as a chemical equation. In a chemical equation, the reactants and products are represented by their chemical formulas. A reaction arrow is used to separate the reactants on the left side of the arrow form products on the right side of the arrow. A "+" sign is used to separate the individual reactants and the individual products when more than one is present. Whole number coefficients are placed before each chemical formula to indicate the ratio of each reactant and product molecule or atom. When no number is shown, the coefficient is assumed to be 1. The physical state of each compound or element is sometimes included in parentheses following the formula of the compound: solid (s), liquid (l), gas (g), or aqueous solution (aq).

Worked Example #1

For the equation below:

$C_6H_{12}O_6$ (aq) → 2 C_2H_6O (aq) + 2 CO_2 (g)

    a. Indicate the reactant(s).

    b. Indicate the product(s).

    c. Indicate the ratio in which the reactants and products react.

    d. If 2 moles of $C_6H_{12}O_6$ react, how many moles of C_2H_6O and how many moles of CO_2 are produced?

    e. Indicate the physical state (s, l, g, aq) of each reactant and product.

Solutions

    a. *The reactant is $C_6H_{12}O_6$. The reactant(s) appear to the left of the reaction arrow.*

    b. *The products are C_2H_6O and CO_2. The products appear to the right of the reaction arrow.*

    c. *One mole of $C_6H_{12}O_6$ reacts to produce 2 moles of C_2H_6O and 2 moles of CO_2.*

    d. *If two moles of $C_6H_{12}O_6$ reacts, then 4 moles of C_2H_6O and 4 moles of CO_2 are produced. There is twice as much C_2H_6O and twice as much CO_2 formed as the number of moles of $C_6H_{12}O_6$, based on the 1, 2, and 2 coefficients.*

    e. *$C_6H_{12}O_6$ and C_2H_6O are in the aqueous state, and CO_2 is in the gas state.*

Try It Yourself #1

For the equation below:

$Ca(OH)_2$ (aq) + 2 HCl (aq) → $CaCl_2$ (aq) + 2 H_2O (l)

    a. Indicate the reactant(s).

    b. Indicate the product(s).

    c. Indicate the ratio in which the reactants and products react.

    d. If 2 moles of $Ca(OH)_2$ react, how many moles of $CaCl_2$ are produced?

    e. Indicate the physical state (s, l, g, aq) of each reactant and product.

Solutions

    a. *The reactants:* _____

    b. *The products:* _____

    c. *Ratio of reactant to products:* _____

    d. *Number of moles of $CaCl_2$ produced:* _____

    e. *Physical state $Ca(OH)_2$:* _____

        Physical state HCl: _____

        Physical state $CaCl_2$: _____

        Physical state H_2O: _____

In a Nutshell: Balancing a Chemical Equation

Applied to a chemical equation, the law of conservation of mass requires that the total number of each type of atom on the reactant side of a chemical equation must always equal the total number of each type of atom on the product side of an equation. A chemical equation that contains an equal number of each type of atom on both sides of the equation is known as a balanced chemical equation. To balance a chemical equation, the correct number of coefficients must be inserted in the equation. To balance a chemical equation follow these guidelines: 1) Assess the equation; 2) balance the equation one atom type at a time by inserting coefficients; and 3) check that the coefficients cannot be divided by a common factor (divisor). To assess the equation, determine the number and type of each atom on the reactant side and product side of the equation. If the two numbers do not match, the equation is not balanced. To balance the equation one atom at a time, systematically place coefficients in front of each reactant and product as necessary to arrive at an equal number of each type of atom on both sides of the equation. Never change the numerical subscripts in a chemical formula because this changes the identity of the compound. Remember that adding a coefficient alters the number of every type of atom in the formula. The balancing process is simplified if you wait to balance last any atom type that appears in more than one compound or element on one side of the equation.

Worked Example #2

Balance the following combustion equation:

    ____$C_7H_6O_2$ (s) + ____O_2 (g) → _____CO_2 (g) + ____H_2O (g)

Assess the equation.

Kind of atom	# on reactant side	# on product side
C	7	1
H	6	2
O	4	3

The equation is unbalanced because the values in the reactant column do not equal the values in the product column for all types of atoms.

Balance the equation one atom type at a time by inserting coefficients.
Balance carbon or hydrogen first. Balance oxygen last because it is present in both compounds on the product side. If you begin with carbon, insert coefficient 7 in front of CO_2.

_____$C_7H_6O_2$ (s) + _____O_2 (g) → 7 CO_2 (g) + _____H_2O (g) *(carbon is now balanced)*
Balance hydrogen by inserting the coefficient 3 in front of H_2O.
_____$C_7H_6O_2$ (s) + _____O_2 (g) → 7 CO_2 (g) + 3 H_2O (g) *(carbon and hydrogen are now balanced)*

To balance oxygen insert the coefficient,15/2 or 7 ½, in front of oxygen as a temporary coefficient. Seventeen oxygen atoms are required on the left hand side to balance the seventeen total oxygen atoms on the right side of the equation (fourteen from 7 CO_2 and three from 3 H_2O). Only fifteen oxygen atoms are used for O_2 because there are two oxygen atoms in $C_7H_6O_2$.

$C_7H_6O_2$ (s) + 15/2 O_2 (g) → 7 CO_2 (g) + 3 H_2O (g) *(oxygen is now balanced)*
Turn the fraction into a whole number by multiplying every coefficient in the equation by 2:
2 $C_7H_6O_2$ (s) + 15 O_2 (g) → 14 CO_2 (g) + 6 H_2O (g)

Check that the coefficients cannot be divided by a common factor (divisor).
There is no common factor for the coefficients 2, 15, 14, and 6. The lowest set of whole number coefficients is in the balanced equation. The values in the second column now equal the values in the third column.

Kind of atom	# on reactant side	# on product side
C	14	14
H	12	12
O	34	34

Try It Yourself #2

Balance the following equation:

_____ H_2O_2 (l) → _____H_2O (l) + _____O_2 (g)

Solution

Assess the equation.

Kind of atom	# on reactant side	# on product side
O		
H		

Is the equation balanced? _____

Balance the equation one atom type at a time by inserting coefficients.

Check that the coefficients cannot be divided by a common factor (divisor).

Extension Topic 8-1 Reaction Stoichiometry Calculations

From a balanced equation, it is possible to calculate the amount (in grams or in moles) of product formed if the mass of one of the reactants is known, provided there is ample supply of the other reactant. This type of calculation is known as a stoichiometry calculation. The coefficients in a chemical equation represent not only the ratio in which the individual molecules react with one another, but also the molar ratio in which they react. In order to perform a stoichiometry calculation, three steps are followed: Step 1) convert the known mass of reactant to moles of reactant; Step 2) convert moles of reactant to moles of product; and Step 3) convert moles of product to mass of product.

Worked Example #3

How many grams of carbon dioxide are produced when 5.0 g of butane reacts with an excess of oxygen?

$$2\ C_4H_{10}\ (g)\ +\ 13\ O_2\ (g)\ \rightarrow\ 8\ CO_2\ (g)\ +\ 10\ H_2O\ (g)$$
Butane

Convert the known mass of reactant to moles of reactant.
Use the molar mass of butane as the conversion factor.

$$5.0\ \cancel{\text{g butane}} \times \frac{1\ \text{mol butane}}{58.11\ \cancel{\text{g butane}}} = 0.086\ \text{mol butane}$$

mass of butane   molar mass of butane

Convert moles of butane to moles of carbon dioxide.
Use the coefficients in the balanced equation to set up a conversion factor between mol butane and mol carbon dioxide. Remember, when no number is shown, the coefficient is 1:

$$0.086\ \cancel{\text{mol butane}} \times \frac{8\ \text{mol carbon dioxide}}{2\ \cancel{\text{mol butane}}} = 0.34\ \text{mol carbon dioxide}$$

Convert moles of carbon dioxide to grams of carbon dioxide.
Use the molar mass of carbon dioxide as a conversion factor.

$$0.34\ \cancel{\text{mol carbon dioxide}} \times \frac{44.01\ \text{g carbon dioxide}}{1\ \cancel{\text{mol carbon dioxide}}} = 15\ \text{g carbon dioxide}$$

mol of carbon dioxide   molar mass of carbon dioxide

Therefore, 5.0 g of butane produces 15 g of carbon dioxide.

Try It Yourself #3

How many grams of water will be produced in Worked Example #3?

Convert the known mass of reactant to moles of reactant.

Molar mass of butane: _____

Convert moles of butane to moles of water.

Ratio of coefficients for butane and water: _____

Convert moles of water to grams of water.

Molar mass of water: _____

Grams of water produced: _____

Practice Problems for Writing and Balancing a Chemical Equation

1. For the following equations, indicate the reactants, the products, the ratio in which the reactants and products react, and the physical state of the reactants and products.

a. $4 \, Al \, (s) \; + \; 3 \, O_2 \, (g) \; \rightarrow \; 2 \, Al_2O_3 \, (s)$

b. $CO_2 \, (g) \; + \; 4 \, H_2 \, (g) \; \rightarrow \; CH_4 \, (g) \; + \; 2 \, H_2O \, (g)$

c. $Zn \, (s) \; + \; 2 \, HCl \, (aq) \; \rightarrow \; ZnCl_2 \, (aq) \; + \; H_2 \, (g)$

2. Balance the following equations.

a. $Fe(OH)_3 \, (aq) \; \rightarrow \; Fe_2O_3 \, (aq) \; + \; H_2O \, (l)$

b. $C_5H_{12} \, (l) \; + \; O_2 \, (g) \; \rightarrow \; CO_2 \, (g) \; + \; H_2O \, (g)$

c. $Mg \, (s) \; + \; Mn_2O_3 \, (aq) \; \rightarrow \; MgO \, (aq) \; + \; Mn \, (s)$

3. Balance the following equation and then determine how many grams of iron oxide are produced when 12.7 g of iron reacts with an excess of oxygen.

$Fe \, (s) \; + \; O_2 \, (g) \; \rightarrow \; Fe_2O_3 \, (s)$

Section 8.2 Energy and Chemical Reactions

Recall that energy is defined as the capacity to do work or transfer heat. Work is the act of moving an object over a distance against an opposing force. Bioenergetics is the field of study concerned with the transfer of energy in reactions occurring in living cells.

In a Nutshell: Units of Energy in Science and Nutrition

Commonly used units of energy include calorie (cal), Joule (J), Calorie (Cal), and kilocalorie (kcal). A calorie (cal) is the amount of heat energy required to raise the temperature of 1 gram of water by 1 °C. One Calorie (with a capital "C") is equal to one kilocalorie, or one thousand calories (with a lowercase "c"). A Joule is defined as the amount of energy required to lift a one kilogram weight a height of 10 centimeters.

Worked Example #4

How many Calories (Cal) of energy are represented by 27.9 J?

Identify the conversion.
The supplied unit is Joule. The requested unit is Calories. Two conversion factors are needed.

4.184 J = 1 cal   1000 cal = 1 Cal

Express the conversion as a conversion factor:

1st conversion: $\dfrac{4.184 \text{ J}}{1 \text{ cal}}$ or $\dfrac{1 \text{ cal}}{4.184 \text{ J}}$

and

2nd conversion: $\dfrac{1000 \text{ cal}}{1 \text{ Cal}}$ or $\dfrac{1 \text{ Cal}}{1000 \text{ cal}}$

Set up the calculation so that the supplied units cancel.
Begin with the supplied unit. Choose the appropriate conversion factors for the numerator and denominator so that the supplied unit (J) cancels and you are left with the answer in the requested unit (Cal). This calculation can be performed in one step as shown below:

$$27.9 \text{ J} \times \frac{1 \text{ cal}}{4.184 \text{ J}} \times \frac{1 \text{ Cal}}{1000 \text{ cal}} = 6.67 \times 10^{-3} \text{ Cal}$$

Try It Yourself #4

How many joules are there in 417 Calories, the amount of energy contained in one glazed donut?

Identify the conversion.

Express the conversion as a conversion factor.

Set up the calculation so that the supplied units cancel.

In a Nutshell: Heat Energy

The energy transferred during a chemical reaction commonly appears in the form of heat energy. In chemistry, this heat energy is known as the change in enthalpy, ΔH. In a chemical reaction, not all of the energy transferred is in the form of heat.

In a Nutshell: Exothermic and Endothermic Reactions

Chemical reactions can either absorb heat or release heat. Whether heat is released or absorbed by a chemical reaction depends on the reaction under consideration. Breaking a chemical bond absorbs energy, while making a chemical bond releases energy. Because not all chemical bonds are identical, they possess different amounts of potential energy. In a chemical reaction—where reactant bonds are broken and new product bonds are formed—heat energy may be released when the sum of the bond energies in the products is greater than the sum of the bond energies in the reactants. Conversely, heat energy is absorbed when the sum of the bond energies in the products is less than the sum of the bond energies in the reactants.

A reaction that releases energy is known as an exothermic reaction. The change in enthalpy is negative ($\Delta H < 0$) because the products are lower in chemical potential energy than the reactants.

When the reactants are lower in energy than the products, energy is absorbed from the surroundings and the surroundings become cooler. This type of reaction is known as an endothermic reaction. The change in enthalpy is positive ($\Delta H > 0$). In an endothermic reaction, energy must be continuously supplied to sustain the reaction.

Exothermic and endothermic reactions can be illustrated in an energy diagram. In an energy diagram, energy appears on the Y-axis and the progress of the reaction as it goes from reactants to products appears on the X-axis. In both endothermic and exothermic reactions, energy is never lost and never created. Energy can only be transformed.

Worked Example #5

For each of the following reactions, indicate whether it is an endothermic or exothermic reaction. Indicate whether the reactants or products are lower in energy.

    a. $C_6H_8O_7(aq) + 3\ NaHCO_3(s) \rightarrow 3\ CO_2(g) + 3\ H_2O(l) + Na_3C_6H_5O_7(aq)$; energy is absorbed

    b. $2\ SO_2\ (g)\ +\ O_2\ (g) \rightarrow\ 2\ SO_3\ (g)\ +$ heat

    a. Energy is absorbed, so the reaction is an endothermic reaction. The reactants are lower in energy than the products.

    b. Energy is released; it is on the right hand side of the chemical equation. The products are lower in energy than the reactants.

Try It Yourself #5

For each of the following reactions, indicate whether it is an endothermic or exothermic reaction. Indicate whether the reactants or products are lower in energy.

    a. $N_2\ (g)\ +\ 2\ O_2\ (g)\ +$ heat $\rightarrow\ 2\ NO_2\ (g)$

    b. $H_2O\ (l) \rightarrow H_2O\ (s)\ +$ heat

    a. Energy is: _____

Reaction is: _____

b. *Energy is:* _____

   Reaction is: _____

In a Nutshell: Calorimetry

Calorimetry is an experimental technique used to measure enthalpy changes (ΔH) in chemical reactions, using an apparatus known as a calorimeter. A calorimeter is designed to measure heat released in a combustion reaction. The caloric content of food can be measured using a calorimeter. The caloric content of the three basic food types are as follows: Carbohydrates have 4 Cal/g, proteins have 4 Cal/g, and fats have 9 Cal/g. From the mass of each biomolecule (carbohydrate, protein, and fat) contained within a particular food item, the total caloric value can be calculated. The number of calories provided by each type of molecule in a particular food are calculated and then summed and rounded to the nearest tens place to obtain the total caloric value of the food.

Worked Example #6

A cup of 1% milk has 2.5 g of fat, 13 g of carbohydrates, and 8 g of protein. How many total Calories does a cup of 1% milk contain?

Determine the Calories supplied by each type of biomolecule.

Calories provided by fat: $2.5 \, \cancel{g} \times \dfrac{9 \, \text{Cal}}{\cancel{g}} = 23 \, \text{Cal}$

Calories provided by carbohydrate: $13 \, \cancel{g} \times \dfrac{4 \, \text{Cal}}{\cancel{g}} = 52 \, \text{Cal}$

Calories provided by protein: $8 \, \cancel{g} \times \dfrac{4 \, \text{Cal}}{\cancel{g}} = 32 \, \text{Cal}$

Sum these values to obtain the total Calories supplied by milk:

23 Cal + 52 Cal + 32 Cal = 107 Cal or 110 Cal

Try It Yourself #6

A serving (1/2 cup) of oatmeal has 3 g of fat, 27 g of carbohydrate, and 5 g of protein. How many total Calories does a serving of oatmeal contain?

Determine the Calories supplied by each type of biomolecule.
Calories provided by fat

Calories provided by carbohydrate:

Calories provided by protein:

Total Calories supplied by oatmeal: _____

In a Nutshell: An Overview of Metabolism and Energy

The chemical reactions of the cell are referred to as biochemical reactions and typically occur as a sequence of reactions. A biochemical pathway is defined by a particular sequence of reactions. There are two basic types of biochemical pathways: catabolic and anabolic. Catabolic and anabolic pathways together are known as metabolism. Catabolic pathways convert large biomolecules, such as carbohydrates, proteins, and fats, to smaller molecules. Anabolic pathways build larger molecules, such as proteins, lipids, and DNA, from smaller molecules. Catabolic pathways release energy overall; while anabolic pathways absorb energy overall. The energy required to build molecules via anabolic pathways comes from the energy released in catabolic processes.

Practice Problems for Energy and Chemical Reactions

1. How many calories are represented by:
   a. 51 J

   b.   175 Calories

2. Are the following reactions endothermic or exothermic?
   a. H_2O (s) + heat → H_2O (l)

   b. C_2H_4 + 3 O_2 → 2 CO_2 + 2 H_2O + heat

3. One egg has 6 g of protein, 4.5 g of fat, and 1 g of carbohydrate. How many total Calories does one egg contain?

4. Which type of metabolic pathways are exothermic, releasing energy overall?

Section 8.3 Kinetics: Reaction Rates

Chemical kinetics is the study of reaction rates—how fast reactants in a chemical reaction are converted into products. How fast a reaction proceeds, known as the reaction rate, is determined by measuring the consumption of reactants or the formation of products over time. The rate of reaction indicates how the concentration of reactants or products changes over time.

In a Nutshell: Activation Energy

The minimum amount of energy that must be attained initially by the reactants for a reaction to proceed is known as the activation energy, E_A. Without this initial input of energy, molecules may collide, but they will only bounce off each other without reacting. The activation energy can be illustrated in an energy diagram. The activation energy, E_A, for a reaction has no effect on the potential energy of either the reactants or the products and therefore no effect on ΔH. The activation energy, E_A, has a profound effect on the rate of a reaction. Reactions that have low activation energies proceed faster than reactions that have high activation energies. An energy diagram with a greater activation energy (E_A) represents a slower reaction.

Worked Example #7

Does this energy diagram represent endothermic or exothermic reactions? Which reaction would you predict to proceed faster, a or b?

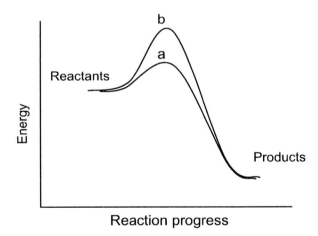

Solution

The energy diagram represents exothermic reactions. The products are lower in energy than the reactants. Reaction a would proceed faster because it has a lower activation energy, E_A.

Try It Yourself #7

Does this energy diagram represent endothermic or exothermic reactions? Which reaction would you predict to proceed faster, a or b?

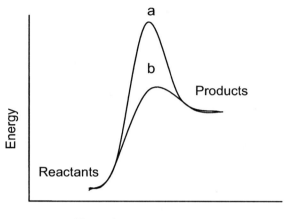

Solution

Products are (higher or lower) in energy than the reactants.

The reaction is: _____

The reaction with lower activation energy, E_A: _____

The faster reaction is: _____

In a Nutshell: Factors Affecting the Rate of a Reaction

The following factors have the most influence on the rate of a chemical reaction: the concentration of reactants, the temperature of the reaction, and whether or not a catalyst is present. The rate of reaction decreases with time as the concentration of reactants decreases because, as reactants are consumed, there are fewer reactant molecules and therefore fewer collisions. For every 10 °C increase in temperature, the rate of reaction approximately doubles because the molecules have more kinetic energy. Conversely, if the temperature of a reaction is lowered, the rate decreases because the molecules have less kinetic energy. A catalyst is a substance that increases the rate of a chemical reaction by lowering the activation energy for the reaction. A catalyst, however, does not influence the change in enthalpy, ΔH, for a reaction. Catalysts facilitate the collision process, making it easier for the reactants to come together. A catalyst does not undergo a chemical change and can be reused over and over. Biological catalysts are molecules are called enzymes.

Worked Example #8

For the following reactions, state whether the indicated change in conditions would increase or decrease the rate of the reaction. Explain why.

  a. $Zn\ (s)\ +\ 2\ HCl\ (aq)\ \rightarrow\ ZnCl_2\ (aq)\ +\ H_2\ (g)$

      More HCl is added to the reaction.

  b. $I_2\ (s)\ +\ H_2\ (g)\ \rightarrow\ 2\ HI\ (g)$

      The temperature is decreased from 130 °C to 25 °C.

  c. $2\ KClO_3\ (s)\ \rightarrow\ 2\ KCl\ (s)\ +\ 3\ O_2(g)$

      A catalyst, manganese dioxide, MnO_2 is added.

  a. *The reaction rate increases because the concentration of one of the reactants was increased.*

  b. *The reaction rate decreases because the temperature is lowered. At low temperatures, atoms and molecules have less kinetic energy, so there are fewer collisions between them.*

  c. *The reaction rate increases because a catalyst has been added. The catalyst lowers the activation energy.*

Try It Yourself #8

For the following reactions, state whether the indicated change in conditions would increase or decrease the rate of the reaction. Explain why.

  a. $2\ N_2O\ (g)\ \rightarrow\ 2\ N_2\ (g)\ +\ O_2\ (g)$

      A catalyst, gold, is added to the reaction.

  b. $Zn\ (s)\ +\ 2\ HCl\ (aq)\ \rightarrow\ ZnCl_2\ (aq)\ +\ H_2\ (g)$

      The temperature of the reaction is increased.

  a. *Reaction rate increases or decreases.*

  b. *Reaction rate increases or decreases.*

Practice Problems for Kinetics: Reaction Rates

1. Which part of an energy diagram determines the rate of a reaction?

2. Name three ways to increase the rate of a reaction.

3. *Hexokinase* is an enzyme that catalyzes one of the steps in the metabolic pathway, glycolysis. Is *hexokinase* chemically altered when it catalyzes the reaction?

Chapter 8 Quiz

1. For the following unbalanced chemical equations, identify the reactants and the products and the physical state of the reactants and the products.

   a. $Mg\ (s)\ +\ O_2\ (g)\ \rightarrow\ MgO\ (s)$

   b. $C_8H_{18}\ (l)\ +\ O_2\ (g)\ \rightarrow\ CO_2\ (g)\ +\ H_2O\ (g)$

   c. $Mg(OH)_2\ (aq)\ +\ HCl\ (aq)\ \rightarrow\ MgCl_2\ (aq)\ +\ H_2O\ (l)$

   d. $P_4O_{10}\ (s)\ +\ H_2O\ (l)\ \rightarrow\ H_3PO_4\ (l)$

2. Balance the chemical equations in Question 1.

   a.

   b.

   c.

   d.

3. Balance the following chemical equation and then determine how many grams of calcium oxide are formed when excess oxygen reacts with 4.35 g of calcium.

Ca (s) + O_2 (g) → CaO (s)

4. How many calories are there in:
   a. 564 J

   b. 0.812 Cal

   c. 4.57 kJ

   d. 21.6 kcal

5. Calculate the total calories in the following foods:
   a. A slice of apple pie containing 19.4 g of fat, 57.5 g of carbohydrate, and 3.7 g of protein.

   b. A slice of pizza containing 9.8 g of fat, 33.6 g of carbohydrate, and 12.3 g of protein.

   c. A blueberry bagel containing 1.5 g of fat, 33.6 g of carbohydrate, and 9 g of protein.

6. Are the following reactions exothermic or endothermic?

   a. H_2O (l) + heat → H_2O (g)

   b. 2 Na (s) + Cl_2 (g) → 2 NaCl (s) + heat

7. Draw energy diagrams for the reactions shown in Questions 6a and 6b, include labels for reactants, products, and ΔH.

8. Which of the following will increase the rate of a reaction?

   a. Adding a catalyst

   b. Decreasing the concentration of a reactant

   c. Decreasing the temperature

9. Reaction "X" is an exothermic reaction that occurs twice as fast in the presence of a catalyst. Draw side by side energy diagrams for reaction "X" occurring with and without a catalyst.

10. Which metabolic pathways create larger molecules?

Chapter 8

Answers to Additional Exercises

8.21   The abbreviation (s) indicates that the substance is in the solid state. The abbreviation (g) indicates that the substance is in the gaseous state. The abbreviation (aq) indicates that the substance is dissolved in water.

8.23   Matter cannot be created nor destroyed in a chemical reaction. It is an application of the law of conservation of matter.

8.25   a. Assess the equation.

Kind of atom	# on reactant side	# on product side
C	2	1
H	6	2
O	3	3

The equation is unbalanced because the values in the reactant column do not equal the values in the product column for all types of atoms.

Balance the equation one atom type at a time by inserting coefficients.
Begin by balancing carbon or hydrogen first. Balance oxygen last because it is present in both compounds on the reactant side and both compounds on the product side. If you begin with carbon, insert the coefficient 2 in front of CO_2.

C_2H_5OH (l) + O_2 (g) → 2 CO_2 (g) + H_2O (l) (carbon is now balanced)
Balance hydrogen by inserting the coefficient 3 in front of H_2O.
C_2H_5OH (l) + O_2 (g) → 2 CO_2 (g) + 3 H_2O (l) (carbon and hydrogen are now balanced)
To balance oxygen, insert the coefficient 3 in front of O_2, oxygen. Seven oxygen atoms are required on the right hand side to balance the seven total oxygen atoms on the left side of the equation (one from C_2H_5OH and 6 from 3 O_2).
C_2H_5OH (l) + 3 O_2 (g) → 2 CO_2 (g) + 3 H_2O (l) (oxygen is now balanced)

Check that the coefficients cannot be divided by a common factor (divisor).

There is no common factor for the coefficients 2, 1, and 3. The lowest set of whole number coefficients is in the balanced equation. The values in the second column now equal the values in the third column.

Kind of atom	# on reactant side	# on product side
C	2	2
H	6	6
O	7	7

b. Assess the equation.

Kind of atom	# on reactant side	# on product side
C	6	1
H	14	2
O	2	3

The equation is unbalanced because the values in the reactant column do not equal the values in the product column for all types of atoms.

Balance the equation one atom type at a time by inserting coefficients.
Begin by balancing carbon or hydrogen first. Balance oxygen last because it is present in both compounds on the product side. If you begin with carbon, insert the coefficient 6 in front of CO_2.

C_6H_{14} (l) + O_2 (g) → 6 CO_2 (g) + H_2O (l) (carbon is now balanced)
Balance hydrogen by inserting the coefficient 7 in front of H_2O.
C_6H_{14} (l) + O_2 (g) → 6 CO_2 (g) + 7 H_2O (l) (carbon and hydrogen are now balanced)
To balance oxygen, insert the coefficient 19/2 in front of oxygen as a temporary coefficient.
C_6H_{14} (l) + 19/2 O_2 (g) → 6 CO_2 (g) + 7 H_2O (l) (oxygen is now balanced)
Turn the fractions into a whole number by multiplying every coefficient in the equation by 2:
2 C_6H_{14} (l) + 19 O_2 (g) → 12 CO_2 (g) + 14 H_2O (l)

Check that the coefficients cannot be divided by a common factor (divisor).

There is no common factor for the coefficients 2, 19, 12, and 14. The lowest set of whole number coefficients is in the balanced equation. The values in the second column now equal the values in the third column.

Kind of atom	# on reactant side	# on product side
C	12	12
H	28	28
O	38	38

c. Assess the equation.

Kind of atom	# on reactant side	# on product side
C	2	1
H	6	2
O	1	3

The equation is unbalanced because the values in the reactant column do not equal the values in the product column for all types of atoms.

Balance the equation one atom type at a time by inserting coefficients.
Begin by balancing carbon or hydrogen first. Balance oxygen last because it is present in both compounds on the product side. If you begin with carbon, insert the coefficient 2 in front of CO_2.

CH_3OCH_3 (g) + O_2 (g) → 2 CO_2 (g) + H_2O (l) (carbon is now balanced)
Balance hydrogen by inserting the coefficient 3 in front of H_2O.
CH_3OCH_3 (g) + O_2 (g) → 2 CO_2 (g) + 3 H_2O (l) (carbon and hydrogen are now balanced)
To balance oxygen, insert the coefficient 3 in front of oxygen.
CH_3OCH_3 (g) + 3 O_2 (g) → 2 CO_2 (g) + 3 H_2O (l) (oxygen is now balanced)

Check that the coefficients cannot be divided by a common factor (divisor).
There is no common factor for the coefficients 1, 3, and 2. The lowest set of whole number coefficients is in the balanced equation. The values in the second column now equal the values in the third column.

Kind of atom	# on reactant side	# on product side
C	2	2
H	6	6
O	7	7

d. Assess the equation.

Kind of atom	# on reactant side	# on product side
C	3	1
H	6	2
O	2	3

The equation is unbalanced because the values in the reactant column do not equal the values in the product column for all types of atoms.

Balance the equation one atom type at a time by inserting coefficients.
Begin by balancing carbon or hydrogen first. Balance oxygen last because it is present in both compounds on the product side. If you begin with carbon, insert the coefficient 3 in front of CO_2.

C_3H_6 (g) + O_2 (g) → 3 CO_2 (g) + H_2O (l) (carbon is now balanced)
Balance hydrogen by inserting the coefficient 3 in front of H_2O.
C_3H_6 (g) + O_2 (g) → 3 CO_2 (g) + 3 H_2O (l) (carbon and hydrogen are now balanced)
To balance oxygen, insert the coefficient 9/2 in front of oxygen as a temporary coefficient.
C_3H_6 (g) + 9/2 O_2 (g) → 3 CO_2 (g) + 3 H_2O (l) (oxygen is now balanced)

Turn the fractions into a whole number by multiplying every coefficient in the equation by 2:
2 C_3H_6 (g) + 9 O_2 (g) → 6 CO_2 (g) + 6 H_2O (l)

Check that the coefficients cannot be divided by a common factor (divisor).
There is no common factor for the coefficients 2, 9, and 6. The lowest set of whole number coefficients is in the balanced equation. The values in the second column now equal the values in the third column.

Kind of atom	# on reactant side	# on product side
C	6	6
H	12	12
O	18	18

8.27   Convert the known mass of reactant to moles of reactant.

Use the molar mass of glucose as the conversion factor.

$$15.0 \; \text{g glucose} \times \frac{1 \; \text{mol glucose}}{180.14 \; \text{g glucose}} = 0.0833 \; \text{mol glucose}$$

Mass of glucose   molar mass of glucose

Convert moles of glucose to moles of carbon dioxide.

Use the coefficients in the balanced equation to set up a conversion factor between mol glucose and mol carbon dioxide. Remember when no number is shown, the coefficient is 1:

$$0.0833 \; \text{mol glucose} \times \frac{6 \; \text{mol carbon dioxide}}{1 \; \text{mol glucose}} = 0.500 \; \text{mol carbon dioxide}$$

Convert moles of carbon dioxide to grams of carbon dioxide.

Use the molar mass of carbon dioxide as a conversion factor.

$$0.500 \; \text{mol carbon dioxide} \times \frac{44.01 \; \text{g carbon dioxide}}{1 \; \text{mol carbon dioxide}} = 22.0 \; \text{g carbon dioxide}$$

Mol of carbon dioxide   molar mass of carbon dioxide

Therefore, 15.0 g of glucose produces 22.0 g of carbon dioxide

8.29   Assess the equation.

Kind of atom	# on reactant side	# on product side
C	4	1
H	10	2
O	2	3

The equation is unbalanced because the values in the reactant column do not equal the values in the product column for all types of atoms.

Balance the equation one atom type at a time by inserting coefficients.

Begin by balancing carbon or hydrogen first. Balance oxygen last because it is present in both compounds on the product side. If you begin with carbon, insert the coefficient 4 in front of CO_2.

C_4H_{10} (g) + O_2 (g) → 4 CO_2 (g) + H_2O (l) (carbon is now balanced)
Balance hydrogen by inserting the coefficient 5 in front of H_2O.
C_4H_{10} (g) + O_2 (g) → 4 CO_2 (g) + 5 H_2O (l) (carbon and hydrogen are now balanced)
To balance oxygen, insert the coefficient 13/2 in front of oxygen as a temporary coefficient.
C_4H_{10} (g) + 13/2 O_2 (g) → 4 CO_2 (g) + 5 H_2O (l) (oxygen is now balanced)

Turn the fractions into a whole number by multiplying every coefficient in the equation by 2:
2 C_4H_{10} (g) + 13 O_2 (g) → 8 CO_2 (g) + 10 H_2O (l)

Check that the coefficients cannot be divided by a common factor (divisor).
There is no common factor for the coefficients 2, 9, and 6. The lowest set of whole number coefficients is in the balanced equation. The values in the second column now equal the values in the third column.

Kind of atom	# on reactant side	# on product side
C	8	8
H	20	20
O	26	26

Convert the known mass of reactant to moles of reactant.
Use the molar mass of butane as the conversion factor.

$$1.05 \ \cancel{g \ butane} \times \frac{1 \ mol \ butane}{58.11 \ \cancel{g \ butane}} = 0.0181 \ mol \ butane$$

Mass of butane   molar mass of butane

Convert moles of butane to moles of water.
Use the coefficients in the balanced equation to set up a conversion factor between mol butane and mol water.

$$0.0181 \; \cancel{\text{mol butane}} \times \frac{10 \; \text{mol water}}{2 \; \cancel{\text{mol butane}}} = 0.905 \; \text{mol water}$$

Convert moles of water to grams of water.

Use the molar mass of water as a conversion factor.

$$0.0905 \; \cancel{\text{mol water}} \times \frac{18.01 \; \text{g water}}{1 \; \cancel{\text{mol water}}} = 1.63 \; \text{g water}$$

Mol of water   molar mass of water

Therefore, 1.05 g of butane produces 1.63 g of water.

8.31   There are 1000 "small c" calories in one "capitol C" Calorie. "Capital C" Calories are normally reported on nutritional food labels.

8.33   a.   Identify the conversion.

The supplied unit is Calorie. The requested unit is calories. One conversion factor is needed.

1000 cal = 1 Cal

Express the conversion as a conversion factor:

$$\frac{1000 \; \text{cal}}{1 \; \text{Cal}} \quad \text{or} \quad \frac{1 \; \text{Cal}}{1000 \; \text{cal}}$$

Set up the calculation so that the supplied units cancel.

Begin with the supplied unit. Choose the appropriate conversion factors for the numerator and denominator so that the supplied unit (Cal) cancels and you are left with the answer in the requested unit (cal).

$$0.234 \; \cancel{\text{Cal}} \times \frac{1000 \; \text{cal}}{1 \; \cancel{\text{Cal}}} = 234 \; \text{cal}$$

b.   Identify the conversion.

The supplied unit is kilocalorie. The requested unit is calories. One conversion factor is needed.

1000 cal = 1 kcal

Express the conversion as a conversion factor:

$$\frac{1000 \text{ cal}}{1 \text{ kcal}} \quad \text{or} \quad \frac{1 \text{ kcal}}{1000 \text{ cal}}$$

Set up the calculation so that the supplied units cancel.

Begin with the supplied unit. Choose the appropriate conversion factors for the numerator and denominator so that the supplied unit (kcal) cancels and you are left with the answer in the requested unit (cal).

$$0.0991 \ \cancel{\text{kcal}} \times \frac{1000 \text{ cal}}{1 \ \cancel{\text{kcal}}} = 99.1 \text{ cal}$$

c. Identify the conversion.

The supplied unit is kilocalorie. The requested unit is calories. One conversion factor is needed.

    1000 cal = 1 kcal

Express the conversion as a conversion factor:

$$\frac{1000 \text{ cal}}{1 \text{ kcal}} \quad \text{or} \quad \frac{1 \text{ kcal}}{1000 \text{ cal}}$$

Set up the calculation so that the supplied units cancel.

Begin with the supplied unit. Choose the appropriate conversion factors for the numerator and denominator so that the supplied unit (kcal) cancels and you are left with the answer in the requested unit (cal).

$$20.7 \ \cancel{\text{kcal}} \times \frac{1000 \text{ cal}}{1 \ \cancel{\text{kcal}}} = 2.07 \times 10^4 \ \text{cal}$$

d. Identify the conversion.

The supplied unit is Calorie. The requested unit is calories. One conversion factor is needed.

    1000 cal = 1 Cal

Express the conversion as a conversion factor:

$$\frac{1000 \text{ cal}}{1 \text{ Cal}} \quad \text{or} \quad \frac{1 \text{ Cal}}{1000 \text{ cal}}$$

Set up the calculation so that the supplied units cancel.

Begin with the supplied unit. Choose the appropriate conversion factors for the numerator and denominator so that the supplied unit (Cal) cancels and you are left with the answer in the requested unit (cal).

$$352 \; \cancel{Cal} \times \frac{1000 \; cal}{1 \; \cancel{Cal}} = 3.52 \times 10^5 \; cal$$

8.35 ΔH is the enthalpy change of a reaction or the amount of heat energy transferred in a reaction.

8.37   In an exothermic reaction, the products are lower in energy than the reactants. If the reaction is reversed, the products become the reactants and the reactants the products. In the reverse reaction the products are higher in energy than the reactants; therefore it is an endothermic reaction.

8.39   a. Exothermic. Heat is a product of the reaction.

   b. Exothermic. Heat is a product of the reaction.

   c. Endothermic. Heat needs to be added to the reaction.

   d. Endothermic. Heat needs to be added to the reaction.

8.41   A calorimeter is the name of the instrument that measures the heat content of various substances.

8.43   a. Determine the Calories supplied by each type of biomolecule.

Calories provided by fat: $71 \; \cancel{g} \times \dfrac{9 \; Cal}{\cancel{g}} = 639 \; Cal$

Calories provided by carbohydrate: $28 \; \cancel{g} \times \dfrac{4 \; Cal}{\cancel{g}} = 112 \; Cal$

Calories provided by protein: $27 \; \cancel{g} \times \dfrac{4 \; Cal}{\cancel{g}} = 108 \; Cal$

Sum these values to obtain the total Calories supplied by almonds:

639 Cal + 112 Cal + 108 Cal = 859 Cal or 860 Cal

   b. Determine the Calories supplied by each type of biomolecule.

Calories provided by fat: $1 \; \cancel{g} \times \dfrac{9 \; Cal}{\cancel{g}} = 9 \; Cal$

Calories provided by carbohydrate: $27 \; \cancel{g} \times \dfrac{4 \; Cal}{\cancel{g}} = 108 \; Cal$

Calories provided by protein: $1 \, \cancel{g} \times \dfrac{4 \text{ Cal}}{\cancel{g}} = 4 \text{ Cal}$

Sum these values to obtain the total Calories supplied by the banana:

9 Cal + 108 Cal + 4 Cal = 121 Cal or 120 Cal

c. Determine the Calories supplied by each type of biomolecule.

Calories provided by fat: $9 \, \cancel{g} \times \dfrac{9 \text{ Cal}}{\cancel{g}} = 81 \text{ Cal}$

Calories provided by carbohydrate: $0 \, \cancel{g} \times \dfrac{4 \text{ Cal}}{\cancel{g}} = 0 \text{ Cal}$

Calories provided by protein: $7 \, \cancel{g} \times \dfrac{4 \text{ Cal}}{\cancel{g}} = 28 \text{ Cal}$

Sum these values to obtain the total Calories supplied by the cheddar cheese:

81 Cal + 0 Cal + 28 Cal = 109 Cal or 110 Cal

d. Determine the Calories supplied by each type of biomolecule.

Calories provided by fat: $10 \, \cancel{g} \times \dfrac{9 \text{ Cal}}{\cancel{g}} = 90 \text{ Cal}$

Calories provided by carbohydrate: $23 \, \cancel{g} \times \dfrac{4 \text{ Cal}}{\cancel{g}} = 92 \text{ Cal}$

Calories provided by protein: $2 \, \cancel{g} \times \dfrac{4 \text{ Cal}}{\cancel{g}} = 8 \text{ Cal}$

Sum these values to obtain the total Calories supplied by the glazed donut:

90 Cal + 92 Cal + 8 Cal = 190 Cal

e. Determine the Calories supplied by each type of biomolecule.

Calories provided by fat: $4 \, \cancel{g} \times \dfrac{9 \text{ Cal}}{\cancel{g}} = 36 \text{ Cal}$

Calories provided by carbohydrate: $0 \, \cancel{g} \times \dfrac{4 \text{ Cal}}{\cancel{g}} = 0 \text{ Cal}$

Calories provided by protein: $22 \, \cancel{g} \times \dfrac{4 \text{ Cal}}{\cancel{g}} = 88 \text{ Cal}$

Sum these values to obtain the total Calories supplied by the swordfish:

36 Cal + 0 Cal + 88 Cal = 124 Cal or 120 Cal

8.45   Adenosine triphosphate, ATP, serves as the important energy carrier molecule between catabolic and anabolic reactions.

8.47   Carbohydrates, fats, and proteins are obtained through the diet and metabolized into smaller molecules.

8.49   The synthesis of proteins is an anabolic pathway.

8.51   Chemical kinetics is the study of reaction rates, how fast reactants in a chemical reaction are converted into products.

8.53   a.   The reaction shown is an endothermic reaction.

b.

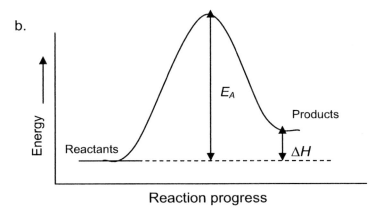

c.   In the presence of a catalyst, the reaction curve will look like this.

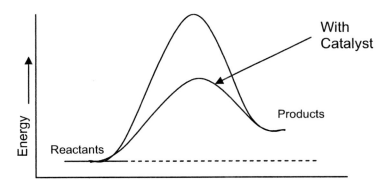

8.55   A catalyst does not affect the value of ΔH. A catalyst affects the value of E_A by making it smaller.

8.57   In the cell, chemical reactions occur at normal body temperature and at a relatively constant concentration. Therefore, to increase the rate of a biochemical reaction, enzymes are used. Enzymes reduce the freedom of motion available to reactants; they lower the activation energy by forcing reactants into a spatial orientation conducive to reaction.

8.59

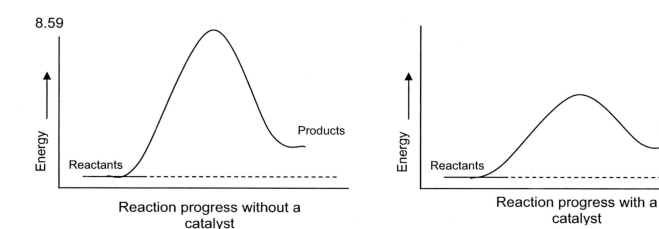

Reaction progress without a
catalyst

Reaction progress with a
catalyst

The reaction without the catalyst has the higher activation energy.

8.61 $\dfrac{2 \text{ miles}}{1 \text{ trip}} \times \dfrac{2 \text{ trips}}{\text{day}} \times 5 \text{ days} \times \dfrac{40 \text{ Cal}}{1 \text{ mile}} = 800 \text{ Cal}$

8.63 100% − 15% = percent of fat = 85%

$0.85 \times 2 \text{ lbs} \times \dfrac{454 \text{ g}}{1 \text{ lb}} \times \dfrac{9 \text{ Cal}}{1 \text{ g}} = 6.95 \times 10^3 \text{ Cal}$

8.65 a. Condensation is an exothermic physical process. Heat energy must be removed.

b. Deposition is an exothermic physical process. Heat energy must be removed.

c. The temperature in the surrounding air would rise because heat energy is removed from the molecules undergoing condensation or deposition and transferred to the surrounding air molecules.

8.67 heat + $ZnCO_3 \rightarrow ZnO + CO_2$

8.69 Yes, the products are the same. Combustion of food in a calorimeter and by the human body both produce carbon dioxide, water, and energy.

8.71 Spirometry measures a patient's oxygen uptake. A patient breathes in 100% oxygen from a pre-filled spirometer. The patient continues to re-breathe oxygen. Gases exhaled from the patient include carbon dioxide and unused oxygen. The exhaled carbon dioxide is removed and the amount of oxygen uptake is determined by the decrease in volume of the spirometer.

Chapter 9

Acids, Bases, pH, and Buffers

Chapter Summary

In this chapter, you learned about the unique properties of acids and bases and their role in health and medicine. You learned about the formation of acids and bases, how to determine the difference between strong acids, weak acids, strong bases, and weak bases, the factors that affect acid-base equilibria, and neutralization reactions. You discovered how to determine the concentrations of hydronium ion and hydroxide ion and how to determine the pH. You gained an understanding of how buffers work and how they control the pH within the body.

Section 9.1 Properties of Acids and Bases

Chemical reactions of acids and bases represent the simplest type of reaction, characterized by the transfer of single protons. There are several different definitions of acids and bases. The first definition states: An acid is a substance that produces protons (H^+) when dissolved in water. A base is a substance that produces hydroxide ions (OH^-) when dissolved in water. The second definition states: An acid is a proton donor and a base is a proton acceptor.

In a Nutshell: Formation of Hydronium Ions

When an acid releases a proton, the proton forms a covalent bond to the oxygen atom in an H_2O molecule to make the hydronium ion, H_3O^+. Therefore, acids form hydronium ions.

In a Nutshell: Formation of Hydroxide Ions

When a base is dissolved in water, hydroxide ions are produced. Hydroxide ions are produced in one of two ways. In the first way, the base accepts a proton from a H_2O molecule—leaving an OH^- ion. In the second way, the base is an ionic compound containing a hydroxyl group (OH^-) that simply dissociates in solution. All bases contain a nonbonding pair of electrons and are either neutral or negatively charged.

Worked Example #1

Write the reaction that represents the reaction of propanonic acid (CH_3CH_2COOH) in water.

$$
\begin{array}{c}
\quad H\ \ H\ \ O \\
\quad |\ \ \ |\ \ \ || \\
H-C-C-C-O-H\ +\ H_2O \\
\quad |\ \ \ | \\
\quad H\ \ H
\end{array}
\rightleftharpoons
\begin{array}{c}
\quad H\ \ H\ \ O \\
\quad |\ \ \ |\ \ \ || \\
H-C-C-C-O^-\ +\ H_3O^+ \\
\quad |\ \ \ | \\
\quad H\ \ H
\end{array}
$$

Propanoic acid releases a proton and forms hydronium ion.

Try It Yourself #1

Write the reaction that represents the reaction of butanoic acid ($CH_3CH_2CH_2COOH$) in water.

In a Nutshell: Conjugate Acids and Bases

In pure water, water molecules react with one another to produce a small amount of hydronium ions and hydroxide ions. This reaction is known as autoionization. Since water can act as either an acid or a base, it is considered an amphoteric molecule.

Two molecules or ions that differ only in the presence or absence of a proton constitute a conjugate acid-base pair. When an acid releases a proton, the part of the acid that is left is known as a conjugate base of the acid. When the base accepts a proton from water, the substance produced is known as the conjugate acid of the base. A base and its corresponding conjugate acid are also known as a conjugate acid-base pair.

Worked Example #2

Label the conjugate acid-base pairs in the reaction that follows. Does water act as an acid or base in this reaction?

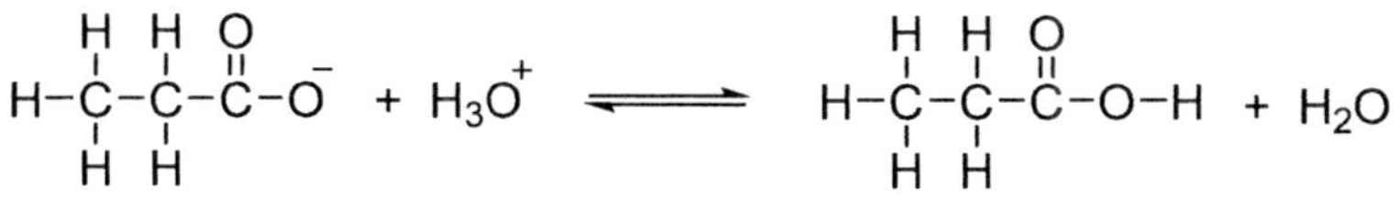

$$
\begin{array}{c}
\overset{\displaystyle H}{\underset{\displaystyle H}{\overset{|}{\underset{|}{H-C}}}}-\overset{\displaystyle H}{\underset{\displaystyle H}{\overset{|}{\underset{|}{C}}}}-\overset{\displaystyle O}{\overset{\|}{C}}-O^{-} \;+\; H_3O^+ \;\rightleftharpoons\; \overset{\displaystyle H}{\underset{\displaystyle H}{\overset{|}{\underset{|}{H-C}}}}-\overset{\displaystyle H}{\underset{\displaystyle H}{\overset{|}{\underset{|}{C}}}}-\overset{\displaystyle O}{\overset{\|}{C}}-O-H \;+\; H_2O
\end{array}
$$

Solution

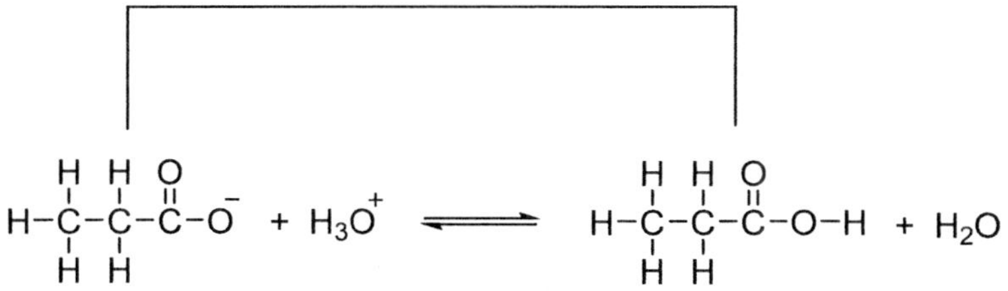

$$
\begin{array}{c}
\overset{\displaystyle H}{\underset{\displaystyle H}{\overset{|}{\underset{|}{H-C}}}}-\overset{\displaystyle H}{\underset{\displaystyle H}{\overset{|}{\underset{|}{C}}}}-\overset{\displaystyle O}{\overset{\|}{C}}-O^{-} \;+\; H_3O^+ \;\rightleftharpoons\; \overset{\displaystyle H}{\underset{\displaystyle H}{\overset{|}{\underset{|}{H-C}}}}-\overset{\displaystyle H}{\underset{\displaystyle H}{\overset{|}{\underset{|}{C}}}}-\overset{\displaystyle O}{\overset{\|}{C}}-O-H \;+\; H_2O
\end{array}
$$

      base   conjugate acid

Water acts as an acid in this reaction, donating a proton to the base.

Try It Yourself #2

Label the conjugate acid-base pairs in the reaction below. Does water act as an acid or base in this reaction?

$$HCl \;+\; H_2O \;\rightleftharpoons\; H_3O^+ \;+\; Cl^-$$

Does water accept or donate a proton? _____

Water is a: _____

Does HCl accept or donate a proton? _____

HCl is a: _____

In a Nutshell: Strengths of Acids and Bases

An acid or base is classified as strong or weak depending on the extent that it dissociates in water. A strong acid completely dissociates in aqueous solution. All the molecules of a strong acid donate a proton to water to form H_3O^+. Complete dissociation is indicated in a chemical reaction by a single forward arrow. Strong acids include HCl, HNO_3, H_2SO_4, $HClO_4$, HBr, and HI. A strong base completely dissociates in aqueous solution. Strong bases consist of hydroxides that completely dissociate in water. Strong bases include KOH, $Ba(OH)_2$, NaOH, $Sr(OH)_2$, $Ca(OH)_2$, and LiOH.

Weak acids are distinguished by the fact that they do not completely dissociate in water; a significant percentage of the undissociated acid is always present in solution. The dissociation of a weak acid in aqueous solution is a reversible reaction, as indicated by the double arrows. In a reversible reaction both the forward and reverse reactions are occurring. Most organic compounds containing carboxylic acids are weak acids. Weak bases are distinguished by the fact that a significant portion of the base is always present in solution; only a small percentage accepts a proton from water, forming a conjugate acid. Ammonia and amines are weak bases.

Amino acids, the building blocks of proteins contain both an amine and a carboxylic acid functional group. In the cell, the amine group is in its conjugate acid form and the acid is in its conjugate base form. Therefore, amino acids are amphoteric molecules.

Worked Example #3

Indicate whether the following is a strong acid, strong base, weak acid, or weak base. For each write the chemical equations that represent the dissociation in water.

   a. $Ca(OH)_2$
   b. CH_3NH_2
   c. HBr

Tools: Use Tables 9-3 and 9-4

   a. $Ca(OH)_2$ is a strong base.

$$Ca(OH)_2 \longrightarrow Ca^{2+}(aq) + 2\,OH^-(aq)$$

   b. CH_3NH_2 is a weak base. It is not on the lists of strong acids or bases.

$$CH_3NH_2 + H_2O \rightleftharpoons CH_3NH_3^+ + OH^-$$

   c. HBr is a strong acid.

$$HBr + H_2O \longrightarrow H_3O^+ + Br^-$$

Try It Yourself #3

Indicate whether the following is a strong acid, strong base, weak acid or weak base. For each write the chemical equations that represent the dissociation in water.

   a. $CH_3CH_2CH_2COO^-$

b. $HClO_4$

c.
$$H-\overset{\overset{\displaystyle H}{|}}{\underset{\underset{\displaystyle H}{|}}{C}}-\overset{\overset{\displaystyle H}{|}}{\underset{\underset{\displaystyle H}{|}}{C}}-\overset{\overset{\displaystyle H}{|}}{\underset{\underset{\displaystyle H}{|}}{C}}-\overset{\overset{\displaystyle H}{|}}{\underset{\underset{\displaystyle H}{|}}{C}}-\overset{\overset{\displaystyle O}{\|}}{C}-OH$$

Tools: Use Tables 9-3 and 9-4

   a. The ion is: _____

   b. The molecule is: _____

   c. The molecule is: _____

In a Nutshell: Acid-Base Equilibria

When reactants of a reversible reaction are mixed together, both the forward and reverse reactions occur simultaneously. When the forward and reverse reactions proceed at the same rate and the concentration of the reactants and products no longer changes, the reaction has reached a state of equilibrium. At equilibrium, the reaction has not stopped. For each set of reactants that is turned into product, another set of products is turned into reactants. Although both reactants and products are present at equilibrium, they are not present in equal concentration. Equilibrium always favors the formation of reactants for weak acids and bases.

In a Nutshell: Le Châtelier's Principle

A reaction at equilibrium can be disturbed by a change in temperature or by a change in the concentration of a reactant or product. Le Châtelier's principle states that, when a reaction at equilibrium is disturbed, the reaction responds by shifting in a direction that restores equilibrium: either the forward direction (to the right) or the reverse direction (to the left.) For example, if more reactant is added to a reversible reaction, the rate of the forward reaction will increase until a new equilibrium is established.

Worked Example #4

Consider the reversible reaction shown below, which occurs in ethylene glycol poisoning:

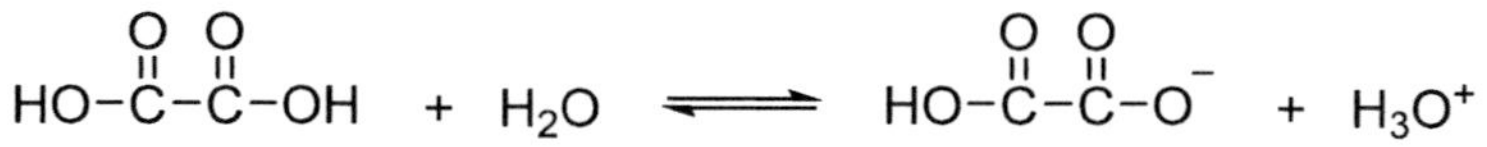

$$HO-\overset{\overset{O}{\|}}{C}-\overset{\overset{O}{\|}}{C}-OH \ + \ H_2O \ \rightleftharpoons \ HO-\overset{\overset{O}{\|}}{C}-\overset{\overset{O}{\|}}{C}-O^- \ + \ H_3O^+$$

oxalic acid   water   oxalate ion   hydronium ion

a. What substances are present at equilibrium?

b. At equilibrium, are the concentrations of oxalic acid and oxalate ion constant or changing?

c. How will the equilibrium shift if oxalic acid is removed from the solution?

d. How the equilibrium shift if more water is added to the solution?

a. *Oxalic acid, water, oxalate ion, and hydronium ion are all present at equilibrium.*

b. *The concentrations of oxalic acid and oxalate ion are constant at equilibrium.*

c. *If oxalic acid is removed, the concentration of one of the reactants decreases; therefore, the equilibrium will shift to the left.*

d. *If more water is added, the concentration of one of the reactants increases; therefore, the equilibrium will shift to the right.*

Try It Yourself #4

Phenylacetic acid builds up in the blood of people with phenylketonuria. In aqueous solution, phenylacetic acid undergoes the following reversible reaction:

$$\bigcirc\!\!\!\!\!-\overset{\overset{H}{|}}{\underset{\underset{H}{|}}{C}}-\overset{\overset{O}{\|}}{C}-OH \ + \ H_2O \ \rightleftharpoons \ \bigcirc\!\!\!\!\!-\overset{\overset{H}{|}}{\underset{\underset{H}{|}}{C}}-\overset{\overset{O}{\|}}{C}-O^- \ + \ H_3O^+$$

phenylacetic   water   phenylacetate   hydronium
   acid   ion   ion

a. What substances are present at equilibrium?

b. At equilibrium, are the concentrations of phenylacetic acid and phenylacetate ion constant or changing?

c. How will the equilibrium shift if phenylacetate ion is removed from the solution?

d. How the equilibrium shift if water is removed from the solution?

a. The substances present at equilibrium are: _____

b. The concentrations are: _____

c. Phenylacetate ion is a reactant or a product? _____

    The concentration of phenylacetate: _____; therefore, the reaction shifts:

    _____.

d. Water is a reactant or a product? _____

    The concentration of water: _____; therefore, the reaction shifts:

    _____.

In a Nutshell: Acid-Base Neutralization Reactions

In a neutralization reaction, a proton is transferred from an acid to a base producing a salt and in some cases water. In a neutralization reaction that produces water, H^+ ions and OH^- ions react in a one to one ratio to form one H_2O molecule.

Worked Example #5

Aluminum hydroxide, $Al(OH)_3$, is one of the active ingredients in Maalox. Write the balanced equation for the neutralization reaction of aluminum hydroxide with stomach acid, HCl.

$$Al(OH)_3 \ + \ 3\,HCl \ \longrightarrow \ 3\,H_2O \ + \ AlCl_3$$
$$\text{base} \qquad\quad \text{acid} \qquad\qquad \text{water} \quad\ \text{salt}$$

Since aluminum hydroxide yields 3 OH^- per formula unit, three HCl molecules are required to neutralize one formula unit of $Al(OH)_3$. Therefore, the coefficient 3 must be placed before HCl (yielding 3 H^+). Three water molecules are produced as well as the salt, $AlCl_3$.

Try It Yourself #5

Write the balanced equation for the neutralization reaction between HBr and $Mg(OH)_2$.

Unbalanced equation for the reaction of $Mg(OH)_2$ and HBr: _____

$Mg(OH)_2$ produces how many OH^- per formula unit? _____

Number of HBr molecules needed to neutral one formula unit of $Mg(OH)_2$: _____

Balanced equation for the reaction of Mg(OH)$_2$ and HBr: _____

Practice Problems for Properties of Acids and Bases

1. Write the reaction that represents the reaction of pentanoic acid, $CH_3CH_2CH_2CH_2COOH$, with water. Label the conjugate acid-base pairs in the reaction. Is pentanoic acid a strong acid, weak acid, weak base, or strong base?

2. Write the products for the following reaction between aniline and water. Label the conjugate acid-base pairs. Is aniline a strong acid, a weak acid, a weak base or a strong base?

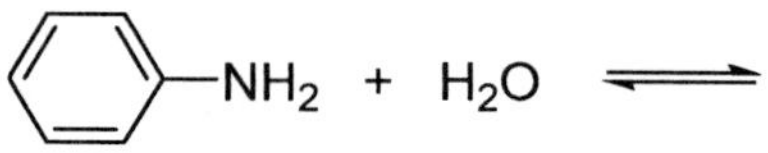

$$\text{aniline} \quad -NH_2 \; + \; H_2O \; \rightleftharpoons$$

   aniline   water

3. Dodecanoic acid or lauric acid is found in coconut oil. In aqueous solution, dodecanoic acid undergoes the following reversible reaction.

$$CH_3(CH_2)_{10}COOH \; + \; H_2O \; \rightleftharpoons \; CH_3(CH_2)_{10}COO^- \; + \; H_3O^+$$

   dodecanoic acid   water   hydronium
                                                                   ion

   a. How will the equilibrium shift if more water is added to the solution?

   b. How will the equilibrium shift if dodecanoic acid is removed from the solution?

   c. How will the equilibrium shift if more hydronium ion is added to the solution?

4. Write the balanced equation for the neutralization reaction between $HClO_4$ and $Ca(OH)_2$.

Section 9.2 pH

The pH of an aqueous solution is a quantitative measure of the concentration of hydronium ions. Aqueous solutions have a pH ranging from 1 to 14. A pH of less than 7 indicates an acidic solution, a pH greater than 7 indicates a basic solution, and a pH equal to 7 indicates a neutral solution.

In a Nutshell: Ion-Product Constant, K_w

In pure water, the concentration of the hydronium ion will be equal to the concentration of the hydroxide ion, which at room temperature (25 °C) is 1.0×10^{-7} M. The product of the hydronium ion concentration and the hydroxide ion concentration is equal to 1.0×10^{-14}, known as the ion-product constant for water, K_w. $K_w = [H_3O^+] \times [OH^-] = 1.0 \times 10^{-14}$. When an acid or a base is present in water, it is no longer pure, the hydronium and hydroxide ion concentrations are no longer equal to 1.0×10^{-7} M. However, K_w is still equal to 1.0×10^{-14}.

Worked Example #6

If an aqueous solution has a $[OH^-] = 4.8 \times 10^{-5}$, what is $[H_3O^+]$?

Solution

Use the equation for the ion-product constant:

$K_w = [H_3O^+] \times [OH^-]$

Set up the equation to solve for $[H_3O^+]$:

$$[H_3O^+] = \frac{K_w}{[HO^-]}$$

Substitute the values for K_w and $[OH^-]$ and solve:

$$[H_3O^+] = \frac{1 \times 10^{-14}}{4.8 \times 10^{-5}} = 2.1 \times 10^{-10} \text{ M}$$

Try It Yourself #6

Complete the following table for an aqueous solution:

$[H_3O^+]$	$[OH^-]$
2.3×10^{-3}	
	1.7×10^{-10}
6.5×10^{-8}	

Use the equation for the ion-product constant.

Set up the equation to solve for $[H_3O^+]$ or $[OH^-]$.

Substitute the values for K_w and $[OH^-]$ or $[H_3O^+]$ and solve.

In a Nutshell: The pH Scale

The pH scale is a simple way to report the hydronium ion concentration without using scientific notation. The pH of a solution is defined as the negative log of the hydronium ion concentration, $pH = -\log[H_3O^+]$. To convert pH to $[H_3O^+]$, use the formula $[H_3O^+] = 10^{-pH}$. Since pH represents a logarithmic scale, one pH unit represents a ten-fold change in hydronium ion concentration.

Worked Example #7

What is the $[HO^-]$ of applesauce, which has a pH of 3.2? Is applesauce acidic or basic?

Applesauce is acidic because the pH is less than 7.
Calculate $[H_3O^+]$ from the pH value.

$[H_3O^+] = 10^{-pH} = 10^{-3.2}$

$[H_3O^+] = 6.3 \times 10^{-4}$ M

Calculate [HO⁻], using the equation for K_W.

$K_W = [H_3O^+] \times [OH^-]$

Set up the equation to solve for [HO⁻]:

$$[HO^-] = \frac{K_W}{[H_3O^+]}$$

Substitute the values for K_w and [H_3O^+] and solve:

$$[HO^-] = \frac{1 \times 10^{-14}}{6.3 \times 10^{-4}} = 1.6 \times 10^{-11} \text{ M}$$

Try It Yourself #7

What is the [HO⁻] of saliva, which has a pH of 6.8? Is this sample of saliva acidic, basic, or neutral?

Saliva is: _____

Calculate [H_3O^+] from the pH value.

Calculate [HO⁻], using the equation for K_W.

Set up the equation to solve for [HO⁻]:

Substitute the values for K_w and [H_3O^+] and solve.

Worked Example #8

What is the pH of a urine sample that has an [HO⁻] of 1.6×10^{-7} M? Is this sample of urine acidic, basic, or neutral?

Calculate [H₃O⁺], using the equation for K$_W$.

$K_W = [H_3O^+] \times [OH^-]$

Set up the equation to solve for [H₃O⁺]:

$$[H_3O^+] = \frac{K_W}{[HO^-]}$$

Substitute the values for K$_w$ and [OH⁻] and solve:

$$[H_3O^+] = \frac{1 \times 10^{-14}}{1.6 \times 10^{-7}} = 6.3 \times 10^{-8} \text{ M}$$

Calculate pH by substituting the value for [H₃O⁺] into the equation for pH:

pH = −log[H₃O⁺] = −log(6.3 × 10⁻⁸) = 7.2

This sample of urine is basic, the pH is greater than 7.

Try It Yourself #8

What is the pH of lemon juice that has an [HO⁻] of 1.27×10^{-12} M? Is lemon juice acidic, basic, or neutral?

Calculate [H₃O⁺], using the equation for K$_W$.

Set up the equation to solve for [H₃O⁺]:

Substitute the values for K$_w$ and [OH⁻] and solve.

Calculate pH by substituting the value for [H₃O⁺] into the equation for pH.

Lemon juice is: _____

In a Nutshell: pH and The Cell

The pH inside a healthy cell ranges from 7.35 to 7.45, a pH referred to as physiological pH. Intracellular pH changes disrupt cell function by causing key molecules in biochemical pathways to change from their neutral form to their ionized form or vice versa. At physiological pH, carboxylic acids and amines are ionized. The charges on the ions in the cell enable the ions to stay within the cell and to bind to enzymes and chemically react as necessary.

In a Nutshell: Measuring pH

The pH of a patient may be measured by taking a sample of blood and using a blood gas analyzer. The pH may also be measured testing the pH of urine by using either paper incorporating pH sensitive dyes or a pH probe.

Practice Problems for pH

1. Indicate whether the following solutions are acidic, basic, or neutral based on their pH.

   a. 12.4

   b. 1.2

   c. 7.0

   d. 6.9

   e. 7.1

2. Complete the following table.

$[H_3O^+]$	$[HO^-]$	pH	Acidic, basic, or neutral
1×10^{-4}			
	1×10^{-2}		
1×10^{-7}			
	1×10^{-11}		

3. What is the pH of tomato juice that has a $[H_3O^+] = 3.98 \times 10^{-5}$? Calculate the $[HO^-]$ as well. Is tomato juice acidic, basic, or neutral?

4. What is the pH of a sodium bicarbonate (baking soda) solution that has a $[H_3O^+] = 6.31 \times 10^{-9}$? Calculate the $[HO^-]$ as well. Is baking soda acidic, basic, or neutral?

Section 9.3 Buffers

A buffer is a solution that resists changes in pH upon addition of small amounts of acid or base. A buffer is a solution composed of a weak acid and its conjugate base, in relatively equal concentration. Adding a small amount of a strong acid to a buffer increases the concentration of H_3O^+. This small amount of H_3O^+ will react with the large amount of the conjugate base present and, according to Le Châtelier's principle, shift the equilibrium back toward the reactants. Adding a small amount of a strong base to a buffer increases the concentration of OH^-. This small amount of OH^- will react with the weak acid present and, according to Le Châtelier's principle, shift the equilibrium toward the products. The greatest amount of acid or base that a buffer can accept while maintaining pH is known as buffer capacity. The higher the concentration of the weak acid and it conjugate base in a buffer system, the higher its buffer capacity.

Worked Example #9

Consider the following buffer system, consisting of the weak acid, formic acid, $HCOOH$, and its conjugate base, $HCOO^-$.

$$HCOOH + H_2O \rightleftharpoons H_3O^+ + HCOO^-$$

a. How would this buffer react if OH^- were added? Show the equation.

b. How would this buffer react if H_3O^+ were added? Show the equation.

a. The equilibrium would respond by shifting to the right:

$$HCOOH + OH^- \longrightarrow HCOO^- + H_2O$$

b. The equilibrium would respond by shifting to the left:

$$HCOOH + H_2O \longleftarrow H_3O^+ + HCOO^-$$

Try It Yourself #9

Consider the following buffer system, consisting of the weak acid, boric acid, H_3BO_3, and its conjugate base, $H_2BO_3^-$.

$$H_3BO_3 + H_2O \rightleftharpoons H_3O^+ + H_2BO_3^-$$

a. How would this buffer react if H_3O^+ were added? Show the equation.

b. How would this buffer react if OH^- were added? Show the equation.

a. The equilibrium shifts: _____

The equation: _____

b. The equilibrium shifts: _____

The equation: _____

In a Nutshell: Buffering Systems in the Body

The primary buffering system in the blood consists of the weak acid, carbonic acid, H_2CO_3, and its conjugate base, the bicarbonate ion, HCO_3^-. Carbonic acid comes from the reaction of carbon dioxide and water. The body's maintenance of the proper balance between acids and bases is known as acid-base homeostasis. The body compensates for short term acid-base imbalances, by regulating breathing rate, which affects the concentration of carbon dioxide in blood and alters the pH.

When blood pH falls below 6.8, acidosis occurs. Respiratory acidosis occurs when carbon dioxide is not removed from the lungs rapidly enough and excess carbon dioxide is dissolved in blood. Metabolic acidosis is caused by a metabolic disorder, such as kidney failure, diabetes, or starvation.

When blood pH rises above 8.0, alkalosis occurs. Respiratory alkalosis occurs when someone hyperventilates and removes carbon dioxide from the lungs faster than it is produced. The concentration of H_3O^+ drops and the pH of the blood rises. Metabolic alkalosis can be caused by excessive vomiting, ingestion of excessive amounts of antacids, and some adrenal gland diseases.

Practice Problems for Buffers

1. Indicate which of the following solutions are buffers.

    a. H_2CO_3 (carbonic acid, a weak acid) and HCO_3Na

    b.   HBr (strong acid) and NaBr

    c.   NaCl and NaOH

    d. CH_3CO_2H and CH_3CO_2Na

2. Consider the following buffer system, consisting of the weak acid, benzoic acid, C_6H_5COOH, and its conjugate base, $C_6H_5COO^-$.

    $$C_6H_5COOH + H_2O \rightleftharpoons H_3O^+ + C_6H_5COO^-$$

    a. How would this buffer react if H_3O^+ were added? Show the equation.

    b. How would this buffer react if OH^- were added? Show the equation.

Chapter 9 Quiz

1. Write the chemical equation for the reaction of HI in water. Label the conjugate acid-base pairs. Is HI a strong acid, weak acid, strong base, or weak base?

2. Write the chemical equation for the reaction of $CH_3CH_2NH_2$ in water. Label the conjugate acid-base pairs. Is $CH_3CH_2NH_2$ a strong acid, weak acid, strong base, or weak base?

3. Which of the following equations represents reactions that are at equilibrium?

    a. $HCl + H_2O \rightarrow H_3O^+ + Cl^-$

    b. $CH_3CH_2COO^- + H_2O \rightleftharpoons CH_3CH_2COOH + OH^-$

    c. $Ba(OH)_2 (s) + H_2O \rightarrow Ba^{2+} (aq) + OH^- (aq)$

    d. $CH_3COOH + H_2O \rightarrow H_3O^+ + CH_3COO^-$

4. The reaction of pyruvic acid in aqueous solution is shown below.

$$\underset{\text{pyruvic acid}}{\text{H-C-C-C-OH}} + \underset{\text{water}}{H_2O} \rightleftharpoons \underset{\text{pyruvate}}{\text{H-C-C-C-O}^-} + \underset{\substack{\text{hydronium}\\\text{ion}}}{H_3O^+}$$

    a. What substances are present at equilibrium?

   b. How will the equilibrium shift if pyruvate is added to the solution?

   c. How will the equilibrium shift if hydronium ion is removed from the solution?

   d. How will the equilibrium shift is pyruvic acid is added to the solution?

5. Write the balanced chemical equation for the neutralization reaction between HNO_3 and $Al(OH)_3$.

6. Complete the following table for an aqueous solution:

$[H_3O^+]$	$[OH^-]$
4.8×10^{-5}	
	9.1×10^{-2}
7.3×10^{-9}	

7. Are the following solutions acidic, basic, or neutral based on their pH.
   a. 7.0

   b. 1.1

   c. 14

   d. 10.6

   e. 2.3

8.   Complete the following table:

$[H_3O^+]$	$[HO^-]$	pH	Acidic, basic, or neutral
1×10^{-7}			
	1×10^{-11}		
1×10^{-12}			

9.   Ketchup has a $[H_3O^+] = 1.29 \times 10^{-4}$. What is the pH of ketchup? Is ketchup neutral, acidic, or basic? Calculate the $[OH^-]$ for ketchup.

10. Consider the following buffer system of phosphoric acid, H_3PO_4, and its conjugate base, dihydrogen phosphate, $H_2PO_4^-$.

$$H_3PO_4 + H_2O \rightleftharpoons H_3O^+ + H_2PO_4^-$$

a. How would this buffer react if H_3O^+ were added? Show the equation.

b. How would this buffer react if OH^- were added? Show the equation.

Chapter 9

Answers to Additional Exercises

9.25

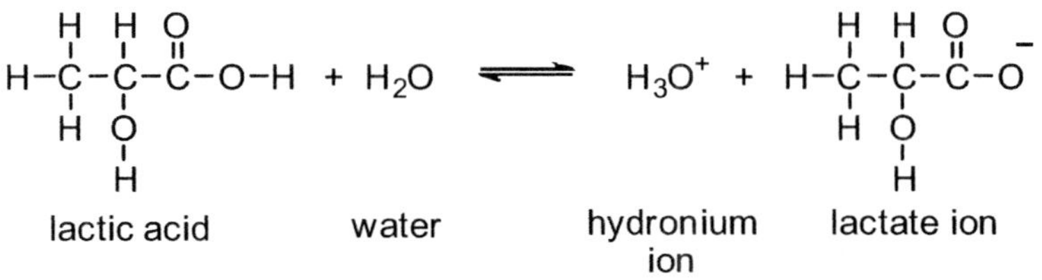

lactic acid   water   hydronium   lactate ion
                                        ion

9.27

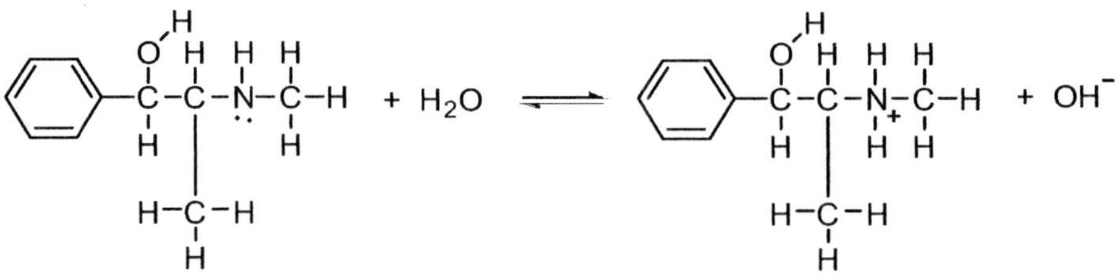

9.29   a. H_3O^+

b. H_2O

c. HI

d. $CH_3CH_2CH_2CH_2COOH$

9.31 $HClO_4 + H_2O \rightarrow ClO_4^- + H_3O^+$

9.33 $H_2SO_4 + H_2O \rightarrow HSO_4^- + H_3O^+$

$HSO_4^- + H_2O \rightarrow SO_4^{2-} + H_3O^+$

9.35   For weak bases, a significant portion of the base is always present in solution. Only a small percentage of the base accepts a proton from water.

9.37   a. Weak acid, most of the acid is still present at equilibrium

b. Weak base

c. Strong acid, it completely dissociates

9.39   a.

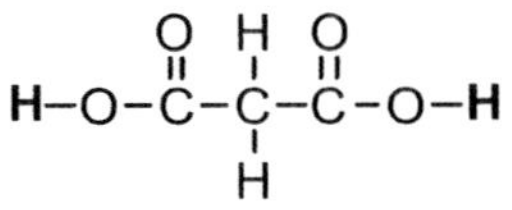

b.

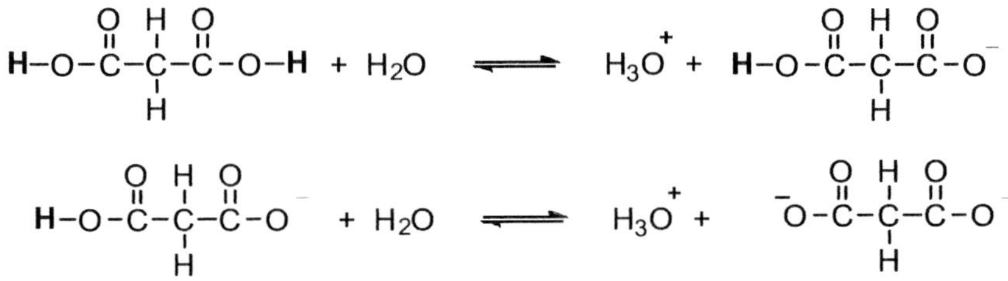

9.41   a. Propanoic acid, water, propanoate, and hydronium ion are present at equilibrium.

        b. The concentrations of propanoic acid and propanoate are constant at equilibrium.

        c. The two opposing arrows indicate that both the forward and reverse reactions occur simultaneously.

9.43   Le Châtelier's principle states that, when a reaction at equilibrium is disturbed, the reaction responds by shifting in the direction that restores equilibrium: either the forward direction (shift to the right) or the reverse direction (shift to the left).

9.45   a. If more ammonia is added, the reaction will shift to the left to consume the excess NH_3 present.

        b. If more NH_4^+ is added, the reaction will shift to the right to consume the excess NH_4^+.

9.47   Hemoglobin will react with the excess oxygen to form the O_2-hemoglobin complex (shifting to the left in the first half of the equation) to restore the equilibrium. The removal of hemoglobin then prevents the reaction to the right that forms CO-hemoglobin and favors the reaction to the left to make more O_2-hemoglobin.

9-49 $2\ HCl\ +\ Ba(OH)_2\ \rightarrow\ 2\ H_2O\ +\ BaCl_2$ Since barium hydroxide yields 2 OH^- per formula unit, two HCl molecules are required to neutralize one formula unit of $Ba(OH)_2$. Therefore, the coefficient 2 must be placed before HCl (yielding 2 H^+). Two water molecules are produced as well as the salt, $BaCl_2$.

9.51 $3\ HCl\ +\ Al(OH)_3\ \rightarrow\ 3\ H_2O\ +\ AlCl_3$ Since aluminum hydroxide yields 3 OH^- per formula unit, three HCl molecules are required to neutralize one formula unit of $Al(OH)_3$. Therefore, the coefficient 3 must be placed before HCl (yielding 3 H^+). Three water molecules are produced as well as the salt, $AlCl_3$.

9.53   a. Basic, pH > 7

        b. Acidic, pH < 7

        c. Neutral, pH = 7

        d. Acidic, pH <7

        e. Basic, pH > 7

        f. Basic, pH > 7

        g. Acidic, pH < 7

9.55

[H_3O^+]	[OH^-]	Is the solution acidic, neutral, or basic?
1.0×10^{-3}	1.0×10^{-11}	**Acidic pH = 3**
1.0×10^{-12}	1.0×10^{-2}	**Basic pH = 12**
1.0×10^{-7}	1.0×10^{-7}	**Neutral pH = 7**
1.0×10^{-5}	1.0×10^{-9}	**Acidic pH = 5**
1.0×10^{-9}	1.0×10^{-5}	**Basic pH = 9**

9.57   Calculate pH by substituting the value for [H_3O^+] into the equation for pH:

$pH = -\log[H_3O^+] = -\log(3.2 \times 10^{-4}) = 3.50$

Apple juice is acidic, pH < 7.

Calculate [OH^-], using the equation for K_W.

$K_W = [H_3O^+] \times [OH^-]$

Set up the equation to solve for [OH^-]:

$$[HO^-] = \frac{K_W}{[H_3O^+]}$$

Substitute the values for K_w and [H_3O^+] and solve:

$$[OH^-] = \frac{1 \times 10^{-14}}{3.2 \times 10^{-4}} = 3.1 \times 10^{-11} \, M$$

9.59   Calculate pH by substituting the value for [H_3O^+] into the equation for pH:

$pH = -\log[H_3O^+] = -\log(3.2 \times 10^{-7}) = 6.49$

Milk is acidic, pH < 7.

Calculate [OH^-], using the equation for K_W.

$K_W = [H_3O^+] \times [OH^-]$

Set up the equation to solve for [OH^-]:

$$[HO^-] = \frac{K_W}{[H_3O^+]}$$

Substitute the values for K_w and [H_3O^+] and solve:

$$[OH^-] = \frac{1 \times 10^{-14}}{3.2 \times 10^{-7}} = 3.1 \times 10^{-8} \, M$$

9.61   A buffer is a solution that resists changes in pH upon addition of small amounts of acid or base.

9.63   When a small amount of acid enters the bloodstream, the bicarbonate ion (HCO_3^-) present reacts with the acid, H_3O^+, forming water and carbonic acid (H_2CO_3). The products do not increase the concentration of H_3O^+, and therefore the pH does not change.

9.65   a. NaOH, a strong base, would react with carbonic acid.

      b. HCl, a strong acid, would react with bicarbonate.

      c. NH_3, a weak base, would react with carbonic acid.

      d. CH_3COO^-, a weak base, would react with carbonic acid.

9.67   The carbonic acid part of the buffer should be given to the patient to decrease the pH.

9.69   Proteins degrade naturally in the gastrointestinal tract due to the reaction of acids and enzymes with proteins. Most intact proteins cannot cross the membrane barrier in the small intestine to proceed to the circulatory system. If the protein could pass through the membrane barrier in the small intestine, it could damage the epithelial cells lining the intestine.

9.71   No, it does not. Protons bind to certain functional groups in a protein, altering its shape and therefore its function.

9.73   Enteric coatings are stable in acidic environment; therefore, the enteric coating control where the in the digestive tract the medication is absorbed.

9.75   The enteric coating is acidic because it reacts in an alkaline environment and is stable in an acidic environment.

Chapter 10

Reactions of Organic Functional Groups in Biochemistry

Chapter Summary

In this chapter, you have seen that biochemistry is organic chemistry. You discovered that functional groups react in the same characteristic way under a given set of conditions. You learned about four types of organic reactions seen in biochemistry. These reactions are oxidation-reduction reactions, dehydration/hydration reactions, acyl group transfer reactions, and phosphoryl group transfer reactions. These reactions will be seen again as you learn about the roles of biomolecules in metabolism.

Section 10.1 The Role of Functional Groups in Biochemical Reactions

Each organic functional group is characterized by a unique chemical reactivity, meaning a predictable set of chemical reactions that the functional group will undergo. The functional groups that undergo chemical reactions include: alkenes, alkynes, primary alcohols, secondary alcohols, phenols, amines, aldehydes, ketones, carboxylic acids, esters, amides, thioesters, and phosphate esters. Their structures are summarized in Table 10-1. Three functional groups are relatively unreactive; they are ethers, tertiary alcohols, and aromatic hydrocarbons. In the cell, there are only six basic reaction types. These six reaction types consist of oxidation-reduction reactions, hydration-dehydration reactions, acyl group transfer reactions (hydrolysis, esterification, and amidation), phosphoryl group transfer reactions, decarboxylation reactions, and reactions that form or break carbon-carbon bonds. Decarboxylation reactions and reactions that form and break carbon-carbon bonds will be covered in a later chapter. When learning these reactions, remember to focus on the change in the functional group and ignore the rest of the molecule, which remains intact.

Section 10.2 Oxidation-Reduction Reactions

The combustion reaction is one of the most common types of oxidation-reduction reactions. Combustion of an organic compound is the reaction between an organic compound and oxygen to produce carbon dioxide, water, and energy. In the cell, glucose undergoes combustion through a series of separate oxidation-reduction reactions to produce carbon dioxide, water, and energy. Cellular respiration describes this sequence of reactions in the cell because oxygen is required.

In a Nutshell: Definitions of Oxidation and Reduction

An oxidation-reduction reaction (sometimes abbreviated redox) is characterized by the transfer of electrons from one reactant to another. The reactant that loses electrons undergoes oxidation; the reactant that gains electrons undergoes reduction. Generally, when a metal cation is formed from a metal, the metal has undergone oxidation; and when a nonmetal anion is formed from a nonmetal, the nonmetal has undergone reduction.

In a Nutshell: Oxidation-Reduction of Organic Molecules

In organic reactions, it is easier to recognize oxidation and reduction reactions by looking for a change in the number of hydrogen atoms or oxygen atoms in the reactants compared to products. An organic molecule that gains oxygen atoms or loses hydrogen atoms or both has undergone oxidation. An organic molecule that does the reverse, loses oxygen atoms or gains hydrogen atoms or both, has undergone reduction. Oxidation always accompanies reduction; the electrons need to be transferred to another reactant. In the cell, the reactants receiving electrons are coenzymes. The term *oxidizing agent* refers to the reactant that gets reduced, and the term *reducing agent* refers to the reactant that gets oxidized. In general, the abbreviation [O] is placed above the reaction arrow to indicate oxidation and [H] is placed above the reaction arrow to indicate reduction.

Worked Example #1

Determine whether the following reactions represent an oxidation or a reduction of the organic substance shown. Explain your reasoning by showing the hydrogen and/or oxygen atoms that have been gained or lost from the reactant.

a.

b.

Solutions

a.

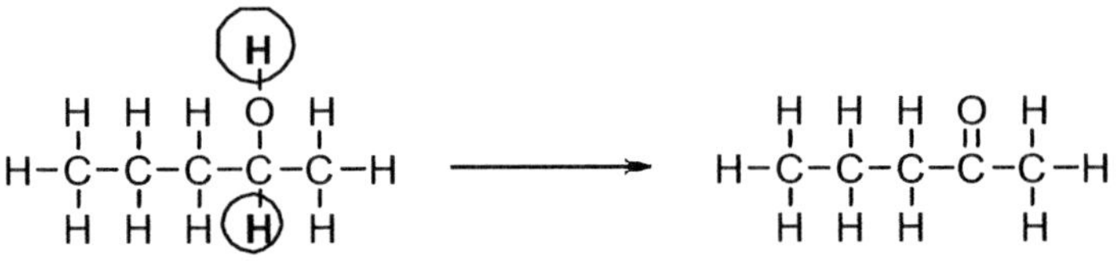

alcohol   ketone

> *The functional group changes from an alcohol to a ketone. There is a decrease in the number of hydrogen atoms and no change in the number of oxygen atoms. Therefore, the reactant has undergone an oxidation.*

b.

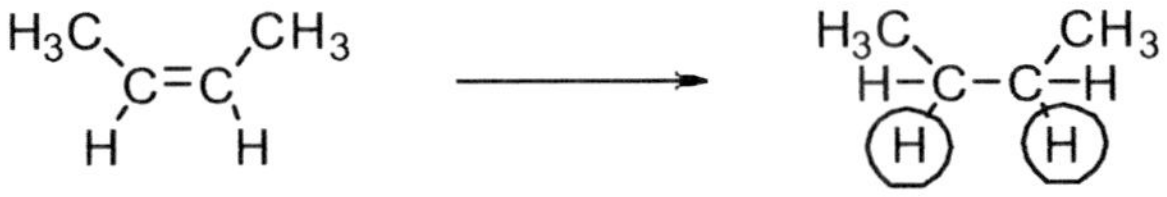

alkene   alkane

> *The functional group changes from an alkene to an alkane. There is an increase in the number of hydrogen atoms and no change in the number of oxygen atoms. Therefore, the reactant has undergone a reduction.*

Try It Yourself #1

Determine whether the following reactions represent an oxidation or a reduction of the organic substance shown. Explain your reasoning by showing the hydrogen and/or oxygen atoms that have been gained or lost from the reactant.

a.

b.

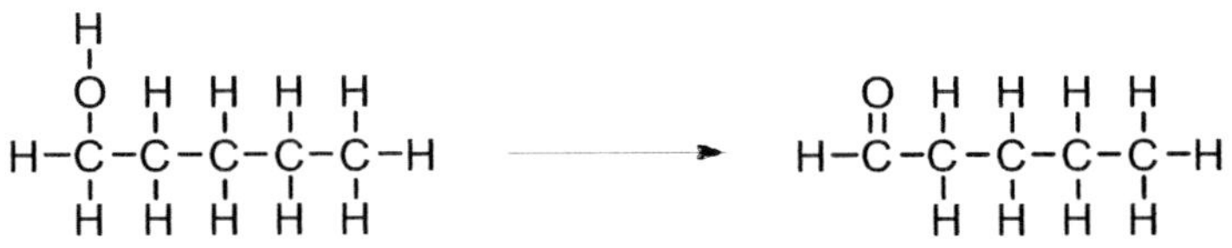

a.   *The functional group changes from* _____ *to* _____.

    The number of hydrogen atoms: _____

    The number of oxygen atoms: _____

    The hydrogen atoms or oxygen atoms that have changed:

*The reaction undergoes:*_____

b. *The functional group changes from* _____ *to* _____.

The number of hydrogen atoms: _____

The number of oxygen atoms: _____

The hydrogen atoms or oxygen atoms that have changed:

The reaction undergoes _____.

In a Nutshell: Hydrocarbon Oxidation-Reductions

Carbon-carbon double bonds undergo reduction to produce carbon-carbon single bonds. The reverse reaction is an oxidation. In the laboratory, the reduction of an unsaturated hydrocarbon is carried out with hydrogen gas as the reducing agent in the presence of a catalyst. This reaction is called catalytic hydrogenation. Catalytic hydrogenation is used by the food industry to prepare shortening from vegetable oils. One outcome of a hydrogenation is that some cis double bonds undergo isomerization to trans double bonds instead of reduction.

In a biological cell, for oxidation-reduction reactions involving carbon-carbon bonds, the coenzyme flavin adenine dinucleotide, FAD, is typically the electron acceptor and is reduced to $FADH_2$ in the process. The reverse process going from $FADH_2$ to FAD is an oxidation reaction. $FADH_2$ is a reducing agent, and FAD is an oxidizing agent.

Worked Example #2

Predict the product formed in the following reaction.

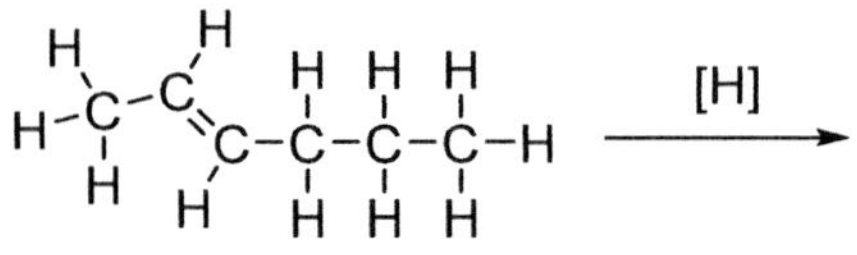

Solution

The [H] indicates that the reaction is a reduction reaction. The functional group is an alkene, which undergoes reduction.

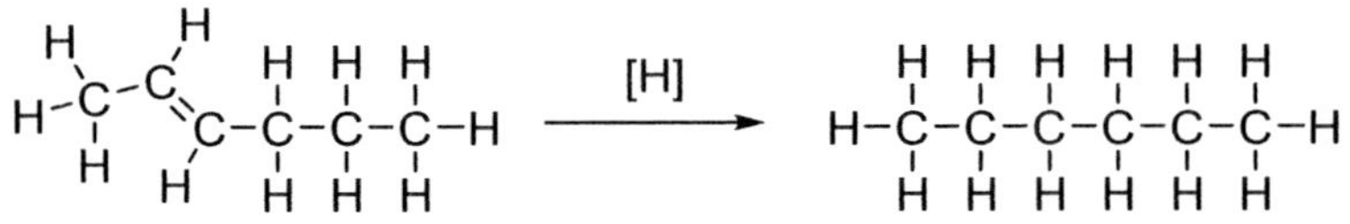

Try It Yourself #2

Predict the product formed in the following reaction.

$$FADH_2 \xrightarrow{[O]}$$

[O] indicates that the reaction is: _____.

The product is: _____.

In a Nutshell: Oxidation-Reductions Involving Carbon-Oxygen Bonds

Some of the most important oxidation-reduction reactions involve functional groups containing carbon-oxygen bonds, such as alcohols and functional groups containing the carbonyl group. To predict the product of an alcohol oxidation, you must first determine whether the alcohol is a primary, secondary, or tertiary alcohol. Primary alcohols undergo oxidation to form aldehydes, secondary alcohols undergo oxidation to form ketones, and tertiary alcohols do not undergo oxidation. Aldehydes can undergo further oxidation in the presence of water to produce carboxylic acids. When an alcohol undergoes oxidation, it loses the hydrogen atom in the O—H group as well as one of the hydrogen atoms on the carbon atom bearing the OH group. When an aldehyde is oxidized to a carboxylic acid, an oxygen atom is added to the molecule.

These oxidation reactions can also proceed in the opposite direction. The reduction of a carboxylic acid produces an aldehyde. The reduction of an aldehyde produces a primary alcohol. The reduction of a ketone produces a secondary alcohol.

In the cell, these oxidation-reduction reactions involving carbon-oxygen bonds use the coenzyme nicotinamide adenine dinucleotide as an electron transfer agent. NAD^+ is an oxidizing agent; it causes the other reactant to undergo oxidation. One hydrogen atom and two electrons are added to the coenzyme to make NADH, and a second hydrogen atom is transferred to solution as a proton. In the reverse process, NADH is a reducing agent; it causes the other reactant to undergo reduction. A proton from solution is also required for the reaction. The convention when writing a biochemical pathway is to show the structures of the main reactant and product on either side of the main reaction arrow. The coenzyme, the oxidizing or reducing agent, is placed next to a curved arrow that intersects the main arrow.

Worked Example #3

Predict the structure of the product formed in each of the following reactions. Write the name of the functional group affected in the reactant and the product.

a.

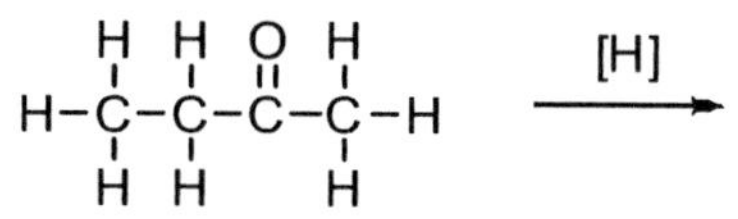

$$H-\underset{\underset{H}{|}}{\overset{\overset{H}{|}}{C}}-\underset{\underset{H}{|}}{\overset{\overset{H}{|}}{C}}-\overset{\overset{O}{\|}}{C}-\underset{\underset{H}{|}}{\overset{\overset{H}{|}}{C}}-H \xrightarrow{[H]}$$

b.

$$H_2O + H-\underset{\underset{H}{|}}{\overset{\overset{H}{|}}{C}}-\underset{\underset{H}{|}}{\overset{\overset{H}{|}}{C}}-\underset{\underset{H}{|}}{\overset{\overset{H}{|}}{C}}-\underset{\underset{H}{|}}{\overset{\overset{H}{|}}{C}}-\overset{\overset{O}{\|}}{C}-H \xrightarrow{[O]}$$

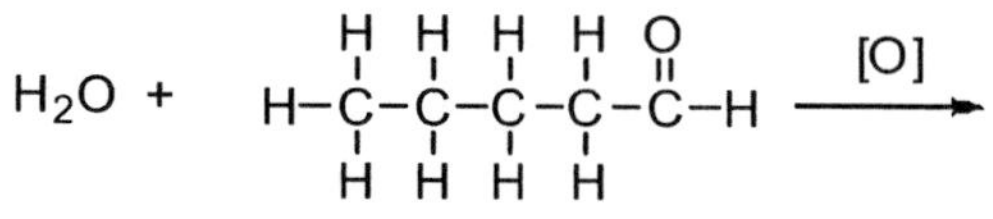

a.   The reactant is a ketone. The [H] above the arrow indicates the ketone is reduced. The reduction of a ketone produces a secondary alcohol.

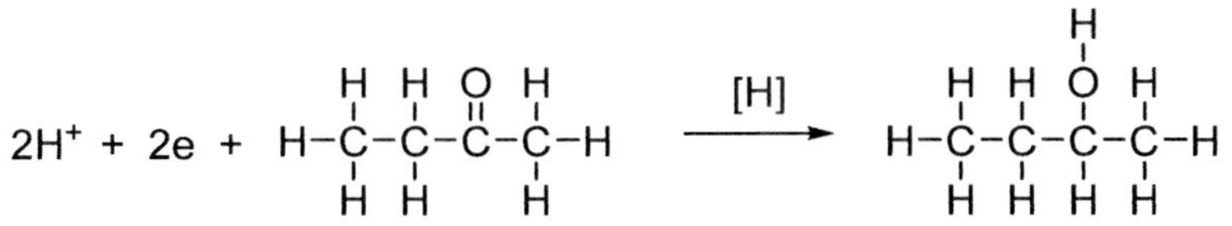

$$2H^+ + 2e + \;\; H-\overset{\overset{\displaystyle H}{|}}{\underset{\underset{\displaystyle H}{|}}{C}}-\overset{\overset{\displaystyle H}{|}}{\underset{\underset{\displaystyle H}{|}}{C}}-\overset{\overset{\displaystyle O}{\|}}{C}-\overset{\overset{\displaystyle H}{|}}{\underset{\underset{\displaystyle H}{|}}{C}}-H \quad \xrightarrow{[H]} \quad H-\overset{\overset{\displaystyle H}{|}}{\underset{\underset{\displaystyle H}{|}}{C}}-\overset{\overset{\displaystyle H}{|}}{\underset{\underset{\displaystyle H}{|}}{C}}-\overset{\overset{\displaystyle O-H}{|}}{\underset{\underset{\displaystyle H}{|}}{C}}-\overset{\overset{\displaystyle H}{|}}{\underset{\underset{\displaystyle H}{|}}{C}}-H$$

ketone   2° alcohol

b. *The reactant is an aldehyde. The [O] above the arrow signifies that the aldehyde undergoes oxidation. The oxidation of an aldehyde produces a carboxylic acid.*

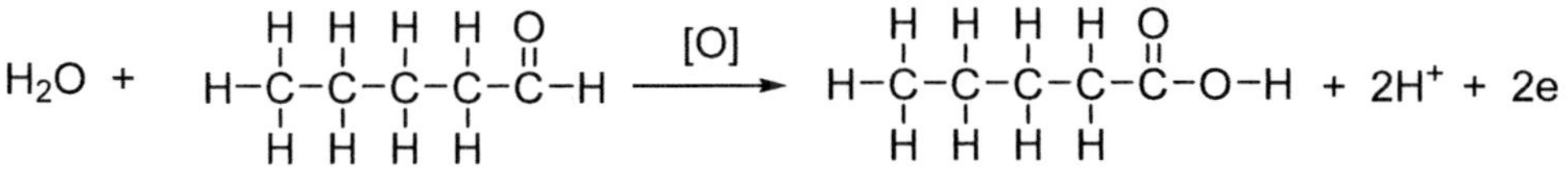

$$H_2O + \;\; H-\overset{\overset{\displaystyle H}{|}}{\underset{\underset{\displaystyle H}{|}}{C}}-\overset{\overset{\displaystyle H}{|}}{\underset{\underset{\displaystyle H}{|}}{C}}-\overset{\overset{\displaystyle H}{|}}{\underset{\underset{\displaystyle H}{|}}{C}}-\overset{\overset{\displaystyle H}{|}}{\underset{\underset{\displaystyle H}{|}}{C}}-\overset{\overset{\displaystyle O}{\|}}{C}-H \quad \xrightarrow{[O]} \quad H-\overset{\overset{\displaystyle H}{|}}{\underset{\underset{\displaystyle H}{|}}{C}}-\overset{\overset{\displaystyle H}{|}}{\underset{\underset{\displaystyle H}{|}}{C}}-\overset{\overset{\displaystyle H}{|}}{\underset{\underset{\displaystyle H}{|}}{C}}-\overset{\overset{\displaystyle H}{|}}{\underset{\underset{\displaystyle H}{|}}{C}}-\overset{\overset{\displaystyle O}{\|}}{C}-O-H + 2H^+ + 2e$$

aldehyde   carboxylic acid

Try It Yourself #3

Predict the structure of the product formed in each of the following reactions. Write the name of the functional group affected in the reactant and the product.

a. 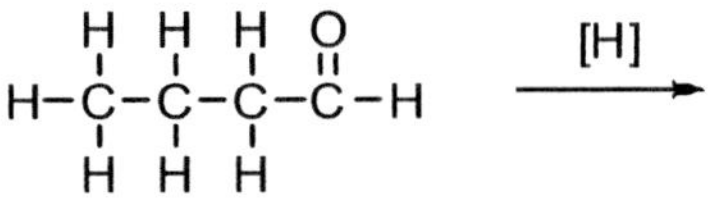 $$H-\overset{\overset{\displaystyle H}{|}}{\underset{\underset{\displaystyle H}{|}}{C}}-\overset{\overset{\displaystyle H}{|}}{\underset{\underset{\displaystyle H}{|}}{C}}-\overset{\overset{\displaystyle H}{|}}{\underset{\underset{\displaystyle H}{|}}{C}}-\overset{\overset{\displaystyle O}{\|}}{C}-H \quad \xrightarrow{[H]}$$

b. 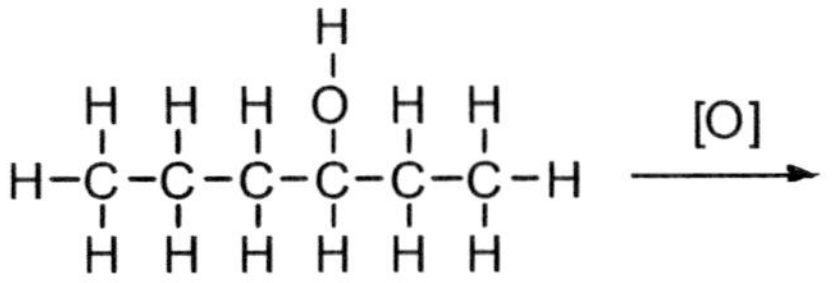 $$H-\overset{\overset{\displaystyle H}{|}}{\underset{\underset{\displaystyle H}{|}}{C}}-\overset{\overset{\displaystyle H}{|}}{\underset{\underset{\displaystyle H}{|}}{C}}-\overset{\overset{\displaystyle H}{|}}{\underset{\underset{\displaystyle H}{|}}{C}}-\overset{\overset{\displaystyle O-H}{|}}{\underset{\underset{\displaystyle H}{|}}{C}}-\overset{\overset{\displaystyle H}{|}}{\underset{\underset{\displaystyle H}{|}}{C}}-\overset{\overset{\displaystyle H}{|}}{\underset{\underset{\displaystyle H}{|}}{C}}-H \quad \xrightarrow{[O]}$$

a. *The reactant is _____.*

    The [H] signifies _____.

    The product is _____.

    The structure of the product is:

b. *The reactant is_____.*

    The [O] signifies _____.

    The product is _____.

The structure of the product is:

Practice Problems for Oxidation-Reduction Reactions

1. Determine whether the following reactions represent an oxidation or a reduction of the organic substance shown. Explain your reasoning by showing the hydrogen and/or oxygen atoms that have been gained or lost from the reactant.

   a.

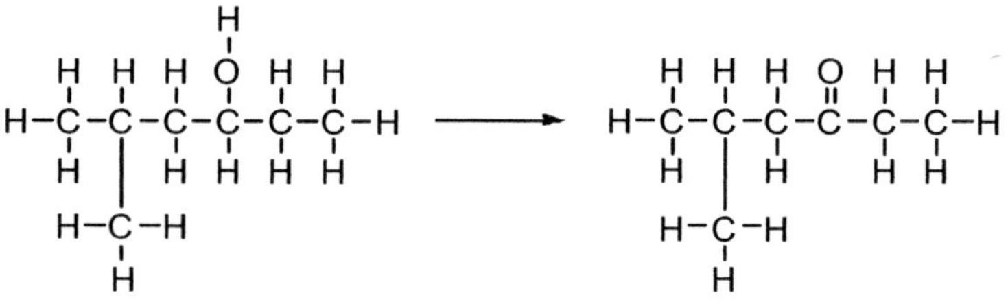

   b.

2. Predict the structure of the product formed in each of the following reactions.

   a.   FAD $\xrightarrow{\text{[H]}}$

   b.   NADH $\xrightarrow{\text{[O]}}$

   c.

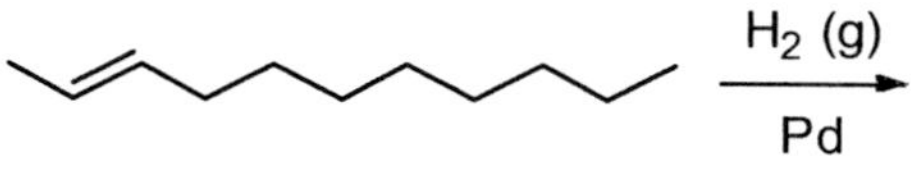

   $\xrightarrow[\text{Pd}]{\text{H}_2 \text{ (g)}}$

3. Predict the structure of the product formed in each of the following reactions. Write the name of the functional group affected in the reactant and the product.

a.

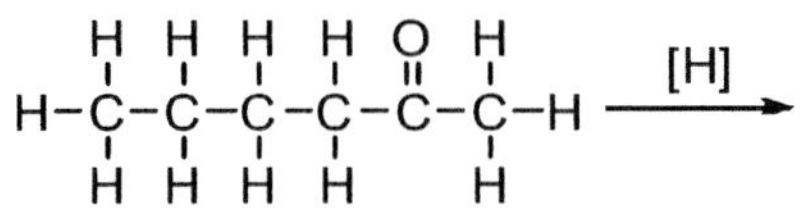

$$
\begin{array}{ccccccc}
& H & H & H & H & O & H \\
& | & | & | & | & || & | \\
H- & C- & C- & C- & C- & C- & C-H \\
& | & | & | & | & & | \\
& H & H & H & H & & H \\
\end{array}
\quad \xrightarrow{\text{[H]}}
$$

b.

$$
\begin{array}{cccc}
& & & H \\
& & & | \\
& H & H & H & O \\
& | & | & | & | \\
H- & C- & C- & C- & C-H \\
& | & | & | & | \\
& H & H & H & H \\
\end{array}
\quad \xrightarrow{\text{[O]}}
$$

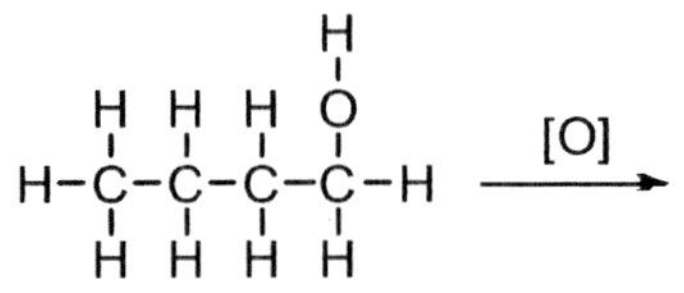

Section 10.3 Hydration-Dehydration Reactions

Hydration and dehydration reactions are common in biochemistry. In a hydration reaction, water is a reactant, and in a dehydration reaction, water is a product. In a hydration reaction, water and an alkene react with each other to produce an alcohol. In the reaction, the H and OH atoms of a water molecule each form a bond to one of the carbon atoms in a carbon-carbon double bond. At the same time, the C=C double bond is converted into a C-C single bond. In the laboratory, the OH group tends to add to the more substituted double-bond carbon atom.

In biochemical applications of the hydration reaction, a carbonyl group usually exists adjacent to the double bond. The H and OH atoms are attached in a specific way: The H is attached to the double-bond carbon atom that is closest to the carbonyl group, and the OH is attached to the double-bond carbon atom that is farthest from the carbonyl group. The carbonyl group is unchanged, but necessary for the reaction.

The reverse reaction, the loss of a water molecule from an alcohol to produce a double bond is known as a dehydration reaction. In a dehydration reaction, OH and H are eliminated from the reactant to form a molecule of H_2O and a carbon–carbon double bond. The loss of OH and H in a dehydration reaction is the reverse of the addition of OH and H in a hydration reaction.

Worked Example #4

Predict the product formed in the following hydration reaction:

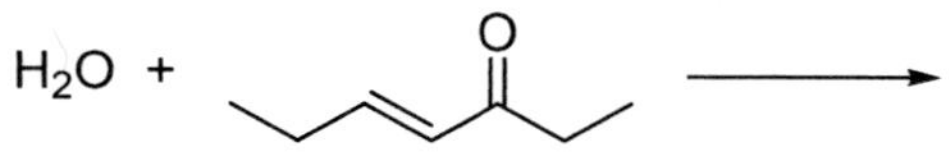

H_2O +

Solution

H_2O +

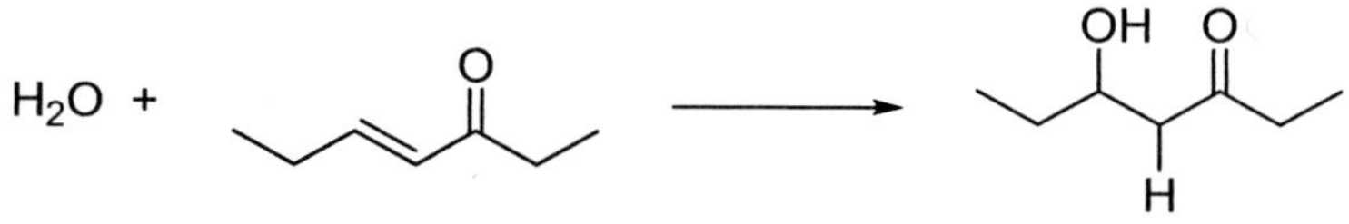

The OH is attached to the double-bond carbon that is farthest away from the carbonyl group.

Try It Yourself #4

Predict the product formed in the following hydration reaction:

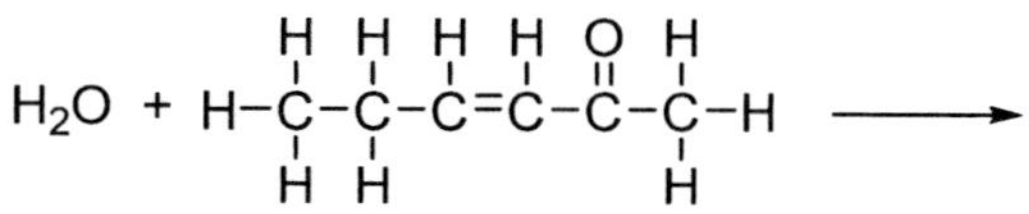

Identify the double bond.

Which double bond carbon atom is farthest from the carbonyl?

Structure of the product:

Worked Example #5

Predict the product formed in the following dehydration reaction:

$$H-\overset{\overset{\displaystyle H}{|}}{\underset{\underset{\displaystyle H}{|}}{C}}-\overset{\overset{\displaystyle H}{|}}{\underset{\underset{\displaystyle OH}{|}}{C}}-\overset{\overset{\displaystyle H}{|}}{\underset{\underset{\displaystyle H}{|}}{C}}-\overset{\overset{\displaystyle O}{\|}}{C}-\overset{\overset{\displaystyle H}{|}}{\underset{\underset{\displaystyle H}{|}}{C}}-H \longrightarrow$$

Solution

$$H-\overset{\overset{\displaystyle H}{|}}{\underset{\underset{\displaystyle H}{|}}{C}}-\overset{\overset{\displaystyle H}{|}}{\underset{\underset{\displaystyle OH}{|}}{C}}-\overset{\overset{\displaystyle H}{|}}{\underset{\underset{\displaystyle H}{|}}{C}}-\overset{\overset{\displaystyle O}{\|}}{C}-\overset{\overset{\displaystyle H}{|}}{\underset{\underset{\displaystyle H}{|}}{C}}-H \longrightarrow H-\overset{\overset{\displaystyle H}{|}}{\underset{\underset{\displaystyle H}{|}}{C}}-\overset{\overset{\displaystyle H}{|}}{C}=\overset{}{\underset{\underset{\displaystyle H}{|}}{C}}-\overset{\overset{\displaystyle O}{\|}}{C}-\overset{\overset{\displaystyle H}{|}}{\underset{\underset{\displaystyle H}{|}}{C}}-H \ + \ H_2O$$

Remember to remove the H atom closest to the carbonyl group when determining where to place the double bond.

Try It Yourself #5

Predict the product formed in the following dehydration reaction:

$$H-\overset{\overset{\displaystyle H}{|}}{\underset{\underset{\displaystyle H}{|}}{C}}-\overset{\overset{\displaystyle H}{|}}{\underset{\underset{\displaystyle OH}{|}}{C}}-\overset{\overset{\displaystyle H}{|}}{\underset{\underset{\displaystyle H}{|}}{C}}-\overset{\overset{\displaystyle O}{\|}}{C}-\overset{\overset{\displaystyle H}{|}}{\underset{\underset{\displaystyle H}{|}}{C}}-\overset{\overset{\displaystyle H}{|}}{\underset{\underset{\displaystyle H}{|}}{C}}-\overset{\overset{\displaystyle H}{|}}{\underset{\underset{\displaystyle H}{|}}{C}}-H \longrightarrow$$

Identify the OH group.

Which H group should be removed?

Structure of the product:

Practice Problems for Hydration/Dehydration Reactions

1. Predict the product formed in the following dehydration reaction.

a.

b.

2. Predict the product formed in the following hydration reaction.

a.

b.

3. For the reactions below, indicate whether they are hydration or dehydration reactions and add a curved arrow showing water to the equation.

a.

b.

Section 10.4 Acyl Group Transfer Reactions

Some of the most common reactions seen in chemistry are reactions that interconvert carboxylic acids and their derivatives: esters, thioesters, and amides. Biochemists refer to this broad class of reactions as acyl group transfer reactions because the group transferred is an acyl group, a carbonyl group, and its attached R group. Water is an essential component of these reactions. Hydrolysis and esterification reactions will be covered in this chapter.

In a Nutshell: Hydrolysis Reactions

Esters, thioesters, and amides react with water in the presence of a catalyst to produce a carboxylic acid and either an alcohol, a thiol, or an amine. In these reactions the acyl group migrates from an oxygen, sulfur, or nitrogen atom to the oxygen atom of water. These particular reactions are known as hydrolysis reactions because water is the agent that breaks the bond between the carbonyl group, and the O, S, or N atom.

To predict the structure of the products formed in a hydrolysis reaction, follow these guidelines: 1) Identify the type of carboxylic acid derivative in the reactant; 2) break the carbonyl carbon-heteroatom bond; and 3) make new bonds. Determine if the carboxylic acid derivative is an ester, thioester, or amide. Determining if the carboxylic acid derivative is an ester, thioester, or amine will enable you to predict what functional group will be in the product structure. One of the products will always be a carboxylic acid (or a carboxylate ion). Break the single bond between the carbonyl carbon and the O, S, or N atom to give two temporary partial structures. Make a new bond between OH and the carbonyl carbon and make a new bond between H and the O, S, or N atom.

When an amide is hydrolyzed, a carboxylic acid and an amine are formed. However, at physiological pH, the amine and the carboxylic acid immediately undergo an acid-base reaction to produce the conjugate base of the acid and the conjugate acid of the amine.

Worked Example #6

Write the structure of the products formed in the following hydrolysis reactions.

a.

b.

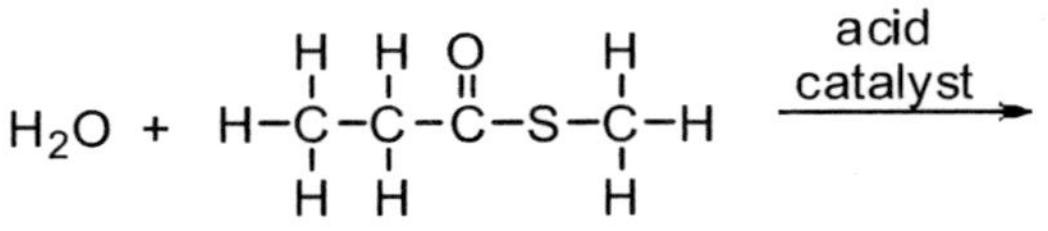

$$H_2O + \text{H-C-C-C-S-C-H} \xrightarrow{\text{acid catalyst}}$$

(structure: H-C(H,H)-C(H,H)-C(=O)-S-C(H,H,H))

c.

(structure with H-C-H branch)

$$H_2O + \text{H-C-C-N-C-C-C-H} \xrightarrow{\text{catalyst}}$$

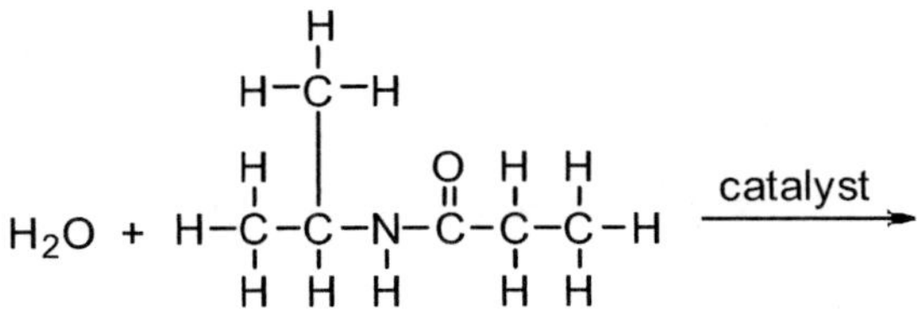

a. *Identify the type of carboxylic acid derivative in the reactant.*

   The reactant is an ester; therefore, the products should be a carboxylic acid and an alcohol.

   Break the carbonyl carbon-heteroatom bond.
   Break the single bond between the carbonyl carbon and the oxygen atom to yield two temporary partial structures:

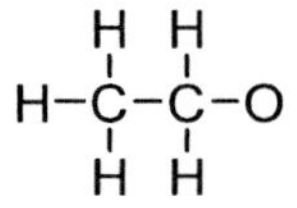

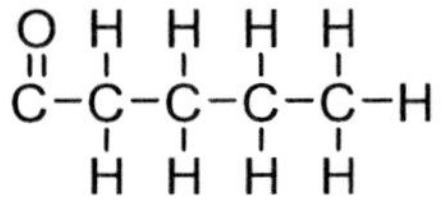

   H-C-C-O   O=C-C-C-C-C-H

   Make new bonds.
   Add OH to the carbonyl carbon to form a carboxylic acid:

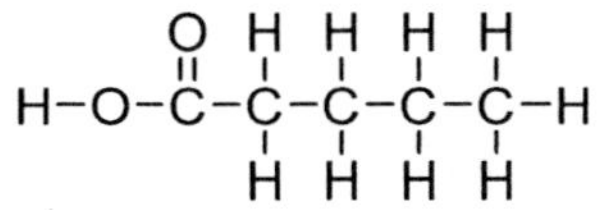

   H-O-C(=O)-C-C-C-C-H

   Add H to the OR partial structure to form an alcohol:

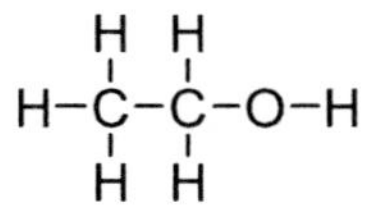

   H-C-C-O-H

b. *Identify the type of carboxylic acid derivative in the reactant.*

   The reactant is a thioester; therefore, the products should be a carboxylic acid and a thiol.

   Break the carbonyl carbon-heteroatom bond.

Break the single bond between the carbonyl carbon and the sulfur atom to yield two temporary partial structures:

$$
\begin{array}{ccc}
\text{H} & \text{H} & \text{O} \\
| & | & \| \\
\text{H}-\text{C}-\text{C}-\text{C} & & \text{S}-\text{C}-\text{H} \\
| & | & | \\
\text{H} & \text{H} & \text{H}
\end{array}
$$

Make new bonds.

Add OH to the carbonyl carbon to form a carboxylic acid:

$$
\begin{array}{ccc}
\text{H} & \text{H} & \text{O} \\
| & | & \| \\
\text{H}-\text{C}-\text{C}-\text{C}-\text{O}-\text{H} \\
| & | \\
\text{H} & \text{H}
\end{array}
$$

Add H to the SR partial structure to form a thiol:

$$
\begin{array}{c}
\text{H} \\
| \\
\text{H}-\text{S}-\text{C}-\text{H} \\
| \\
\text{H}
\end{array}
$$

c. *Identify the type of carboxylic acid derivative in the reactant.*

   The reactant is an amide; therefore, the products are a carboxylic acid and an amine.

Break the carbonyl carbon-heteroatom bond.

Break the single bond between the carbonyl carbon and the nitrogen atom to yield two temporary partial structures:

$$
\begin{array}{cc}
\text{H} & \\
| & \\
\text{H}-\text{C}-\text{H} & \\
\text{H} \; | & \text{O} \; \text{H} \; \text{H} \\
| \;\; | & \| \;\; | \;\; | \\
\text{H}-\text{C}-\text{C}-\text{N} & \text{C}-\text{C}-\text{C}-\text{H} \\
| \;\; | \;\; | & | \;\; | \\
\text{H} \; \text{H} \; \text{H} & \text{H} \; \text{H}
\end{array}
$$

Make new bonds.

Add OH to the carbonyl carbon to form a carboxylic acid:

$$
\begin{array}{ccc}
\text{O} & \text{H} & \text{H} \\
\| & | & | \\
\text{H}-\text{O}-\text{C}-\text{C}-\text{C}-\text{H} \\
& | & | \\
& \text{H} & \text{H}
\end{array}
$$

Add H to the NRR' partial structure to form an amine:

H-C-H structure (2-aminopropane with methyl branch):

```
        H
        |
     H-C-H
     H  |
     |  |
  H-C-C-N-H
     |  | |
     H  H H
```

Remember that the initial amine and carboxylic acid products undergo a subsequent acid-base reaction to form the conjugate acid of the amine and the conjugate base of the carboxylic acid. Therefore, the following products are the result:

```
        H
        |
     H-C-H
     H  | H           O  H  H
     |  | |+          ||  |  |
  H-C-C-N-H   +   ⁻O-C-C-C-H
     |  | |            |  |
     H  H H            H  H
```

Try It Yourself #6

Write the structure of the products formed in the following hydrolysis reactions.

a.

```
              H  O     H  H  H
              |  ||    |  |  |          catalyst
  H₂O  +  H-C-C-O-C-C-C-H      ――――――→
              |       |  |  |
              H       H  H  H
```

b.

```
              H  H  O      H  H  H
              |  |  ||     |  |  |      catalyst
  H₂O  +  H-C-C-C-S-C-C-C-H   ――――――→
              |  |        |  |  |
              H  H        H  H  H
```

c.

```
              H  H  H     O  H  H
              |  |  |     ||  |  |      catalyst
  H₂O  +  H-C-C-C-N-C-C-C-H   ――――――→
              |  |  |  |     |  |
              H  H  H  H     H  H
```

a. *Identify the type of carboxylic acid derivative in the reactant.*

   The reactant is:

Break the carbonyl carbon-heteroatom bond.

The two temporary partial structures are:

Make new bonds.
Add OH to the carbonyl carbon.
Add H to the O, S, or N atom.
If the reaction is the hydrolysis of an amide, remember the acid-base reaction between to the products.
The products are:

b. *Identify the type of carboxylic acid derivative in the reactant.*
    The reactant is

Break the carbonyl carbon-heteroatom bond.
The two temporary partial structures are:

Make new bonds.
Add OH to the carbonyl carbon.
Add H to the O, S, or N atom.

If the reaction is the hydrolysis of an amide, remember the acid-base reaction between to the products.

The products are:

c. *Identify the type of carboxylic acid derivative in the reactant.*

   The reactant is

   Break the carbonyl carbon-heteroatom bond.
   The two temporary partial structures are:

   Make new bonds.
   Add OH to the carbonyl carbon.
   Add H to the O, S, or N atom.

   If the reaction is the hydrolysis of an amide, remember the acid-base reaction between to the products.
   The products are:

In a Nutshell: Esterification Reactions

A carboxylic acid and an alcohol can react to form an ester in an esterification reaction, which is the reverse reaction of the hydrolysis of an ester. When sulfur replaces oxygen in an alcohol, the functional group is known as a thiol. When a thiol is used instead of an alcohol in an esterification reaction, a thioester is formed instead of an ester. The reaction is the reverse of the hydrolysis of a thioester.

To predict the product formed in an esterification reaction that produces either an ester or a thioester, use the following guidelines: 1) Remove the OH group and the H atom; and 2) connect the two resulting partial structures. Remove the OH group on the carboxylic acid and the H atoms on the oxygen atom of the alcohol or the sulfur atom of the thiol to form two incomplete product structures and a water molecule. Form a carbon-oxygen single bond between the carbonyl carbon and the OR group to produce an ester. Form a carbon-sulfur single bond between the carbonyl carbon and the SR group to form a thioester.

Worked Example #7

Write the products formed in the following esterification reactions.

a.

b.

a. *Remove the OH group and the H atom.*

Remove the OH group on the carboxylic acid and the H on the oxygen atom of the alcohol to form two incomplete structures and a water molecule:

Connect the two resulting partial structures.

Form a carbon-oxygen single bond between the carbonyl carbon and the OR group to produce an ester:

$$\begin{array}{ccccccc} & H & H & H & O & H & H \\ & | & | & | & \| & | & | \\ H- & C- & C- & C- & O-C- & C- & C-H \\ & | & | & | & | & | \\ & H & H & H & & H & H \end{array}$$

b   Remove the OH group and the H atom.

Remove the OH group on the carboxylic acid and the H on the sulfur atom of the thiol to form two incomplete structures and a water molecule:

$$\begin{array}{cccc} H\ H\ H & & O\ H & \\ |\ |\ | & & \|\ | & \\ H-C-C-C-S & & C-C-H & + H_2O \\ |\ |\ | & & | & \\ H\ H\ H & & H & \end{array}$$

Connect the two resulting partial structures.

Form a carbon-sulfur single bond between the carbonyl carbon and the SR group to produce a thioester:

$$\begin{array}{cccccc} H\ H\ H & & O\ H & \\ |\ |\ | & & \|\ | & \\ H-C-C-C-S-C-C-H & \\ |\ |\ | & & | & \\ H\ H\ H & & H & \end{array}$$

Try It Yourself #7

Write the products formed in the following esterification reactions.

a.

$$\begin{array}{ccc} H\ H & & H\ O \\ |\ | & & |\ \| \\ H-C-C-O-H\ +\ H-C-C-O-H \\ |\ | & & | \\ H\ H & & H \end{array} \xrightarrow{\text{catalyst}}$$

b.

$$\begin{array}{cc} H\ H\ H\ O & H\ H\ H \\ |\ |\ |\ \| & |\ |\ | \\ H-C-C-C-C-O-H\ \ +\ \ H-S-C-C-C-H \\ |\ |\ | & |\ |\ | \\ H\ H\ H & H\ H\ H \end{array} \xrightarrow{\text{catalyst}}$$

a.   Remove the OH group and the H atom.
The two partial structures are:

The other product is:_____

Connect the two resulting partial structures.

The structure of the product is:

b. *Remove the OH group and the H atom.*

 The two partial structures are:

 The other product is:_____

 Connect the two resulting partial structures.

 The structure of the product is:

Practice Problems for Acyl Group Transfer Reactions

1. Predict the products formed in the following hydrolysis reactions.

 a.

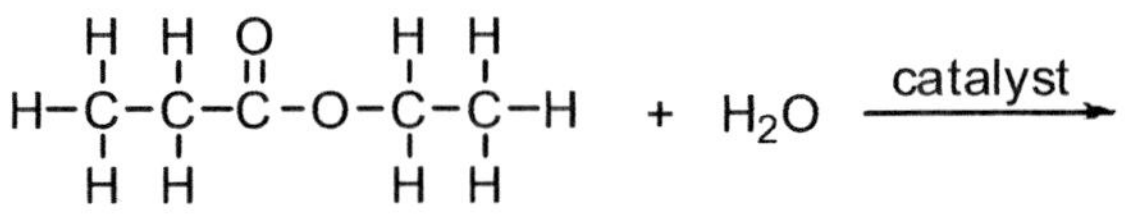

$$
\begin{array}{ccc}
H & H & O \qquad H \ H \\
| & | & || \qquad | \ | \\
H-C-C-C-O-C-C-H & + & H_2O \xrightarrow{\text{catalyst}} \\
| & | & \qquad | \ | \\
H & H & \qquad H \ H
\end{array}
$$

 b.

$$
\begin{array}{c}
H \\
| \\
H-C-H \\
H \ | \qquad O \ H \ H \ H \\
| \ | \qquad || \ | \ | \ | \\
H-C-C-S-C-C-C-C-H \quad + \quad H_2O \xrightarrow{\text{catalyst}} \\
| \ | \qquad | \ | \ | \\
H \ H \qquad H \ H \ H
\end{array}
$$

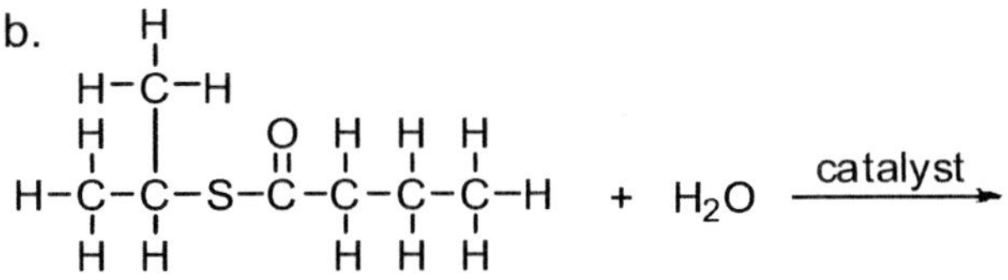

c.

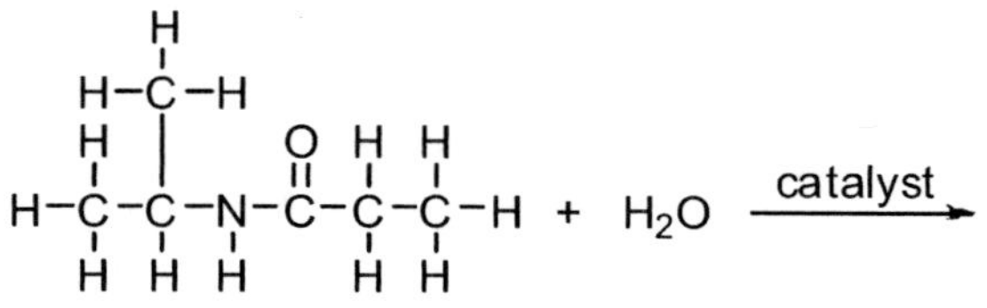

$$\text{H}\underset{\underset{\displaystyle \text{H}}{\displaystyle |}}{\overset{\overset{\displaystyle \text{H}}{\displaystyle |}}{-\text{C}-}}\text{H}$$

H–C–C–N–C–C–C–H + H$_2$O $\xrightarrow{\text{catalyst}}$

(with the substituent chain: H–C–C–N–C(=O)–C–C–H, methyl branch at top, H H H on lower positions, O H H above)

2. Predict the product formed in the following esterification reactions.

a.

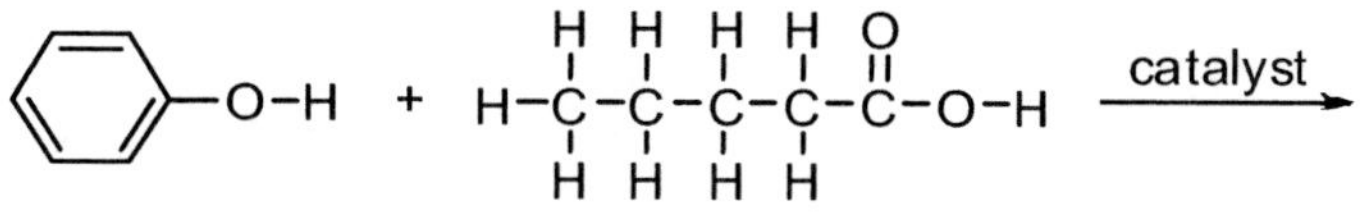

⬡–O–H + H–C–C–C–C–C(=O)–O–H $\xrightarrow{\text{catalyst}}$

(pentanoic acid: H H H H O above, H H H H below)

b.

H–C–C–C–C–C(=O)–O–H + H–SCoA $\xrightarrow{\text{catalyst}}$

(H H H H O above, H H H H below)

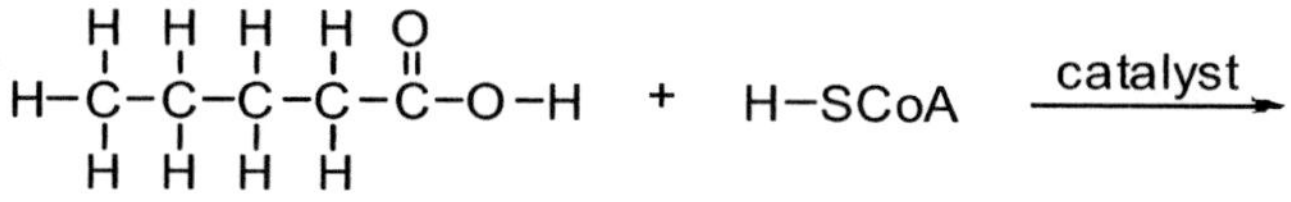

3. Predict the product(s) formed in the following reactions.

a.

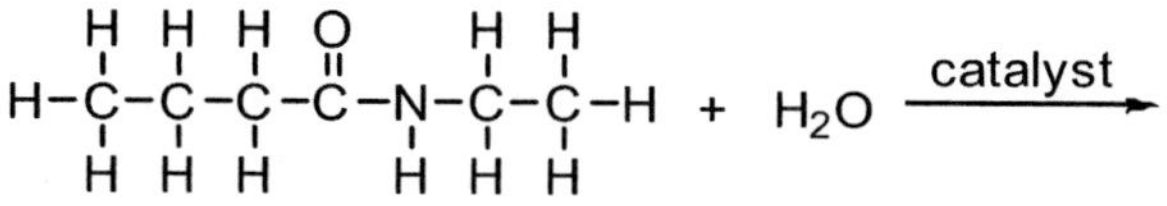

H–C–C–C–C(=O)–N–C–C–H + H$_2$O $\xrightarrow{\text{catalyst}}$

(H H H O above left, H H above right; H H H below left, H H H below right)

b.

H$_2$O + H–C–C–C–C(=O)–O–C–H $\xrightarrow{\text{catalyst}}$

(H H H O above, H above; H H H below, H below)

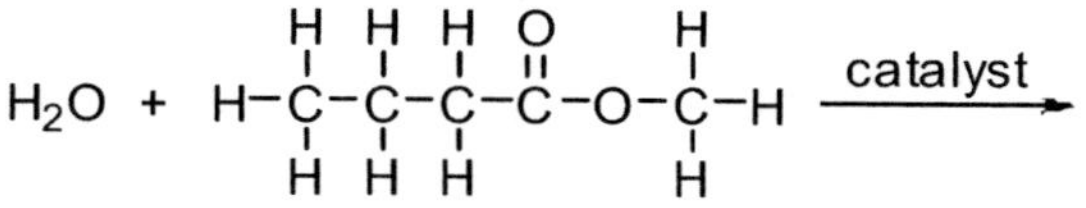

Section 10.5 Phosphoryl Group Transfer Reactions

Reactions that transfer phosphoryl groups, known as phosphoryl group transfer reactions, play a central role in the way energy is transferred in biochemical reactions. A phosphoryl group is comprised of three oxygen atoms attached to a central phosphorus atom, by one P-O double bond and two P-O single bonds. The phosphorus atom is attached to the oxygen atom of another functional group creating a phosphate ester.

In a Nutshell: The Products of Phosphoryl Group Transfer Reactions

The phosphoryl group can be transferred to another phosphoryl group to produce a di or triphosphate. Phosphoryl groups are unique in their ability to join together via a phosphoanhydride bond. The phosphoryl group can be transferred to the oxygen atom of a water, alcohol, or carboxylic acid molecule.

In a Nutshell: Energy Transfer in the Cell

The phosphoanhydride bond is sometimes called a "high energy" bond because its hydrolysis releases energy and its formation absorbs energy; the former drive the latter reactions. ATP is called the "energy currency" of the cell because it is able to store chemical energy and transfer energy when needed through a phosphoryl group transfer.

Worked Example #8

Answer the following questions about the hydrolysis of ADP to **A**denosine **M**ono**P**hosphate, AMP.

a. Circle the phosphoryl group transferred.
b. Circle the phosphoanhydride bond that is broken.

a. *The dashed lines show the phosphoryl group that is transferred from ADP to water to produce AMP and inorganic phosphate.*

b.

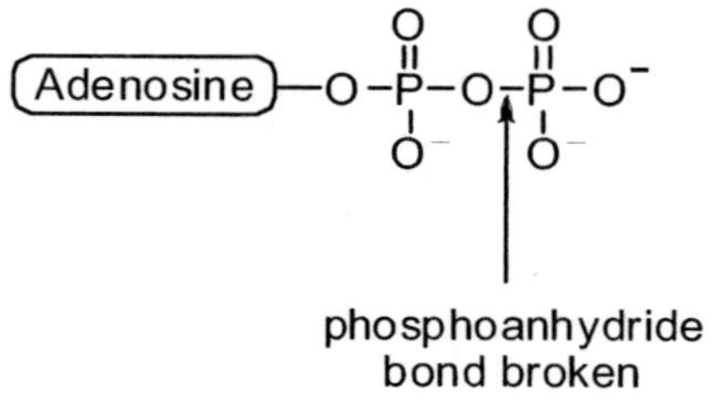

phosphoanhydride
bond broken

Try It Yourself #8

The enzyme, *glycerol kinase*, catalyzes the following phosphoryl transfer reaction.

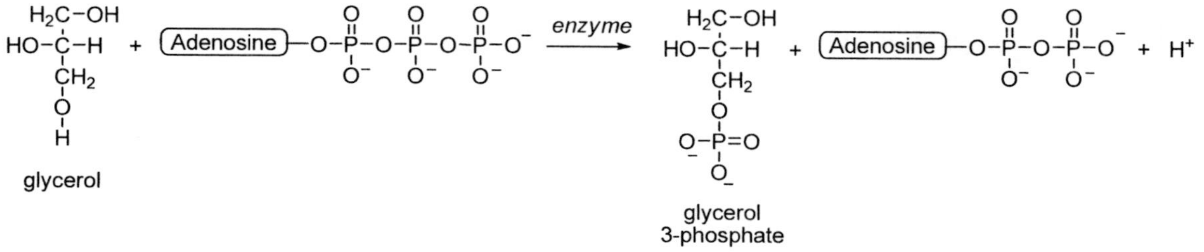

glycerol   glycerol
3-phosphate

    a.   Circle the phosphoryl group transferred.

    b.   Circle the phosphoanhydride bond that is broken.

Solution

    a-b.

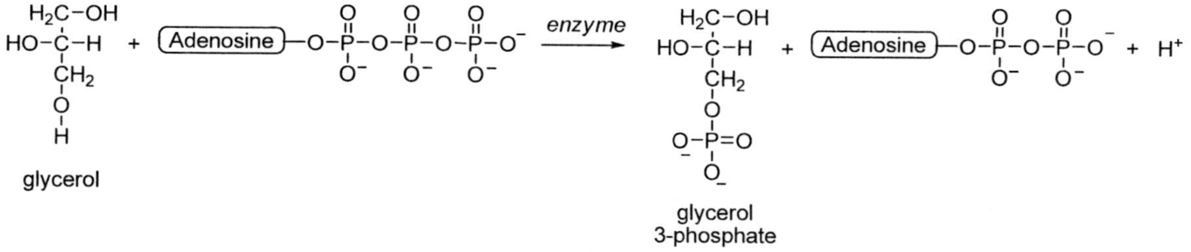

glycerol   glycerol
3-phosphate

Practice Problems for Phosphoryl Group Transfer Reactions

1. **G**uanosine **T**ri**P**hosphate, GTP, is very similar to ATP. GTP is produced in the citric acid cycle. Write the products for the hydrolysis of GTP.

$$\boxed{\text{Guanosine}} - O - \overset{\overset{O}{\|}}{\underset{\underset{O^-}{|}}{P}} - O - \overset{\overset{O}{\|}}{\underset{\underset{O^-}{|}}{P}} - O - \overset{\overset{O}{\|}}{\underset{\underset{O^-}{|}}{P}} - O^- \ + \ H_2O \longrightarrow$$

GTP

2. Answer the following questions about phosphoryl group transfer is catalyzed by an enzyme.

3-Phosphoglycerate phosphate + $\boxed{\text{Adenosine}} - O - \overset{\overset{O}{\|}}{\underset{\underset{O^-}{|}}{P}} - O - \overset{\overset{O}{\|}}{\underset{\underset{O^-}{|}}{P}} - O^-$ (ADP) $\longrightarrow$ 3-Phosphoglycerate + $\boxed{\text{Adenosine}} - O - \overset{\overset{O}{\|}}{\underset{\underset{O^-}{|}}{P}} - O - \overset{\overset{O}{\|}}{\underset{\underset{O^-}{|}}{P}} - O - \overset{\overset{O}{\|}}{\underset{\underset{O^-}{|}}{P}} - O^-$ (ATP)

3-Phosphoglycerate phosphate:
$$\overset{\overset{\overset{O^-}{|}}{O - P = O}}{\underset{\underset{\overset{|}{O}}{|}}{}}$$
$$O = C$$
$$HO - C - H$$
$$CH_2$$
$$\underset{\underset{O^-}{}}{O - P = O}$$

3-Phosphoglycerate:
$$\overset{\overset{O^-}{|}}{O = C}$$
$$HO - C - H$$
$$CH_2$$
$$O$$
$$O - P = O$$
$$O^-$$

   a. Circle the phosphoryl group transferred.

   b. What functional group is the phosphoryl group transferred from?

   c. What functional group is the phosphoryl group transferred to?

Chapter 10 Quiz

1. Predict the product for the following oxidation-reduction reactions.

    a.   NAD$^+$ $\xrightarrow{\text{[H]}}$

    b.

    c.

2. For the oxidation-reduction reactions below, determine whether the organic compound has undergone an oxidation or a reduction by placing an "H" or "O" in the brackets above the arrow.

    a.

    b.

    c.

3.   Predict the products for the following dehydration–hydration reactions.

a.

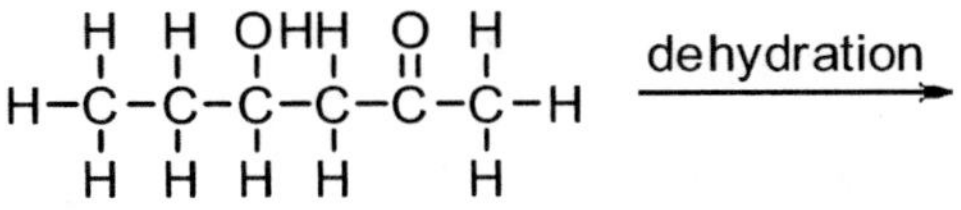

```
      H  H OHH  O  H
      |  |  | |  ||  |
   H-C-C-C-C-C-C-H        dehydration
      |  |  | |     |       ───────────→
      H  H  H H     H
```

b.

```
      H  O           H
      |  ||           |
   H-C-C-C=C-C-H  +  H₂O  ───────────→
      |      | | |
      H      H H H
```

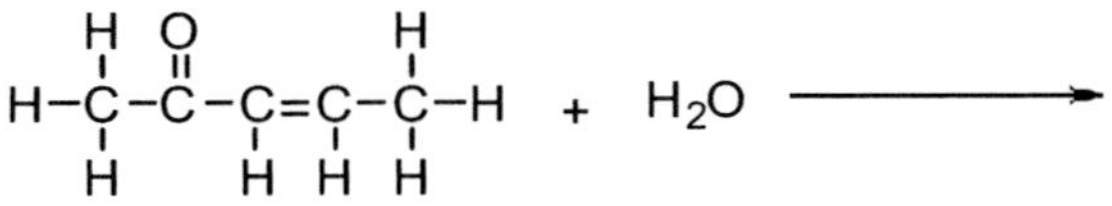

4.   Determine whether the reactions shown below are hydration or dehydration reactions and add a curved arrow showing water to the equation.

a.

b.

```
      H  O           H                    H  O  H  H  H
      |  ||           |                    |  ||  |  |  |
   H-C-C-C=C-C-H  ─────→  H-C-C-C-C-C-H
      |      | | |                         |      |  |  |
      H      H H H                         H      H OHH
```

5.   Predict the products for the following hydrolysis reactions. Identify the functional groups in the reactants and products.

a.

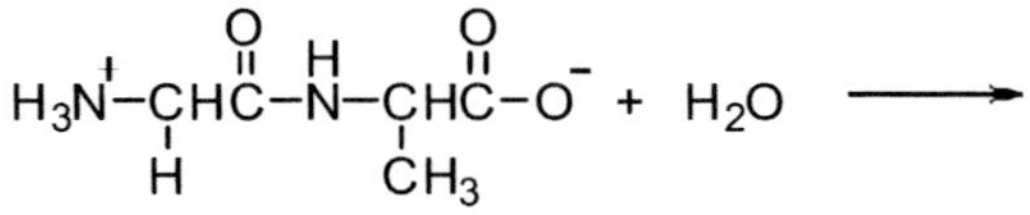

```
         O   H    O
         ||   |    ||
   H₃N⁺-CHC-N-CHC-O⁻  +  H₂O  ───────────→
         |        |
         H        CH₃
```

b.

```
      H  H  H      O  H
      |  |  |      ||  |
   H-C-C-C-S-C-C-H  +  H₂O  ───────────→
      |  |  |          |
      H  H  H          H
```

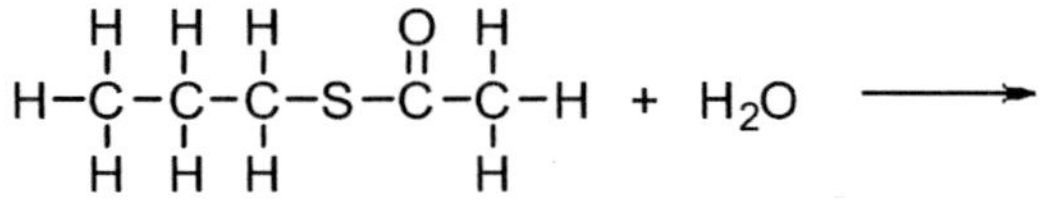

c.

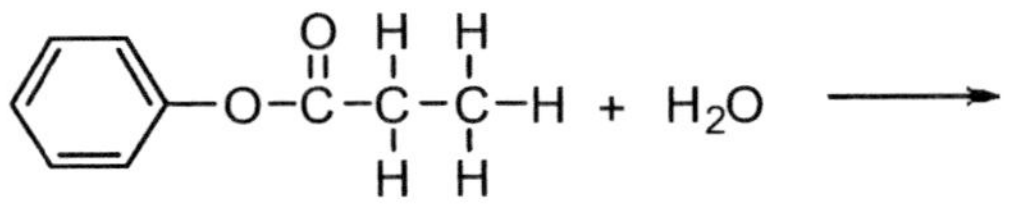

$$\text{(phenyl)}-O-\overset{O}{\underset{}{\overset{||}{C}}}-\overset{H}{\underset{H}{\overset{|}{C}}}-\overset{H}{\underset{H}{\overset{|}{C}}}-H \;+\; H_2O \longrightarrow$$

6. Write the products formed in the following esterification reactions.

a.

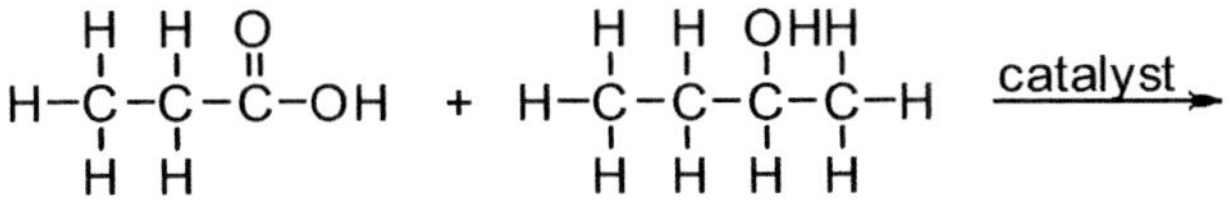

$$H-\overset{H}{\underset{H}{\overset{|}{C}}}-\overset{H}{\underset{H}{\overset{|}{C}}}-\overset{O}{\overset{||}{C}}-OH \;+\; H-\overset{H}{\underset{H}{\overset{|}{C}}}-\overset{H}{\underset{H}{\overset{|}{C}}}-\overset{OH}{\underset{H}{\overset{|}{C}}}-\overset{H}{\underset{H}{\overset{|}{C}}}-H \xrightarrow{\text{catalyst}}$$

b.

$$HSCoA \;+\; H-\overset{H}{\underset{H}{\overset{|}{C}}}-\overset{H}{\underset{H}{\overset{|}{C}}}-\overset{H}{\underset{H}{\overset{|}{C}}}-\overset{O}{\overset{||}{C}}-OH \xrightarrow{\text{catalyst}}$$

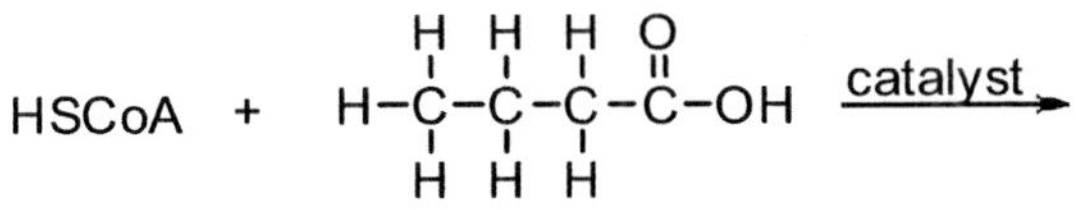

7. Write the product for the following reaction.

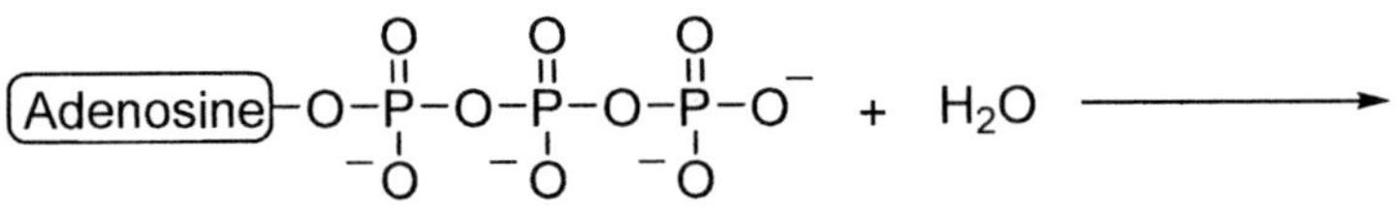

$$\boxed{\text{Adenosine}}-O-\overset{O}{\underset{-O}{\overset{||}{P}}}-O-\overset{O}{\underset{-O}{\overset{||}{P}}}-O-\overset{O}{\underset{-O}{\overset{||}{P}}}-O^- \;+\; H_2O \longrightarrow$$

ATP

8.   For the reaction shown below, highlight the phosphoryl group transferred. What functional group is the phosphate group transferred from?

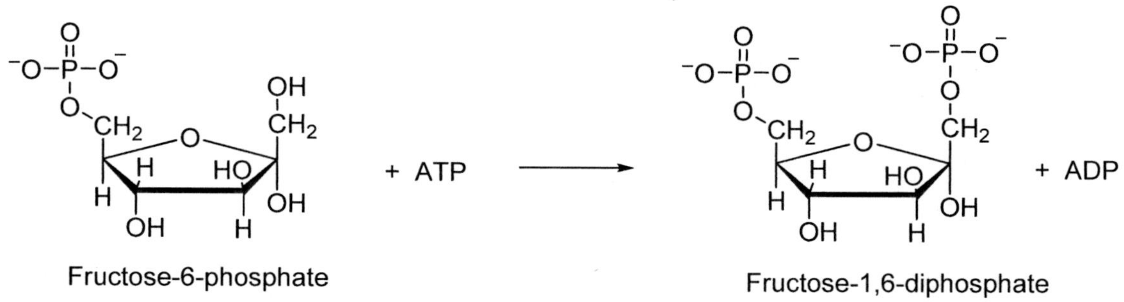

Fructose-6-phosphate   + ATP ⟶   Fructose-1,6-diphosphate   + ADP

9.   Identify the following reactions as oxidation-reduction reactions, hydration/dehydration, hydrolysis reactions, acyl group transfer reactions, or phosphoryl group transfer reactions.

a.

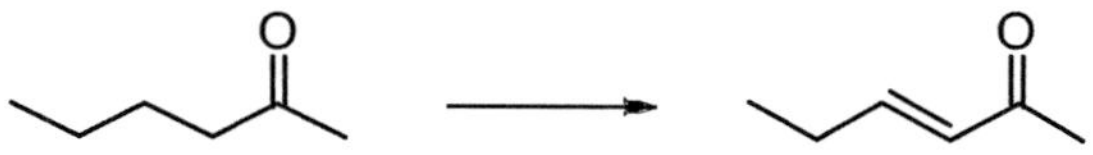

b.

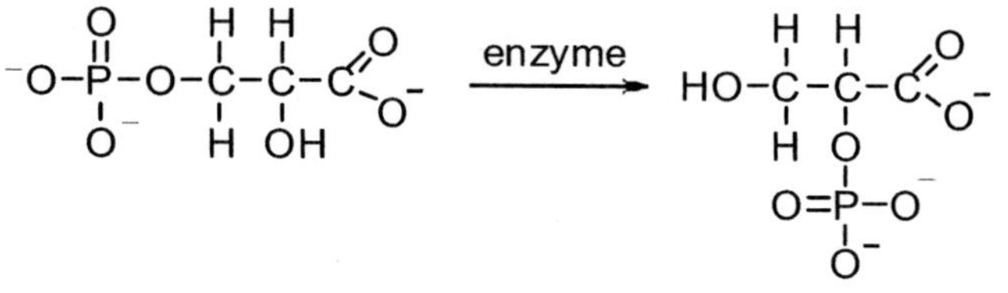

10. Identify the following reactions as oxidation-reduction reactions, hydration/dehydration, hydrolysis reactions, acyl group transfer reactions, or phosphoryl group transfer reactions.

a.

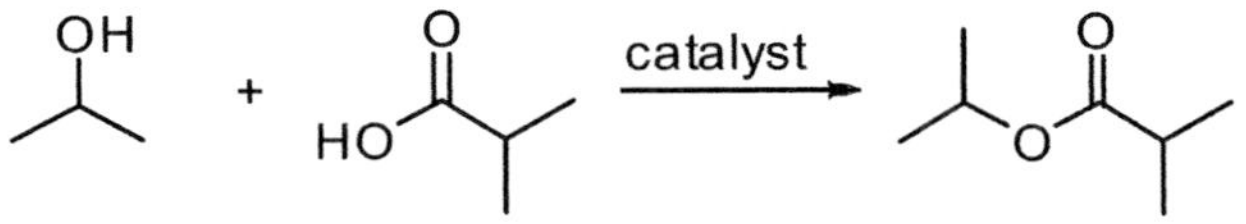

b.

Chapter 10

Answers to Additional Exercises

10.23   Alcohols, thiols, amines, carboxylic acids, esters, thioesters, and amides are involved in acyl transfers.

10.25   a.   A ketone has a carbonyl group attached to two R groups.

   b.   An ester has a carbonyl group attached to an R group and an OR group.

   c.   A thioester has a carbonyl group attached to an R group and an SR group.

10.27   The reactant that undergoes loss of electrons in an oxidation-reduction reaction is said to undergo *oxidation*.

10.29   An organic reactant that gains hydrogen atoms has undergone reduction.

10.31   An organic reactant that gains oxygen atoms has undergone oxidation.

10.33   a.   Oxidation

$$Zn: \longrightarrow Zn^{2+} + 2\,e^-$$

Reduction

$$2\,H^+ + 2\,e^- \longrightarrow H-H$$

Cl^- is a spectator ion.

   b.   Oxidation

$$Na\cdot \longrightarrow Na^+ + e^-$$

Reduction

$$:\!\overset{..}{\underset{..}{Cl}}\!-\!\overset{..}{\underset{..}{Cl}}\!: + 2\,e^- \longrightarrow 2\;:\!\overset{..}{\underset{..}{Cl}}\!:^-$$

10.35   a.   Acetaldehyde has undergone an oxidation reaction; it has gained oxygen atoms. NAD^+ has undergone a reduction reaction; it has gained a hydrogen atom to become NADH.

   b.   The reactant has undergone an oxidation reaction; the reactant has lost hydrogen atoms. FAD has undergone a reduction reaction; it has gained hydrogen atoms to become $FADH_2$.

10.37

This reaction is a type of reduction reaction known as a catalytic hydrogenation. The product has an increase in the number of hydrogen atoms.

10.39   An unsaturated fat is more likely to be a liquid.

10.41 $FADH_2$ is most likely used in the biochemical reduction of an alkene.

10.43 a.

b.

c.

d.

10.45   a. $C_3H_8 + 5\ O_2 \rightarrow 3\ CO_2 + 4\ H_2O$

b.

c.

d. $FADH_2 + NAD^+ \rightarrow FAD + NADH + H^+$

10.47   a.

b.

c. $FADH_2$

d.

10.49 d. A free radical does not contain an octet of electrons.

10.51 a. hydration

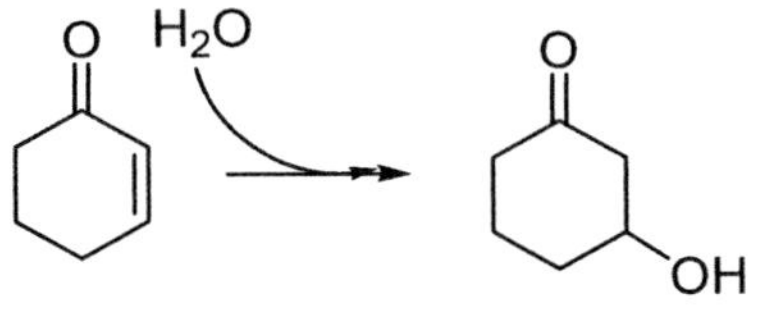

b. dehydration

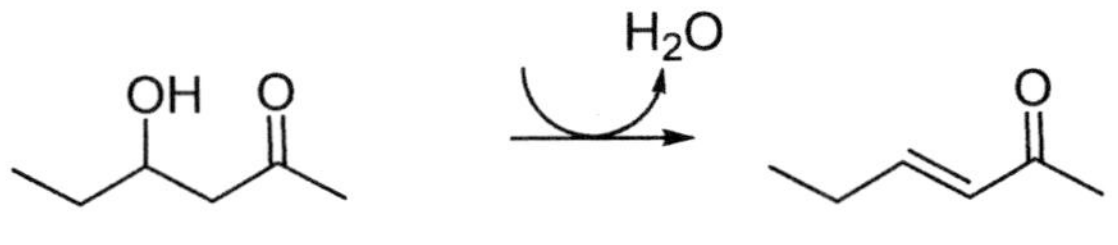

c. hydration

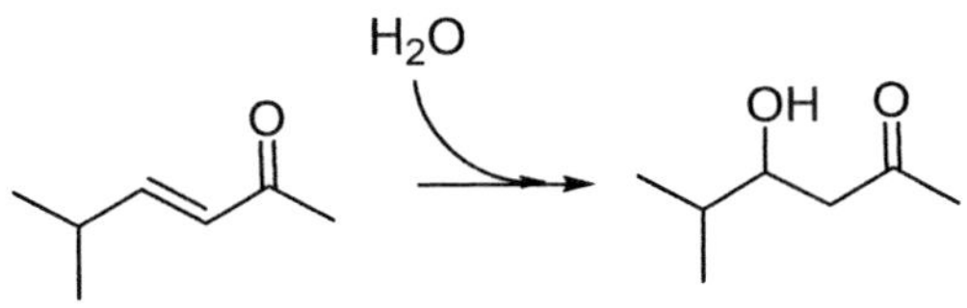

10.53 a.

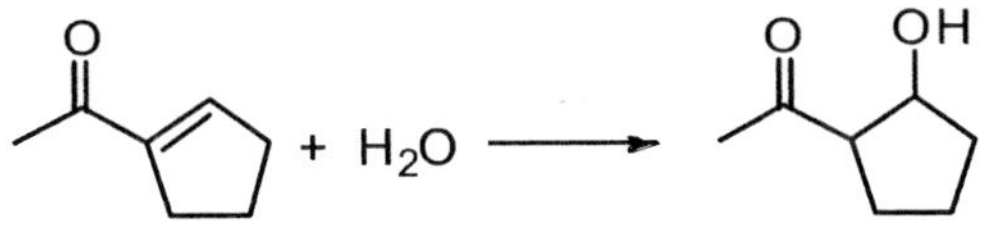

b.

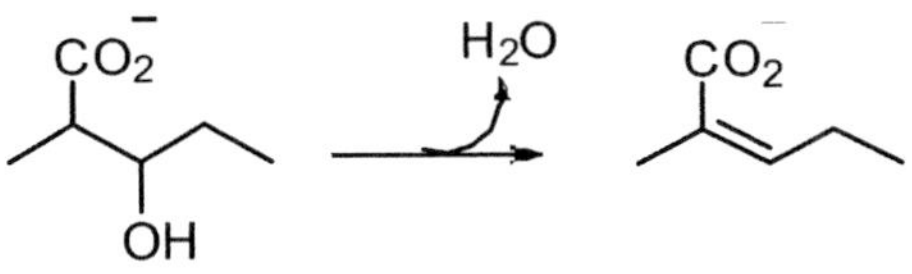

10.55 a.

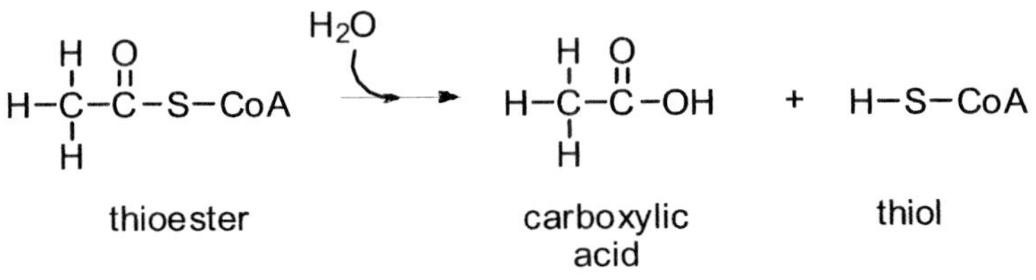
thioester
carboxylic acid
thiol

b.

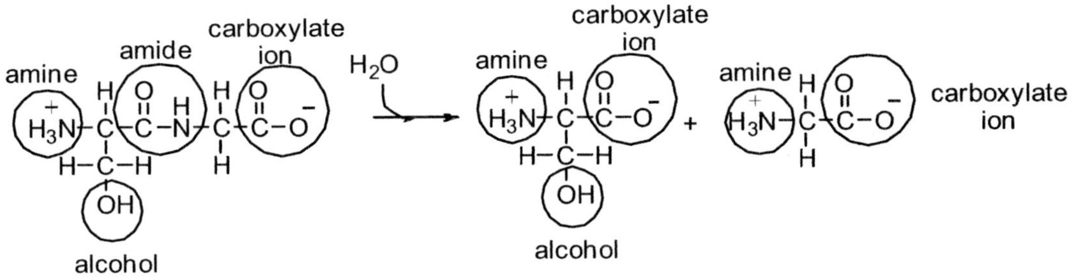
amine
amide
carboxylate ion
alcohol
amine
carboxylate ion
alcohol
amine
carboxylate ion

10.57   a.

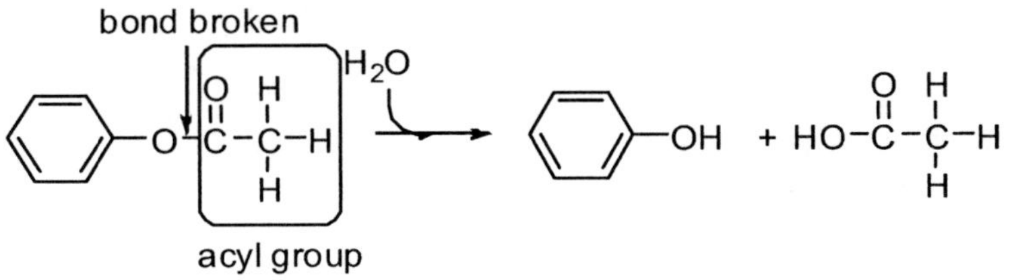

bond broken

H₂O

acyl group

b.

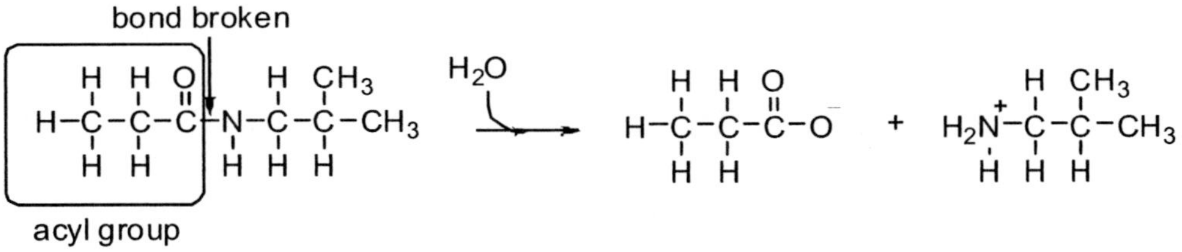

bond broken

H₂O

acyl group

c.

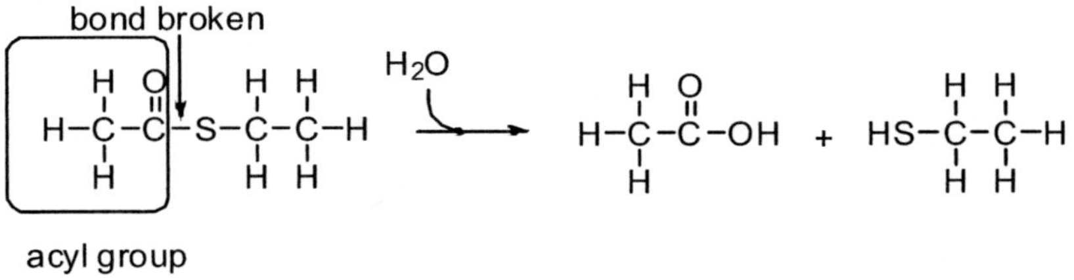

bond broken

H₂O

acyl group

d.

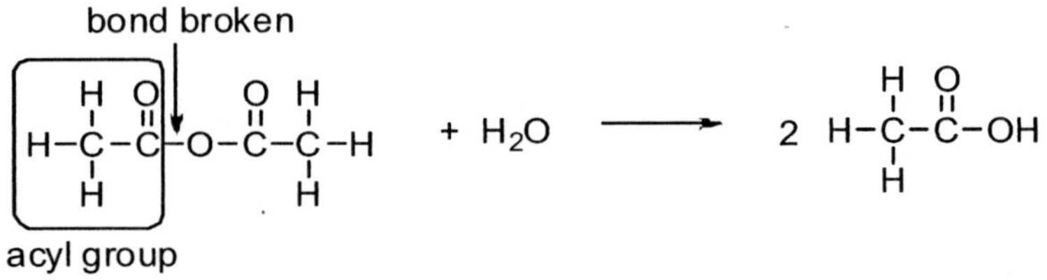

bond broken

+ H₂O

acyl group

10.59   a.

catalyst

+ H₂O

b.

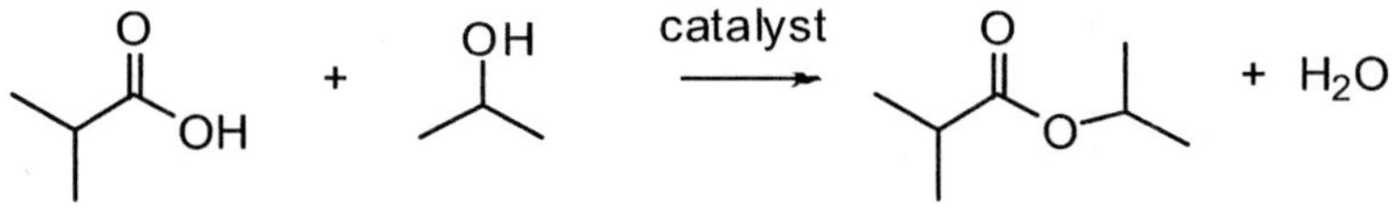

catalyst

+ H₂O

c.

catalyst

+ H₂O

10.61   A phosphoryl group can be transferred to the oxygen atom of water or the oxygen atom of an alcohol group, just like an acyl group.

10.63   a.

ATP   glucose   glucose 6-phosphate   ADP

b.   The phosphate group is being transferred from a triphosphate.

c.   The phosphate group is being transferred from a phosphate group.

d.   The phosphate group is being transferred to an alcohol group.

10.65   a.   hydrolysis, a type of acyl group transfer reaction

b.   oxidation-reduction

c.   hydration reaction

10.67   a.

b.   In the reaction that occurs in yeast, NADH is the reactant.

c.   In the reaction that occurs in yeast, acetaldehyde is reduced.

10.69   The reaction is a reduction reaction. The number of hydrogen atoms increases; FAD gains two hydrogen atoms. FAD is derived from riboflavin, vitamin B_2.

10.71   The four fat soluble vitamins are vitamin A, vitamin D, vitamin E, and vitamin K. These vitamins are soluble in fat and not water because they are hydrophobic, and they are stored in fat tissue in the body.

10.73   A vitamin deficiency occurs when the body does not have enough of a particular vitamin. It can be caused by insufficient vitamin intake in the diet or poor vitamin absorption. The symptoms of riboflavin deficiency include inflammation of the tongue and skin, lesions in the mouth, and cataracts.

Chapter 11

Proteins: Structure and Function

Chapter Summary

In this chapter, you were introduced to one of the most important biomolecules, proteins. You have seen how proteins are built from the basic building blocks, the amino acids. You examined the structure of amino acids and learned how they form peptide chains. You studied how these peptide chains fold into more complex structures through different interactions based on the chemical structure of the amino acids in the peptide chain. You learned how enzymes work and how molecules can inhibit the function of the enzymes.

Section 11.1 Amino Acids

Proteins have the most diverse functions of all biomolecules. They can act as enzymes, receptors, structural proteins, immunoglobins, transport proteins, and dietary proteins. The basic building blocks of all proteins are the amino acids. An amino acid contains an amine functional group and a carboxylic acid functional group, which are attached to the same carbon known as the α-carbon. Also attached to the α-carbon is an R group or side chain. There are 20 different natural amino acids, each with a different R group.

In a Nutshell: Amino Acid Equilibria and pH

Since the amine group of an amino acid is basic and the carboxylic acid group is acidic, the charge on these groups depends on the pH of the solution. At physiological pH, about 7.3, the amine is in its conjugate acid form and the carboxylic acid is in its conjugate base form. A compound containing both a positive and negative charge is known as a zwitterion.

If the pH of the solution changes, a proton is accepted or donated from one of these two functional groups. At low pH, the carboxylate group gains a proton to become a neutral carboxylic acid. At high pH, the NH_3^+ group loses a proton to become a neutral amine.

In a Nutshell: Amino Acid Side Chains

The various side chains of the amino acids can be classified as either nonpolar or polar. The nonpolar side chains consist of alkanes, aromatic hydrocarbons, or contain nitrogen and sulfur atoms in nonpolar covalent bonds. The amino acids with nonpolar side chains are glysine, alanine, valine, leucine, isoleucine, proline, tryptophan, phenylalanine, and methionine.

The polar side chains can be further subdivided into acidic, basic, and neutral side chains. The two amino acids with acidic side chains are aspartic acid and glutamic acid. At physiological pH, the side chains on these amino acids exist in their conjugate base form. The three amino acids with basic side chains are lysine, arginine, and histidine. At physiological pH, the side chains of these amino acids have a positive charge because they accept a proton. The six amino acids with neutral polar side chains are serine, threonine, tyrosine, cysteine, glutamine, and asparagine. The side chains on these amino acids are not ionizable—they do not lose or gain a proton.

In a Nutshell: Essential Amino Acids

There are ten amino acids that cannot be synthesized in the body and must be obtained through diet. These amino acids, known as the essential amino acids, are as follows: arginine, histidine, isoleucine, leucine, lysine, methionine, phenylalanine, threonine, tryptophan, and valine. The essential amino acids can be found in both animal and vegetable sources. Animal sources contain all 10 of the essential amino acids and are known as complete proteins. Single vegetable sources of amino acids are usually missing one or more essential amino acid.

Worked Example #1

One of the 20 natural amino acids is shown below.

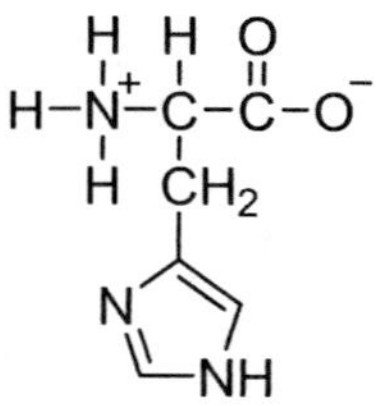

a. Circle and label the amine. In what form is this functional group at physiological pH?

b. Circle and label the carboxylic acid functional group that is present in all amino acids. In what form is this functional group at physiological pH?

c. Circle and label the side chain. What is the name of this amino acid?

d. Is the side chain polar or nonpolar? If it is polar, is it neutral, acidic, or basic?

Solutions

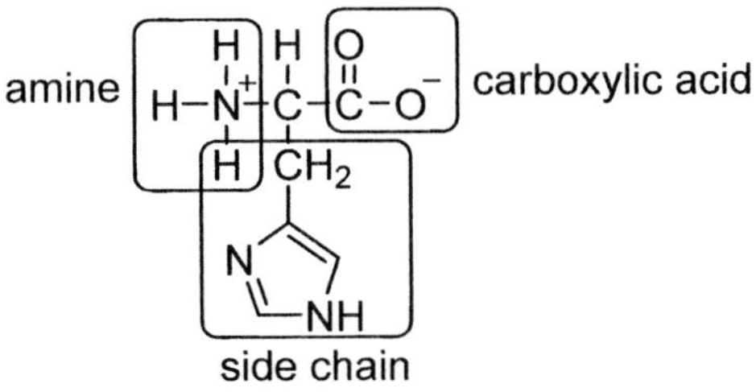

a. *Normally the amine is its conjugate acid form, –NH₃⁺ at physiological pH.*

b. *Normally the carboxylic acid is in its conjugate base form –CO₂⁻ at physiological pH.*

c. *The R group is shown above. This amino acid is histidine.*

d. *The side chain is polar and it is basic.*

Try It Yourself #1

One of the 20 natural amino acids is shown below.

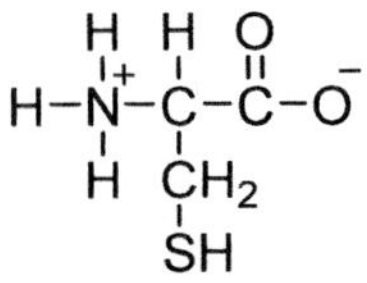

a. Circle and label the amine. Is it in its neutral or ionized form?

b. Circle and label the carboxylic acid functional group that is present in all amino acids. Is it in its neutral or ionized form?

c. Circle and label the side chain. What is the name of this amino acid?

d. Is the side chain polar or nonpolar? If it is polar, is it neutral, acidic, or basic?

Solutions

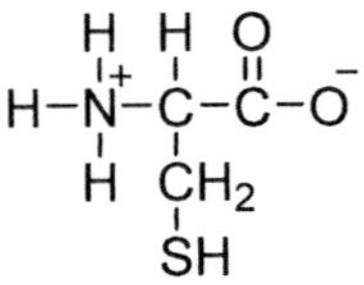

a. *The amine is in its: _____*

b. *The carboxylic acid is in its: _____*

c. *The name of the amino acid is: _____*

d. The side chain is:

Practice Problems for Amino Acids

1. For each of the amino acids place an "X" in the boxes that apply.

Amino acid	Nonpolar	Polar	Acidic side chain	Basic side chain	Neutral side chain	Essential amino acid
Tryptophan						
Threonine						
Aspartic acid						
Arginine						

2. Draw the structure of lysine at pH 10.

3. Which amino acids contain an alcohol functional group?

Section 11.2 Chirality and Amino Acids

Chirality is a symmetry property present not only in molecules but in many objects. An object is chiral if it is nonsuperposable on its mirror reflection. Two objects are nonsuperposable when you can never get all the components to perfectly overlay. An object is achiral if its mirror reflection is superposable.

In a Nutshell: Enantiomers

Stereoisomers are molecules with the same chemical formula and connectivity of atoms but a different spatial arrangement of atoms. Two nonsuperposable mirror image stereoisomers are known as enantiomers. When evaluating whether or not a molecule is superposable on its mirror image, consider all conformations of the molecule, but do not break any bonds. In the IUPAC naming system, the prefixes "D" and "L" are sometimes used to distinguish two enantiomers. Other prefixes in common use are "R" and "S", + and −, and d and l.

In a Nutshell: Fischer Projections

A simplified way of writing enantiomers is to show them as Fischer projections. In a Fischer projection, a tetrahedral carbon atom with four different atoms or groups attached to it is always shown as a crosshair and the main carbon chain is written vertically. The carboxylic acid or most oxidized functional group is placed at the top of the structure. Therefore when writing an amino acid, the α-carbon appears at the crosshair. In the Fischer projection for an L-amino acid, the amine group is on the left of the crosshair and a hydrogen atom is on the right.

In a Nutshell: Properties of Enantiomers

Enantiomers have identical physical and chemical properties when placed in an achiral environment, such as most laboratory conditions. Consequently, enantiomers are very difficult to separate. However, when enantiomers are placed in a chiral environment, such as the body, they can exhibit profoundly different chemical properties. Often, particularly in man-made drugs, a chiral substance will be produced in a 50:50 mixture of enantiomers, known as a racemic mixture.

Worked Example #2

The Fischer projection of one of the natural amino acids is shown below.

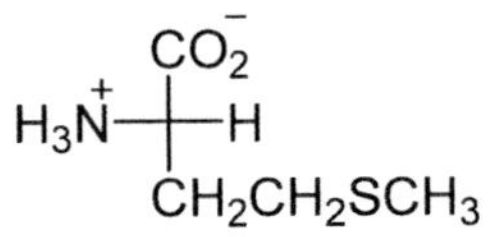

$$H_3\overset{+}{N}-\!\!\!\begin{array}{c}CO_2^- \\ | \\ \hline \\ | \\ CH_2CH_2SCH_3\end{array}\!\!\!-H$$

    a. Which amino acid is shown?

    b. Is this amino acid D- or L-? Explain.

    c. Place an arrow pointing to the α-carbon.

d. Is this molecule chiral?

e. Write the Fischer projection of the enantiomer of this amino acid. How would it be named?

a. *The side chain of the amino acid is $CH_2CH_2SCH_3$. The amino acid is L-methionine.*

b. *The amino acid is the L-amino acid. The amine group is on the left of the α-carbon.*

c.

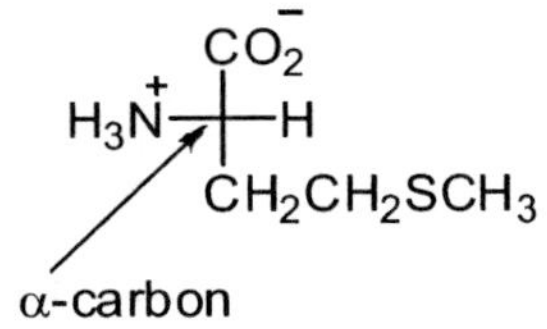

$$H_3\overset{+}{N} \overset{\displaystyle CO_2^-}{\underset{\displaystyle CH_2CH_2SCH_3}{\rule[0.5ex]{2em}{0.4pt}}} H$$

α-carbon

d. *The molecule is chiral because it is nonsuperposable on its mirror image.*

e. *D-methionine*

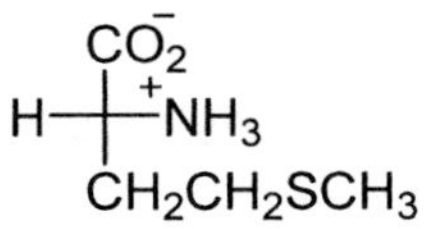

$$H \overset{\displaystyle CO_2^-}{\underset{\displaystyle CH_2CH_2SCH_3}{\rule[0.5ex]{2em}{0.4pt}}} \overset{+}{N}H_3$$

Try It Yourself #2

The Fischer projection of one of the natural amino acids is shown below.

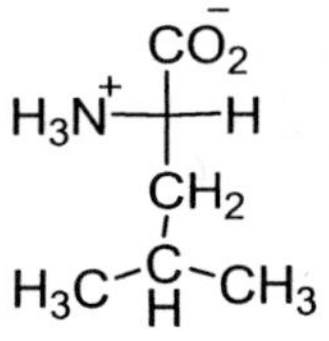

$$H_3\overset{+}{N} \overset{\displaystyle CO_2^-}{\rule[0.5ex]{2em}{0.4pt}} H$$
$$CH_2$$
$$H_3C - \overset{\displaystyle}{\underset{\displaystyle H}{C}} - CH_3$$

a. Which amino acid is shown?

b. Is this amino acid D- or L-? Explain.

c. Place an arrow pointing to the α-carbon.

d. Is this molecule chiral?

e. Write the Fischer projection of the enantiomer of this amino acid. How would it be named?

a. *The side chain for the amino acid is:*

The name of the amino acid is: _____

b.   The amine group is on the (left or right) of the α-carbon.

The amino acid is: _____

c.

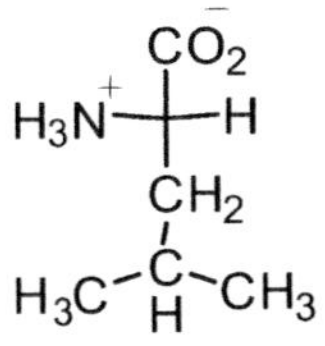

$$\begin{array}{c} \overset{\scriptstyle -}{CO_2} \\ | \\ H_3\overset{+}{N}\!\!-\!\!\!\!\begin{array}{c}|\\|\\|\end{array}\!\!\!\!-H \\ | \\ CH_2 \\ | \\ H_3C\!\!-\!\!\underset{H}{\overset{|}{C}}\!\!-\!\!CH_3 \end{array}$$

d.   The mirror image of this amino acid is (superposable or nonsuperposable).   The
amino acid is:_____

e.   The name of the enantiomer is:_____

Worked Example #3

Indicate whether the following objects are chiral or achiral.

   a. a square

   b. a pair of scissors

   c. a mitten

   d. a golf ball

   a. *A square is achiral. Its mirror image is superposable.*

   b. *A pair of scissors is chiral.*

   c. *A mitten is chiral.*

   d. *A golf ball is achiral. Its mirror image is superposable.*

Try It Yourself #3

Indicate whether the following objects are chiral or achiral.

   a. your left foot

   b. a circle

c. a triangle

d. a hammer

a. *Is the mirror image superposable?* _____

    Your left foot is: _____ .

b. *Is the mirror image superposable?* _____

    A circle is: _____ .

c. *Is the mirror image superposable?* _____

    A triangle is: _____ .

d. *Is the mirror image superposable?* _____

    A hammer is: _____ .

Practice Problems for Chirality and Amino Acids

1. The Fischer projection for an amino acid is shown below.

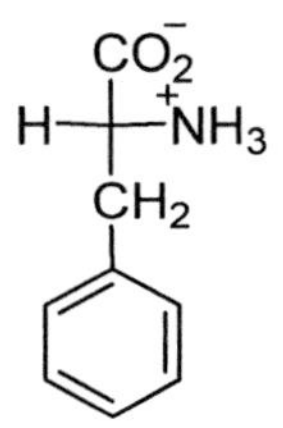

   a. Is the amino acid D- or L-? Explain.

   b. Write the Fischer projection of the enantiomer of this amino acid.

2. The Fischer projection for an amino acid is shown below.

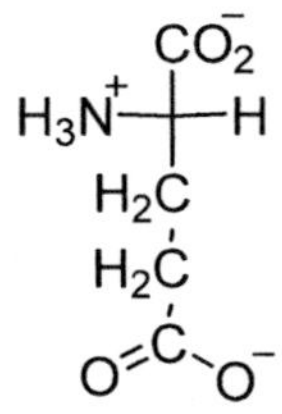

   a. Which amino acid is shown?

b. Is this amino acid D- or L-? Explain.

c. Place an arrow pointing to the α-carbon.

d. Write the Fischer projection of the enantiomer of this amino acid. How would it be named?

3. Are the following objects achiral or chiral?
    a. your right ear

    b. a ping pong ball

    c. a shoe

    d. a corkscrew

Section 11.3 Peptides

Peptides are molecules composed of two or more amino acids. A peptide containing two amino acids is also known as a dipeptide; one containing three amino acids, a tripeptide; and so forth. Peptides containing more than ~12 amino acids are known as polypetides. A polypeptide with more than about 50 amino acids folded into its active three-dimensional structure is referred to as a protein.

In a Nutshell: The Peptide Bond

The carboxylic acid of one amino acid and the amine of another amino acid react to form an amide functional group. The new covalent bond formed is known as a peptide bond. This reaction also forms a molecule of water. The reverse reaction forming two amino acids from a dipeptide and water is known as a hydrolysis reaction.

In a Nutshell: Small Peptides

The reaction of a dipeptide with a third amino acid produces a tripeptide. Every amino acid is capable of forming two peptide bonds, one with its amine functional group and one with is carboxylic acid functional group. The first and last amino acids in a peptide have a free amine or carboxylate group. The end of a peptide containing the free amine group is known as the N-terminus; the end containing the free carboxylate group is known as the C-terminus. The convention for writing peptides is to list the sequence of amino acids starting from the N-terminus and continuing in order to the C-terminus. Each amino acid has a three-letter abbreviation and a one-letter abbreviation, which are used when giving the amino acid sequence for a peptide or protein.

Worked Example #4

A skeletal line structure for a tetrapeptide, Tuftsin, which is used in immune system function, is shown below:

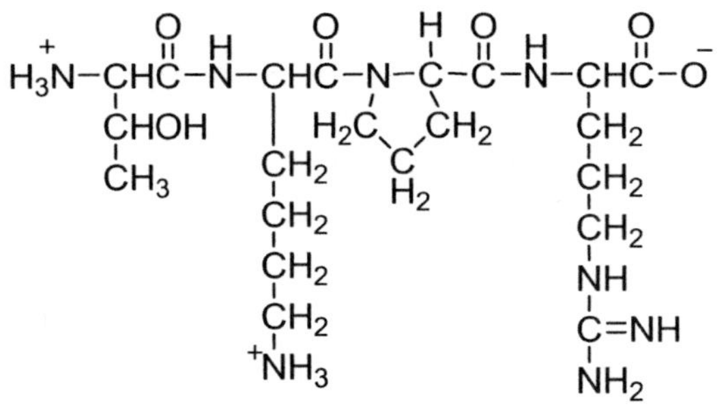

a. Circle each amino acid in the peptide and identify it by its three-letter abbreviation.

b. Label the N-terminus and the C-terminus.

c. What is the amino acid sequence for this peptide?

d. Identify each peptide bond.

Solutions

a.

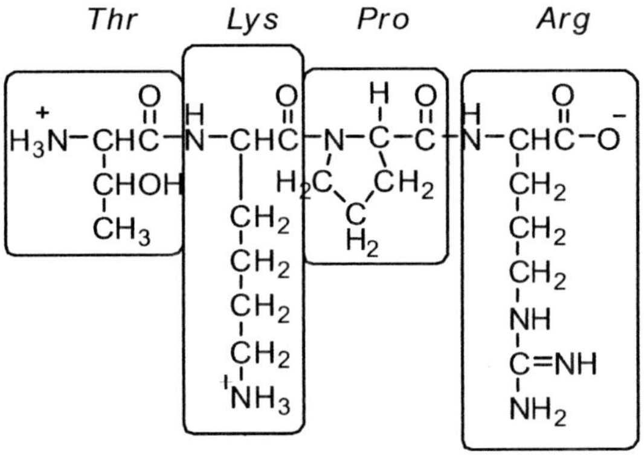

Thr   Lys   Pro   Arg

b.

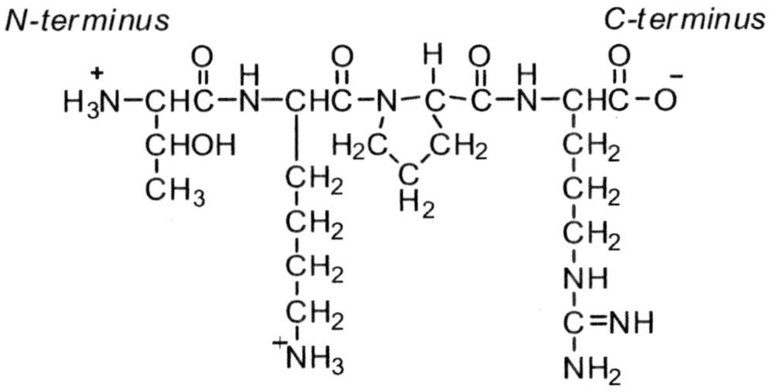

N-terminus   C-terminus

c.   Thr-Lys-Pro-Arg

d.

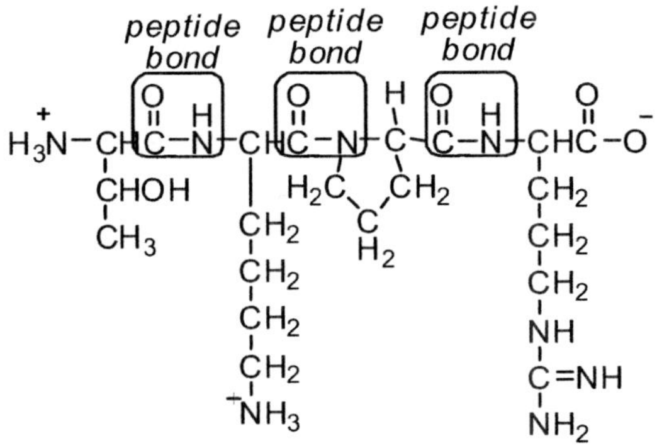

peptide   peptide   peptide
bond   bond   bond

Try It Yourself #4

The structure of a peptide is shown below.

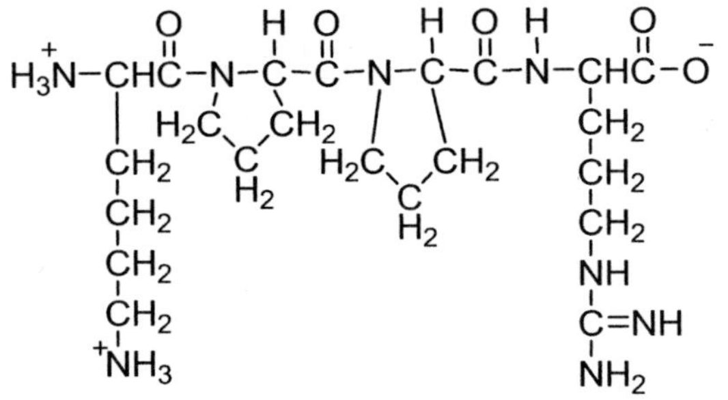

a. How many amino acids are in this peptide? Is this peptide a di-, tri-, tetra-, or pentapeptide?

b. Circle each amino acid in the peptide and identify it by its three letter-abbreviation.

c. What amino acid is the N-terminus?

d. What amino acid is the C-terminus?

e. How many peptide bonds are in this peptide?

f. Write the amino acid sequence for this peptide.

a. *There are _____ amino acids in this peptide. This peptide is a:*

_____

b.

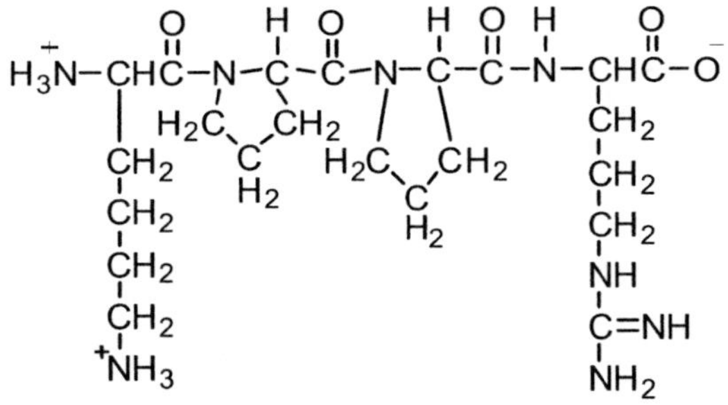

c. *The N-terminus is: _____*

d. *The C-terminus is: _____*

e. *There are _____ peptide bonds in this peptide.*

f. *The amino acid sequence is: _____*

Practice Problems for Peptides

1. Write the structural formula for the tripeptide Gly-His-Lys.

2. Angiotensin II plays an important role in the biochemical system that controls blood volume and blood pressure in the body. Its structure is shown below:

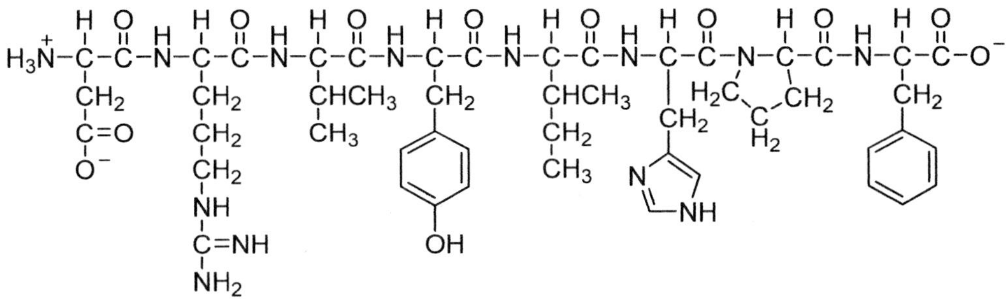

   a. Circle each amino acid in the peptide and identify it by its three-letter abbreviation.

   b. Label the N-terminus and the C-terminus.

   c. What is the amino acid sequence for this peptide?

3. How many amino acids are found in a pentapeptide? How many peptide bonds are in a pentapetide?

Section 11.4 Protein Architecture

In a Nutshell: The Three-Dimensional Shape of Proteins

The three-dimensional shape of a protein is not linear; the polypeptide chain folds into a unique three-dimensional shape, known as the native conformation. The chain of amino acids folds into its specific three-dimensional shape in order for the protein to perform its unique function. Electrostatic interactions cause the proteins to fold. Electrostatic interactions occur between atoms within the protein as well as with atoms in the environment in which the protein is suspended. When part of a protein, amino acids are often referred to as residues to emphasize that the only difference between amino acids in a protein is their side chains. Most enzymes exist in an aqueous medium; therefore, polar residues will tend to be on the exterior of the protein where they can interact with water molecules. Nonpolar residues will be on the interior. There are four levels of architecture for protein folding: 1) primary structure, 2) secondary structure, 3) tertiary structure, and 4) quaternary structure.

In a Nutshell: The Primary Structure of Proteins

The primary structure of a protein is its amino acid sequence. Errors in the primary structure can have profound effects on its overall shape.

Worked Example #5

Thyroid stimulating hormone (TSH) is composed of two polypeptide chains. How many N-termini does it have? How many C-termini does it have?

Solution

There are two N-termini and two C-termini because there are two polypeptide chains.

Try It Yourself #5

What could happen to the shape of a protein if one of the amino acids in the protein sequence is changed?

Solution:

In a Nutshell: Secondary Structure

The secondary through quaternary structures of a protein describe the folding patterns and three-dimensional structure of the protein based on its primary sequence. Secondary structure refers to regular folding patterns in localized regions of the polypeptide backbone. These folds are held together by hydrogen bonds between the N-H and the carbonyl groups (C=O) in the polypeptide backbone. The most common secondary structures are the α-helix and the β-pleated sheet. Proteins differ in the amount and type of secondary structure that they contain.

An α-helix is a long or short section of polypeptide that coils in the shape of a helix. The helix is held in shape by hydrogen bonds between N-H and C=O groups on the polypeptide backbone. The carbonyl group forms a hydrogen bond with the N-H group of an amino acid that is situated about four amino acids farther down the sequence. Amino acid side chains extend outward from the α-helix and do not participate in the hydrogen bonding that creates the α-helix. Proteins are frequently depicted using ribbon drawings, which trace the polypeptide backbone depicting it as a ribbon. In a ribbon drawing an α-helix appears as a coil.

The β-pleated sheet, also known as a β-sheet, is formed when two or more polypeptide strands stack on top of one another in a folding pattern analogous to a pleated skirt. Hydrogen bonding between N-H and C=O groups on parallel (or antiparallel) strands stabilizes this secondary structure. In a ribbon drawing a β–pleated sheet is drawn as a wide flat arrow, with the arrow pointing in the direction of the C-terminus.

Worked Example #6

In a ribbon drawing, does a coil or a flat arrow represent a β-pleated sheet? What type of intermolecular force holds the β-pleated sheet together?

Solution

A wide flat arrow is used to represent a β-pleated sheet in a ribbon drawing. Hydrogen bonding between the N-H and C=O groups holds the β-pleated sheet together.

Try It Yourself #6

In a ribbon drawing, does a wide flat arrow or a coil represent an α-helix? Where are the side chains of an amino acid located in the α-helix?

An α-helix is represented by a: _____

The side chains are located: _____

In a Nutshell: Tertiary Structure

The tertiary structure of a protein describes the complex and irregular folding of the polypeptide beyond its secondary structure, showing the three-dimensional picture of the entire protein. Tertiary structure is determined by interactions that include distant amino acid residues as well as the surrounding environment. For some proteins, tertiary structure also includes prosthetic groups, non-peptide organic molecules or metal ions that are strongly bound to the protein and essential to its function.

The fundamental driving force behind protein folding is energy. A folded protein and its surrounding environment are lower in potential energy than the unfolded protein. There are four main electrostatic interactions between amino acid side chains that are responsible for the tertiary structure of a protein:1) disulfide bridges (covalent bond, hydrophobic), 2) salt bridges (ionic bond, hydrophilic), 3) hydrogen bonding (hydrophilic), and 4) dispersion forces (hydrophobic).

Disulfide bridges are covalent bonds formed between two cysteine residues. The two thiol functional groups on each cysteine residue react to form a disulfide bond. Disulfide bridges are also known as disulfide bonds. The oxidation of two thiols produces the disulfide bond; the reduction of a disulfide bond produces two thiols. Since a disulfide bridge is a covalent bond, the interaction is strong, and can only be made or broken through a chemical reaction.

Ionic bonds are often called salt bridges in the context of proteins. Ionic bonds form in proteins between positive and negatively charged residues, such as a basic residue and an acidic residue. These interactions are stronger than hydrogen bonds.

Hydrogen bonding occurs between polar residues when one residue contains either an N-H or an O-H bond. Although hydrogen bonding is weaker than disulfide or ionic bonds, there are often so many hydrogen bonds in a protein that they contribute significantly to the conformational integrity of the protein. Tertiary structure is also influenced by hydrogen bonding between polar residues and the aqueous environment of some proteins. In these proteins, polar residues tend to be located on the outside of the protein, so that they can interact with the aqueous surroundings through hydrogen bonding with the water molecules.

Nonpolar residues interact with other nonpolar residues through dispersion forces. Nonpolar residues avoid interactions with polar residues and water. Thus, in an aqueous environment, proteins tend to fold so that the nonpolar residues are on the interior.

Prosthetic groups are part of the tertiary structure of some proteins. Prosthetic groups are non-peptide organic molecules or metal ions, or a combination of both, that are essential to a protein's function. In proteins that function as enzymes, the prosthetic group is called a cofactor or a coenzyme.

Worked Example #7

Label the following interactions as a disulfide bond, a salt bridge, hydrogen bonding, or dispersion forces.

a.

b.

c.

d.

Solutions

    a. *Hydrogen bonding. The bond is between an O-H group and an N-H group.*

    b. *Dispersion forces. The interaction is between two nonpolar residues.*

    c. *Disulfide bond.*

    d. *Salt bridge. The interaction is between a negative charge, COO^-, and NH_3^+.*

Try It Yourself #7

Label the following interactions as a disulfide bond, a salt bridge, hydrogen bonding, or dispersion forces.

    a.

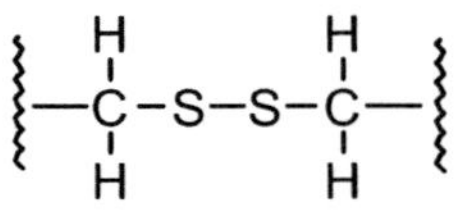

    b.

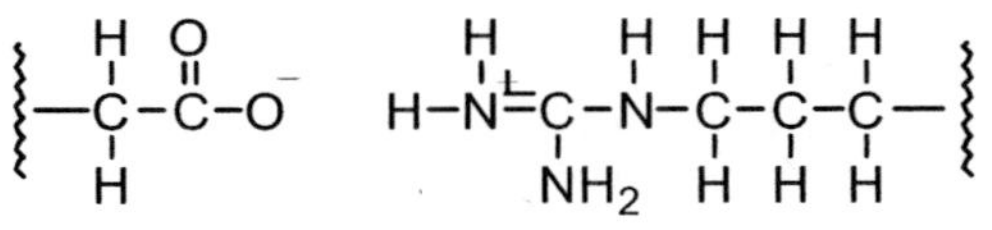

    c.

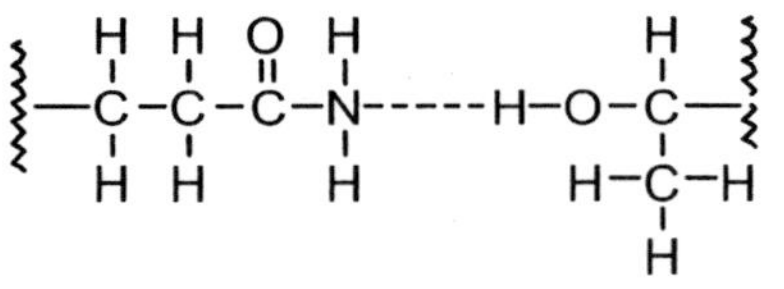

    d.

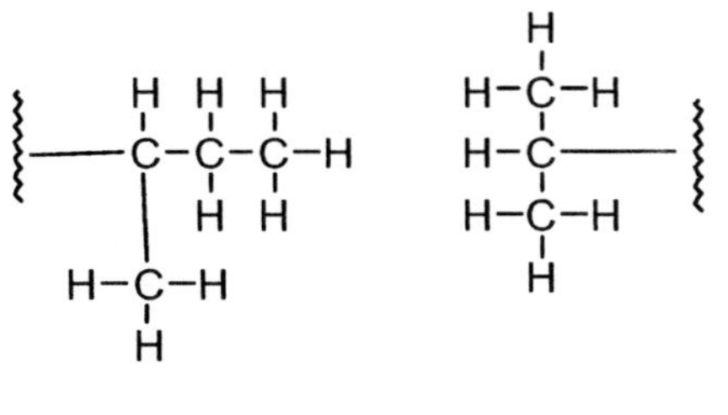

Solutions

    a. *The interaction is between: _____.*

        The interaction is: _____

    b. *The interaction is between: _____.*

        The interaction is: _____

    c. *The interaction is between: _____.*

        The interaction is: _____

    d. *The interaction is between: _____.*

The interaction is: _____

In a Nutshell: Quaternary Structure

Some proteins are composed of more than one polypeptide chain, also called a subunit. The quaternary structure of a protein is the relative arrangement and position of the subunits within the overall structure of the protein. The different subunits interact with one another and are stabilized by the same interactions that determine the tertiary structure of a protein: disulfide bridges, salt bridges, hydrogen bonding, and dispersion forces.

In a Nutshell: Denaturation of a Protein

Denaturation of a protein refers to any process that disrupts the secondary, tertiary, or quaternary structure of a protein so that it can no longer perform its function. Denaturing a protein does not alter its primary structure. Agents that denature a protein include heat, pH changes, mechanical agitation, detergents, and some metals.

Worked Example #8

For the following pairs of amino acids, indicate how they might interact to contribute to the tertiary or quaternary structure of a protein. Choose from the following interactions: disulfide bond, salt bridge, hydrogen bonding, or dispersion forces.

- a. proline and alanine
- b. threonine and glutamine
- c. two cysteines
- d. arginine and aspartic acid

- a. *Dispersion forces. The two residues are nonpolar, so the interaction might be dispersion forces.*
- b. *Hydrogen bonding. There is an O-H group in threonine and an N-H group in glutamine.*
- c. *Disulfide bond. There is an S-H group on each cysteine.*
- d. *Salt bridge. Arginine will have a positive charge on one of the nitrogens in the side chain, and aspartic acid will have a negative charge on the carbocylate group in the side chain.*

Try It Yourself #8

For the following pairs of amino acids, indicate how they might interact to contribute to the tertiary or quaternary structure of a protein. Choose from the following interactions: disulfide bond, salt bridge, hydrogen bonding, or dispersion forces.

a. two cysteines

b. phenylalanine and valine

c. glutamic acid and histidine

d. asparagine and serine

a. *The interaction is between:* _____.

The interaction is: _____

b. *The interaction is between:* _____.

The interaction is: _____

c. *The interaction is between:* _____.

The interaction is: _____

d. *The interaction is between:* _____.

The interaction is: _____

In a Nutshell: Types of Proteins

There are three general classes of proteins, distinguished by their tertiary and quaternary structure as well as their solubility. The three classes are fibrous proteins, globular proteins, and membrane proteins. **Fibrous proteins** contain parallel peptide chains, resulting in long fibers or sheets. These proteins are strong and insoluble in water and have a structural role in nature. There are three basic types of fibrous proteins: keratins, elastins, and collagens. **Globular proteins** are soluble polypetides folded into complex overall spherical shapes. **Membrane proteins** are specialized proteins that span the cell membrane and often serve as receptors of biochemical signals, or as ion channels that shuttle ions in and out of the cell.

Worked Example #9

A globular protein is found in an aqueous environment. What types of residues would you expect to find on the interior of the protein? What types of residues would you expect to find on the exterior of the protein?

Solution

Because the aqueous environment is polar, polar, hydrophilic residues should be on the exterior of the protein. The hydrophobic, nonpolar residues should be on the inside of the protein.

Try It Yourself #9

A membrane protein is found in the hydrophobic environment of a cell membrane. What types of residues would you expect to find on the interior of the protein? What types of residues would you expect to find on the exterior of the protein?

Solution

The cell membrane environment is: _____

The residues on the exterior of a membrane protein are: _____

The residues on the interior of a membrane protein are: _____

Practice Problems for Protein Architecture

1. In which levels of protein architecture are hydrogen bonding found?

2. What types of interactions are broken when a protein is denatured? What levels of protein architecture are affected when a protein is denatured?

3. A genetic mutation changed an aspartic acid residue, that formed a salt bridge to a leucine residue in a protein. How might this change affect the secondary or tertiary structure of the protein?

Section 11.5 Enzymes

In a Nutshell: How do Enzymes Work?

One of the most important roles that proteins serve is their function as enzymes, the globular proteins that catalyze most of the chemical reactions occurring a cell. Enzymes typically act on only one particular reactant, referred to as a substrate. As with all catalysts, enzymes work by lowering the energy of activation, E_A, for the reaction. The enzyme itself is unchanged at the end of a chemical reaction; thus, the same enzyme molecule may be used over and over again.

The enzyme name is usually drawn over the reaction arrow. The enzymes are named after the substrate or type of reaction they catalyze with the addition of the suffix "ase." Their names are usually italicized. Enzymes are classified by the types of reactions that they catalyze.

In a Nutshell: The Enzyme-Substrate Complex

The first step in an enzyme-catalyzed reaction is the binding of the substrate to the enzyme. The substrate binds to the enzyme at a pocket or cleft within the protein, known as the active site or binding site. The active site is where the chemical reaction actually takes place. The shape of the active site is complementary to the shape of the substrate and is largely responsible for the selectivity that an enzyme has for its substrate. The substrate binds to the active site through ionic interactions and intermolecular forces of attraction (dispersion forces, dipole-dipole forces, and hydrogen bonding). The binding interactions together with the complementary shape create a good fit between the enzyme and its substrate, called the enzyme-substrate complex, ES.

The binding of a substrate to an enzyme is described by the lock-and-key model. In this model the enzyme and substrate have complementary shapes which allow them to fit together like a key fits into a lock.

In the enzyme-substrate complex, the substrate is held in a position and orientation that holds the reacting functional group(s) near one another, thereby facilitating the chemical reaction. It is the placement of the substrate in an optimal geometry and proximity to the other reactant that lowers the energy of activation, E_A, for the reaction.

In a Nutshell: Cofactors and Coenzymes

Some enzymes require a cofactor to achieve their catalytic effect. Cofactors can be metal ions, such as Fe^{2+}, Mg^{2+}, and Zn^{2+}. Cofactors can also be organic compounds such as $NAD^+/NADH$ or $FAD/FADH_2$ and coenzyme A. Cofactors that are organic compounds are called coenzymes. A coenzyme undergoes a chemical change during the reaction, but it is regenerated in a subsequent reaction so that it can be employed by the enzyme again.

In a Nutshell: pH and Temperature Dependence of Enzymes

An enzyme must be in its native conformation to achieve its catalytic activity, which requires that both pH and temperature be in the optimal range for the enzyme. Changes in pH can alter the charge on acidic and basic residues, which in turn affects the shape of the protein. For most enzymes, the optimal pH is physiological pH (pH = 7.3). Too high temperatures will denature enzymes and decrease reaction rates. Most enzymes function best when around body temperature, 37°C.

Worked Example #10

How does an enzyme catalyze a reaction?

Solution

An enzyme catalyzes a reaction by holding the substrate in a position and orientation so that the reacting functional group(s) are near one another, thereby facilitating the chemical reaction. It is the placement of the substrate in an optimal geometry and proximity to the other reactant that lowers the energy of activation, E_A.

Try It Yourself #10

What interactions are responsible for binding an enzyme to a substrate?

The interactions are: _____

In a Nutshell: Enzyme Inhibitors

Enzyme inhibitors are compounds that prevent an enzyme from performing its function. pH and temperature are examples of nonspecific inhibitors because they can denature many different types of enzymes at the same time. Specific enzyme inhibitors target only one enzyme. There are two types of specific enzyme inhibitors: competitive inhibitors and noncompetitive inhibitors.

A competitive inhibitor competes with the substrate for the active site of the enzyme, because it too has a structure that is complementary to the active site. By binding to the active side, an inhibitor blocks the substrate from binding to the enzyme. The effectiveness of a competitive inhibitor depends on the relative concentrations of substrate and inhibitor. The more inhibitor molecules present, the less likely the substrate will find a free enzyme to which to bind.

Noncompetitive inhibitors bind at a location on the enzyme other than the active site of the enzyme. Binding of a noncompetitive inhibitor causes the shape of the enzyme to change in such a way that the active site can no longer bind the substrate. Therefore, no ES complex is formed. In contrast to competitive inhibitors, increasing the concentration of substrate will not restore enzyme activity when a noncompetitive inhibitor is bound to the enzyme. Enzyme activity can only be restored when the concentration of noncompetitive inhibitor is low.

Worked Example #11

Celebrex® is an NSAID (nonsteroidal anti-inflammatory drug) that is a competitive inhibitor for the enzyme *cyclooxygenase 2,* COX-2. COX-2 is an enzyme that produces prostaglandins, which can cause pain and inflammation. Does Celebrex® bind to the active site of COX-2 or does it change the shape of COX-2?

Solution

Because Celebrex® is a competitive inhibitor for the COX-2 enzyme, it binds to the active site of the enzyme.

Try It Yourself #11

Cholesterol is a noncompetitive inhibitor of an enzyme required for its own synthesis, known as *HMG-CoA Reductase.* Does cholesterol bind to the active site of *HMG-CoA Reductase* or does it change the shape of *HMG-CoA Reductase*?

Solution:

Practice Problems for Enzymes

1. How is the rate of an enzyme-catalyzed reaction affected by the following changes?

    a. An increase in temperature

    b. An increase in pH

2. NAD^+/NADH is a coenzyme. How does this coenzyme change during a reaction? Why is it considered a coenzyme?

3. Do competitive inhibitors of an enzyme have the same shape as the substrate for that enzyme?

Chapter 11 Quiz

1. For each of the amino acids place an "X" in the boxes that apply.

Amino acid	Nonpolar	Polar	Acidic side chain	Basic side chain	Neutral side chain	Essential amino acid
Glycine						
Serine						
Glutamic acid						
Histidine						

2. Which of the amino acids in Question 1 are chiral?

3. A peptide chain has the following sequence Thr-Lys-Pro-Arg.

   a. What amino acid is the N-terminus?

   b. What amino acid is the C-terminus?

   c. How many peptide bonds are in this peptide?

   d. Which amino acids in this peptide are polar?

   e. How many amino acids in this peptide are chiral?

   f. Write the structure formula of this peptide.

4. Would you expect the amino acid aspartic acid to participate in hydrogen bonding, disulfide bridges, salt bridges, or dispersion forces in protein architecture? What level (primary, secondary, tertiary, or quaternary) of protein architecture would be involved?

5. Would you expect the amino acid serine to participate in hydrogen bonding, disulfide bridges, salt bridges, or dispersion forces in protein architecture? What level (primary, secondary, tertiary, or quaternary) of protein architecture would be involved?

6. What amino acid forms disulfide bridges? What functional group does this amino acid contain?

7. How does an enzyme catalyze a reaction?

8. Does a competitive inhibitor of an enzyme have the same shape as the substrate for the enzyme? How does a competitive inhibitor prevent an enzyme from performing its function?

9. Does a noncompetitive inhibitor of an enzyme have the same shape as the substrate for the enzyme? How does a noncompetitive inhibitor prevent an enzyme from performing its function?

10. Draw the structure of aspartic acid at pH 1. Explain why low pH would denature a protein.

Chapter 11

Answers to Additional Exercises

11.45   a. Glycine

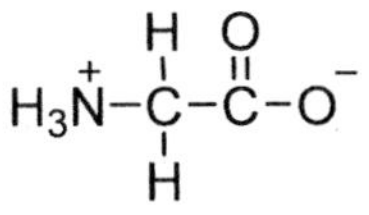

b. Aspartic acid

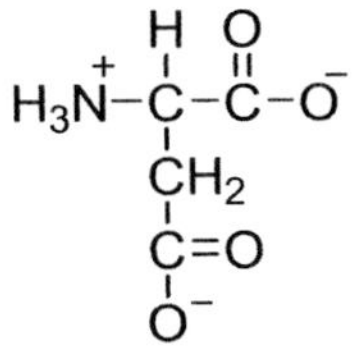

c. Tyrosine

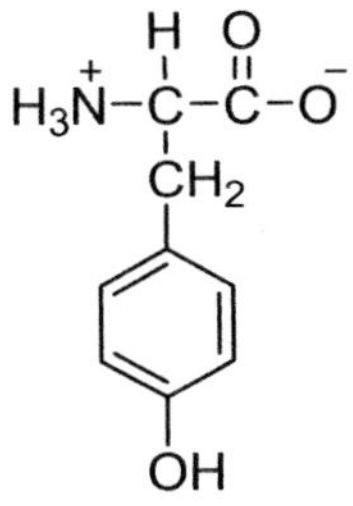

d. Cysteine

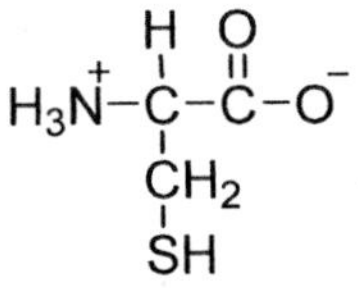

e. Glutamic acid

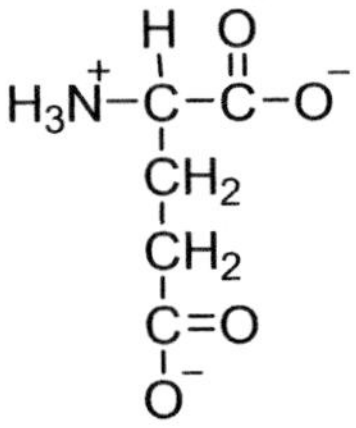

11.47   The amino acids with nonpolar side chains are glycine, alanine, valine, leucine, isoleucine, proline, tryptophan, phenylalanine, and methionine. The amino acids with basic side chains are lysine, arginine, and histidine. The amino acids with acidic side chains are aspartic acid and glutamic acid. The amino acids with polar but neutral side chains are serine, threonine, tyrosine, cysteine, glutamine, and asparagine.

11.49   Essential amino acids are amino acids that are not synthesized by the body and must be supplied through the diet.

11.51   a.   The amino acid with a thiol functional group in its side chain is cysteine.

b.   The amino acids with an amide functional group in their side chains are asparagine and glutamine.

11.53   Cysteine has a thiol functional group.

$$\underset{\underset{SH}{|}}{\underset{\underset{CH_2}{|}}{H_3\overset{+}{N}-\overset{\overset{H}{|}}{C}-\overset{\overset{O}{\parallel}}{C}-O^-}}$$

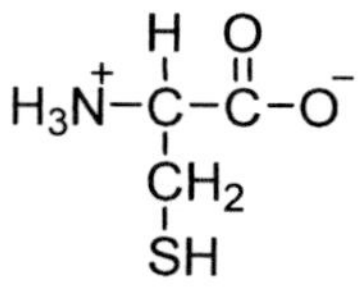

11.55

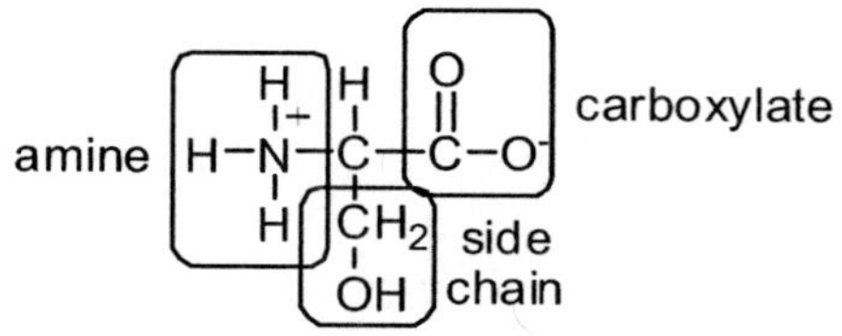

a.   The amine is in its ionized form.

b.   The proton has been lost from the carboxylic acid, forming a carboxylate ion.

c.   Yes, the amino acid shown is a zwitterion. At approximately physiological pH, most amino acids exist as a zwitterion.

d.   This amino acid is serine.

e.   The side chain is polar.

f.   At pH = 12 serine has the structure shown below. The net charge is −2.

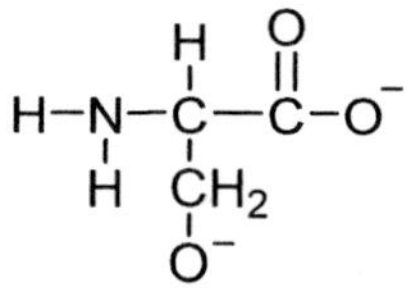

$$H-\overset{\overset{H}{|}}{\underset{\underset{H}{|}}{N}}-\overset{\overset{H}{|}}{\underset{\underset{CH_2}{|}}{C}}-\overset{\overset{O}{\parallel}}{C}-O^-$$
$$\underset{O^-}{|}$$

11.57

Amino acid	Nonpolar	Polar	Acidic side chain	Basic side chain	Neutral side chain	Essential amino acid
Phe	x					x
Asn		x			x	
Met	x					x

11.59   Meat contains all the essential amino acids.

11.61   a.   A corkscrew is chiral.

b.   An orange is achiral.

c.   A car is chiral.

d.   A nail is achiral.

11.63   A racemic mixture of alanine would contain a 50:50 mixture of L-alanine and D-alanine.

11.65   a.

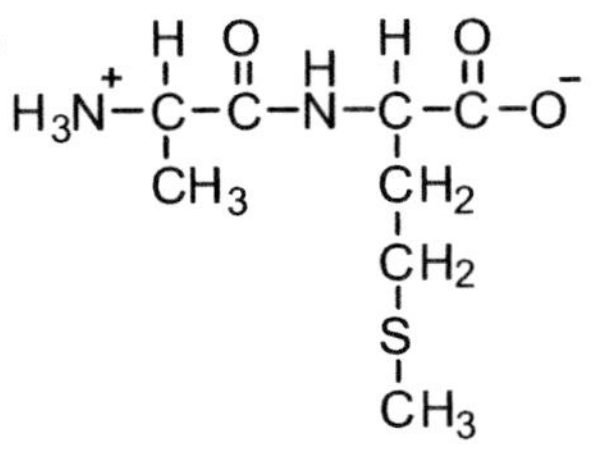

Ala-Met

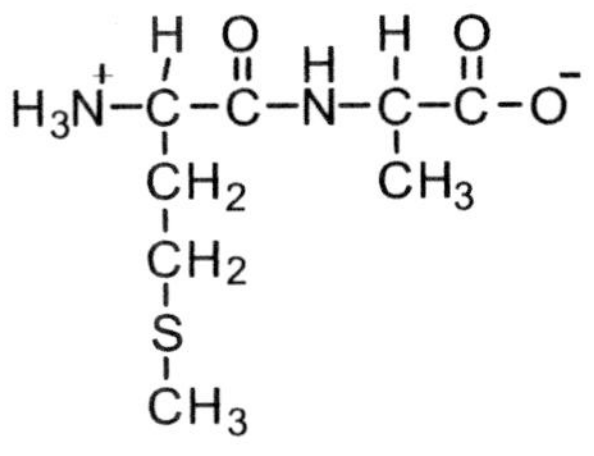

Met-Ala

b.

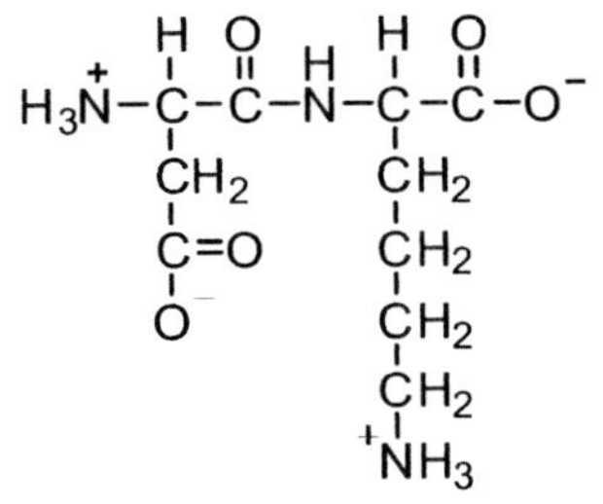

Asp-Lys

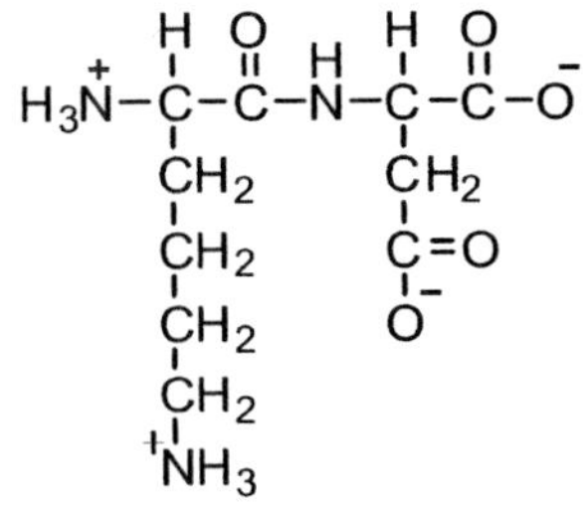

Lys-Asp

c.

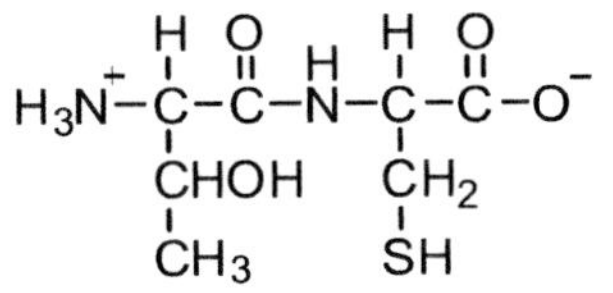

Thr-Cys

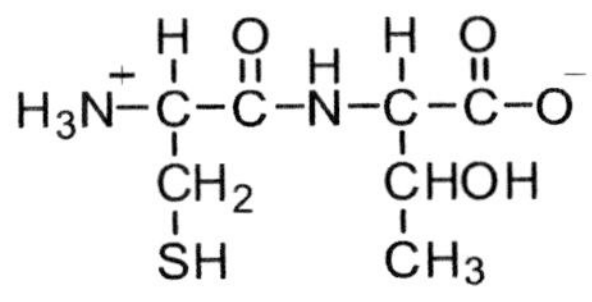

Cys-Thr

d.

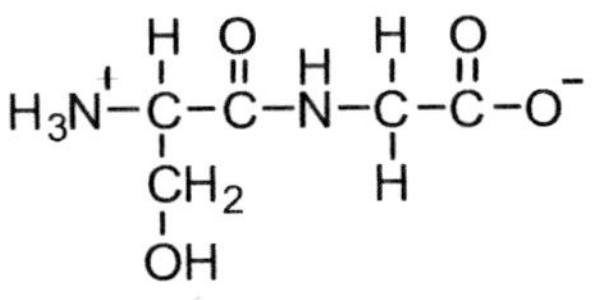

Ser-Gly

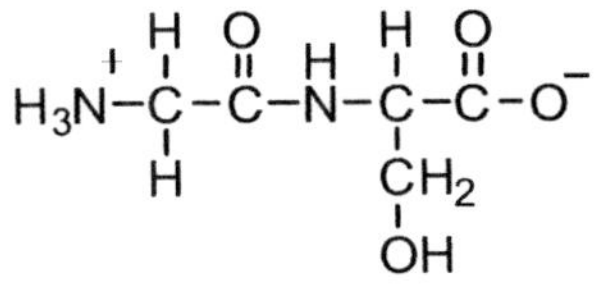

Gly-Ser

11.67   Leu-Glu-His

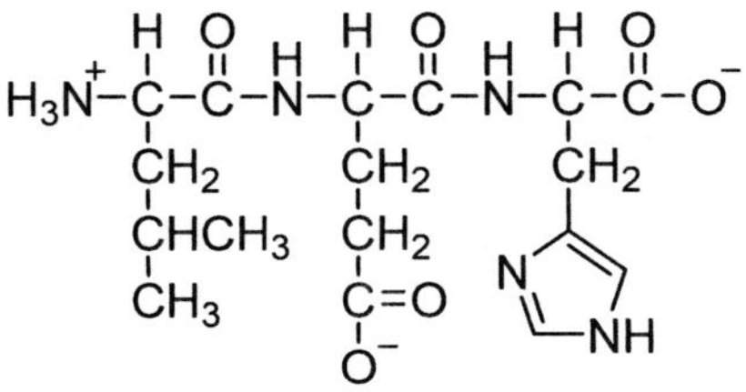

11.69   a. Lys-Thr-Thr-Lys-Ser

b. The N-terminal amino acid is lysine.

c. The C-terminal amino acid is serine.

d. Lysine has an amine functional group and serine and threonine have alcohol functional groups.

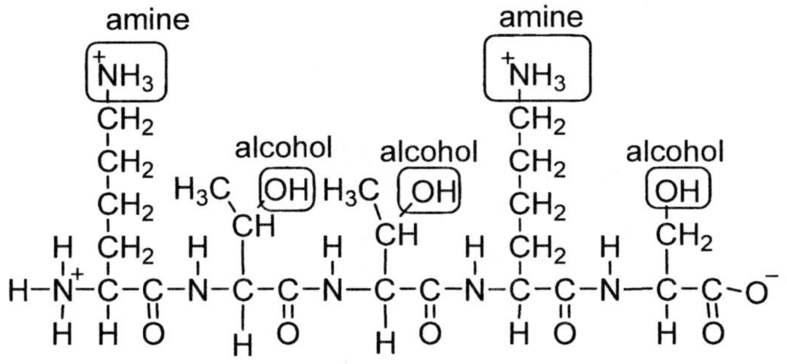

11.71   Vasopressin should be more polar because it contains Arg, an amino acid with a polar side chain, instead of Leu.

11.73   The DNA of the cell contains the code for the amino acid sequence.

11.75   The two most common forms of secondary structure are the α-helix and the β-pleated sheet. Secondary structure is formed by hydrogen bonding between the amide carbonyl and the amide nitrogen of the peptide backbone.

11.77   The four types of electrostatic interactions that are responsible for the tertiary structure of a protein are: disulfide bridges, salt bridges, hydrogen bonding and dispersion forces.

A disulfide bridge

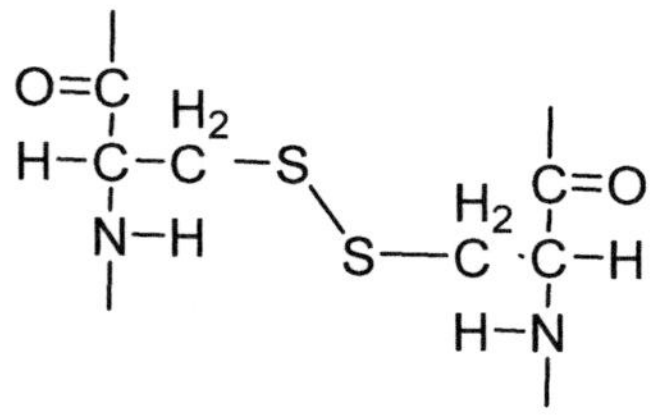

Salt bridge

O=C   O⁻   H–N⁺–CH₂–CH₂–CH₂–CH₂–C–H
H–C–CH₂–C   H   N–H
        N–H

(structure with +NH and C=O groups, N-H)

Hydrogen bonding

H–N
O=C H
H–C–C–O–H------N–C–C–C–
H–N H   H   H H

Dispersion forces

(–C–C with CH₃ groups; H–C– structure)

11.79 Aspartic acid, glutamic acid, lysine, arginine, and histidine have side chains that can form salt bridges.

11.81 The three general classes of proteins distinguished by their quaternary structure and solubility are fibrous proteins, globular proteins, and membrane proteins.

11.83

$$H–\overset{H}{\underset{H}{C}}–\overset{H}{\underset{H}{C}}–S–S–\overset{H}{\underset{H}{C}}–H \xrightarrow{[H]} H–\overset{H}{\underset{H}{C}}–\overset{H}{\underset{H}{C}}–SH + HS–\overset{H}{\underset{H}{C}}–H$$

thiol   thiol

11.85 Most fibrous proteins provide structure to the body and are found in hair, skin, fingernails (keratins); skin, blood vessels, heart, lung, intestines, tendons, ligaments (elastins); and connective tissue (collagens). The quaternary structure provides elasticity and rigidity of connective tissue.

11.87 Membrane proteins are found in the cell membrane.

11.89 Heat, pH changes, mechanical agitation, detergents, and some metals may denature a protein. When a protein is denatured it loses its secondary, tertiary, and quaternary structure. Since the denatured protein has lost its shape, it has also lost its function.

11.91 $E + S \rightleftharpoons ES \rightleftharpoons E + P$

E is the enzyme, S is the substrate, *ES* is the enzyme-substrate complex, and P is the product.

11.93   Some cofactors include Fe^{2+}, Mg^{2+}, and Zn^{2+}.

11.95   The substrate binds to the active site through ionic interactions and intermolecular forces of attraction (dispersion forces, dipole-dipole forces, and hydrogen bonding).

11.97   The activation energy, E_A, is lower for the enzyme-catalyzed reaction compared to the uncatalyzed reaction.

11.99   A competitive inhibitor competes with the substrate for the active site of the enzyme because it too has a structure that is complementary to the active site. By binding to the active site, the competitive inhibitor blocks the substrate from binding to the enzyme and thus prevents the reaction from occurring.

11.101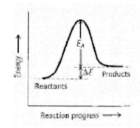

11.103   Angiotension II initiates a number of physiological events that raise blood pressure. Therefore, stopping the production of angiotension II should not increase blood pressure.

11.105   The *angiotension converting enzyme* breaks the peptide bond between Phe and His in angiotension I.

11.107   When valine replaces glutamic acid in hemoglobin, an acidic polar side chain has been replaced with a nonpolar side chain and a potential salt bridge has been removed, changing the shape of the protein. The tertiary and quaternary shape of the protein has been affected.

11.109   At low pH, the carboxylic acids are in their unionized form. They cannot form salt bridges with positively charged side chains. Therefore, the tertiary and quaternary structure of the protein is disrupted.

Chapter 12

Carbohydrates: Structure and Function

Chapter Summary

In this chapter, you learned about carbohydrates and how they provide fuel for the cell. You first studied the structure of simple carbohydrates and then learned how the simple carbohydrates form more complex carbohydrates. You observed the role that complex carbohydrates play in living things. You also examined the biochemical pathway that the body uses to produce energy from carbohydrates. Finally, you learned that carbohydrates serve an important function as cell markers.

Section 12.1 Role of Carbohydrates

Carbohydrates are the most abundant of the biomolecules. Half of the earth's carbon exists in the form of two types of carbohydrates, which are found in plant material: cellulose and starch. Cellulose provides structure and starch provides energy for a plant. Plants synthesize carbohydrates from carbon dioxide and water during photosynthesis. The energy to drive photosynthesis comes from the sun. In the body, carbohydrates are metabolized to carbon dioxide, water, and energy through another series of biochemical pathways.

In a Nutshell: Overview of Catabolism

Energy can be extracted from carbohydrates, fats, and when necessary, proteins, through biochemical pathways that are collectively referred to as catabolism. In the first stage of catabolism, each biomolecule is hydrolyzed into its smaller building blocks; proteins are hydrolyzed into amino acids; fats are hydrolyzed into fatty acids and glycerol; and carbohydrates are hydrolyzed into monosaccharides.

In the second stage of catabolism, each of these small intermediates is broken down further by separate biochemical pathways. All these pathways end in the same final product, acetyl CoA.

In the third and final stage of catabolism, acetyl CoA enters the citric acid cycle which extracts energy in the form of electrons from acetyl CoA. In the process, acetyl CoA is converted to two molecules of carbon dioxide and coenzyme A. The electrons harvested in the citric acid cycle are carried by the coenzymes NADH and $FADH_2$ into the electron-transport chain. In the electron-transport chain these electrons, together with oxygen, drive the phosphorylation of ADP to ATP.

Section 12.2 Monosaccharides

Monosaccharides are carbohydrates that cannot be hydrolyzed into simpler carbohydrates. They are also known as simple sugars because they cannot be hydrolyzed into anything simpler.

In a Nutshell: Structure of Monosaccharides

Monosaccharides consist of a carbon chain containing an aldehyde or a ketone and two or more OH groups. The most common monosaccharides are glucose and fructose. Monosaccharides are soluble in water because their hydroxyl groups are able to form hydrogen bonds with water molecules.

In a Nutshell: Chirality and the D-sugars

Most monosaccharides are chiral. The Fischer projection of a carbohydrate has several carbon atoms that are represented by crosshairs, each containing a hydroxyl group. To draw the enantiomer of a monosaccharide, every group at a crosshair must be exchanged. D-glucose and L-glucose are enantiomers—nonsuperposable mirror image stereoisomers. In a Fischer projection, a pair of enantiomers can be readily identified because every set of horizontal groups—OH and H—at a crosshair has the opposite orientation.

The two enantiomeric forms of a carbohydrate each have the same name, but are distinguished by the prefix D- or L- before the name. A D-sugar has the OH group at the crosshair farthest from the carbonyl group pointing to the right. Conversely, in an L-sugar the OH group on the crosshair farthest from the carbonyl group points to the left. Nature produces only D-sugars with only a few exceptions.

Diastereomers are non-superposable stereoisomers that are not mirror images. In a Fischer projection, a pair of diastereomers will have at least one, but not all of the groups/atoms at the crosshairs interchanged.

Worked Example #1

Fischer projections of three different monosaccharides are shown below:

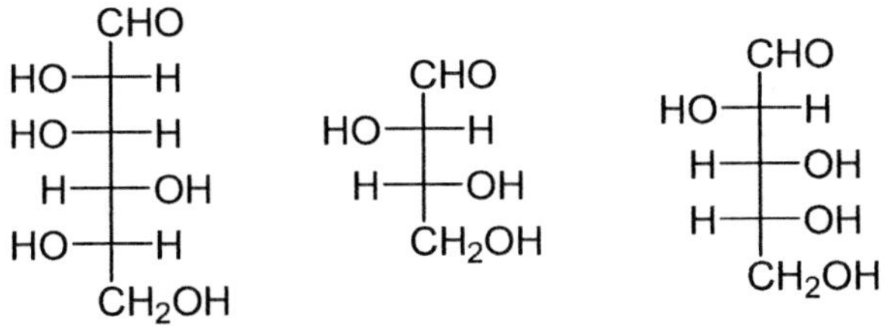

Gulose   Threose   Arabinose

a. Are these structures D-sugars or L-sugars? How can you tell?

b. Write the structure of the enantiomer of gulose shown. What is the name of the enantiomer?

a. *L-Gulose—the OH group on the crosshair farthest from the CHO group points to the left. D-Threose—the OH group on the crosshair farthest from the CHO group points to the right. D-Arabinose—the OH group on the crosshair farthest from the CHO group points to the right.*

b.

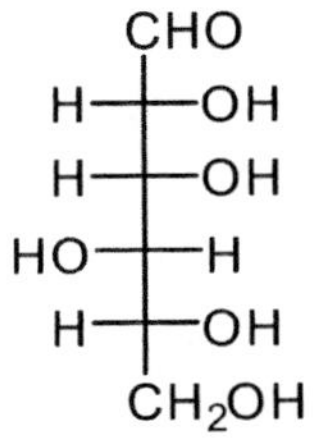

D-Gulose

Try It Yourself #1

Fischer projections of three different monosaccharides are shown below:

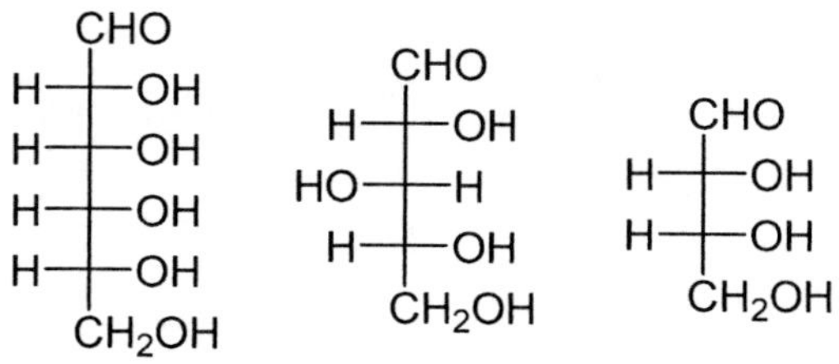

CHO
H—OH
H—OH
H—OH
H—OH
CH₂OH
Allose

CHO
H—OH
HO—H
H—OH
CH₂OH
Xylose

CHO
H—OH
H—OH
CH₂OH
Erythrose

a. Are these structures D-sugars or L-sugars? How can you tell?

b. Write the structure of the enantiomer of erythrose shown. What is the name of the enantiomer?

a. *Allose—The OH group on the crosshair farthest from the aldehyde points to:*

_____. *It is a _____-sugar.*

Xylose—The OH group on the crosshair farthest from the aldehyde points to:

_____. *It is a _____-sugar.*

Erythrose—The OH group on the crosshair farthest from the aldehyde points to:

_____. *It is a _____-sugar.*

b.

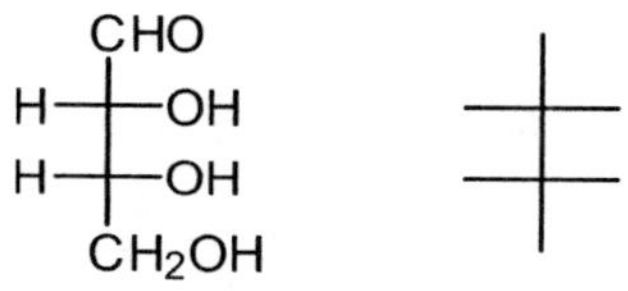

CHO
H—OH
H—OH
CH₂OH
Erythrose

Erythrose
enantiomer

The name of the enantiomer is: _____

Worked Example #2

The structures of D-idose, L-idose, and D-altrose are shown below:

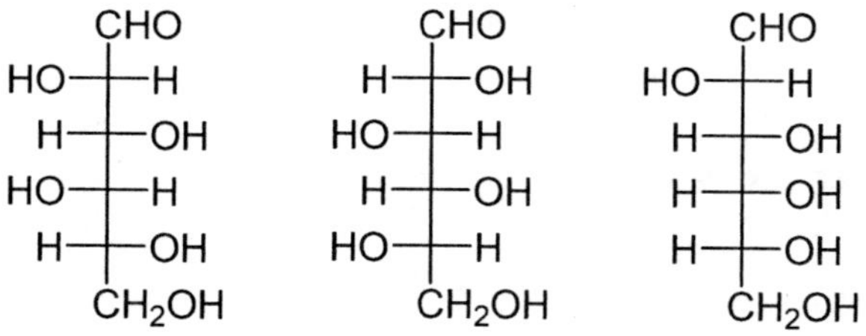

CHO
HO—H
H—OH
HO—H
H—OH
CH₂OH
D-idose

CHO
H—OH
HO—H
H—OH
HO—H
CH₂OH
L-idose

CHO
HO—H
H—OH
H—OH
H—OH
CH₂OH
D-altrose

a. Are D-idose and L-idose enantiomers or diastereomers?

b. Are D-ildose and D-altose enantiomers or diastereomers?

a. D-idose and L-idose are non-superposable mirror images of each other; therefore, they are enantiomers.

b. D-idose and D-altose are diastereomers. They are nonsuperposable stereoisomers. The OH and H groups are changed at only one cross hair.

Try It Yourself #2

The structure of D-ribose and D-lyxose are shown below.

```
      CHO                    CHO
   H——OH              HO——H
   H——OH              HO——H
   H——OH               H——OH
      CH2OH                CH2OH

   D-ribose            D-lyxose
```

a. Write the structure for L-lyxose.

b. Are D-lyxose and L-lyxose enantiomers or diastereomers?

c. Are D-ribose and D-lyxose enantiomers or diastereomers?

a.

L-lyxose

b. Are D-lyxose and L-lyxose mirror images of each other? _____

They are: _____

c. Are D-ribose and D-lyxose mirror images of each other? _____

They are: _____

In a Nutshell: Haworth Drawings

Most monosaccharides react to form a five-membered ring containing an oxygen atom, called a furanose, or a six-membered ring containing an oxygen atom, called a pyranose. The convention for drawing monosaccharide rings is to place the oxygen atom in the ring at the top of the ring in a furanose structure and on the top right of the ring in a pyranose structure.

Monosaccharides contain a hydroxyl group on almost every carbon atom in the ring, which is oriented either above or below the ring. To illustrate this three-dimensional characteristic of monosaccharides, the ring form is written as a Haworth projection, which is a flattened representation of the ring that emphasizes the orientation of the hydroxyl group with respect to the ring. The orientation of each hydroxyl group in a monosaccharide is important because these orientations describe the stereoisomer and hence the identity of the monosaccharide. In the pyranose and furanose form of a monosaccharide, a D-sugar always has the C6 CH_2OH group positioned above the ring.

The carbon atom bonded to both the ring oxygen atom and a hydroxyl group is known as the anomeric carbon. The hydroxyl group on the anomeric carbon can be oriented either above or below the ring. When it is below the ring, the sugar is known as the α-anomer, and when it is above the ring, the sugar is known as the β-anomer. A pyranose is numbered beginning with C1 at the anomeric carbon, whereas a furanose has its anomeric carbon at C2. It is conventional when writing Haworth projections to place the anomeric carbon to the right of the oxygen atom in the ring.

In a Nutshell: Mutarotation

Pyranoses and furanoses are unstable in aqueous solutions and readily undergo reactions that cause the ring to open and re-close. The ring opens when the bond between the ring oxygen and C1 or C2 breaks, producing the open chain form of the monosaccharide. In going to the open chain form, the anomeric carbon becomes either an aldehyde or a ketone and the oxygen atom in the ring becomes a hydroxyl group at C5.

When the open-chain form closes again it can close to become either the α- or β- anomer. When the pure α- or β- form of a sugar is placed in an aqueous solution, it undergoes a process called mutarotation. All three forms of a monosaccharide—α-, β- and open-chain form—are in equilibrium in aqueous solution. Equilibrium favors both ring forms over the open-chain form.

Worked Example #3

Two forms of mannose are shown below.

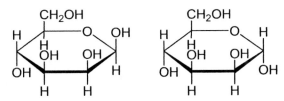

a. Are these sugars D- sugars or L-sugars? How can you tell?

b. Name each from of mannose: ___-___mannose and ___-___mannose.

c. Do the Haworth projections show these monosaccharides to be pyranoses or furanoses? Explain.

d. Are these two forms of mannose enantiomers or diastereomers?

a. These sugars are D-sugars because the C6 CH₂OH group lies above the ring.

b. In the structure on the left the OH group attached to the anomeric carbon is above the ring; therefore, it is β-D-mannose. In the structure on the right the OH group attached to the anomeric carbon lies below the ring; therefore, it is α-D-mannose.

c. These sugars are pyranoses. They are represented by a six-membered ring with an oxygen atom in it.

d. These two forms of mannose are diastereomers. Only the positions of the H and OH group on the anomeric have changed.

Try It Yourself #3

Two forms of allose are shown below.

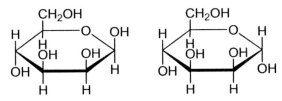

a. Are these sugars D-sugars or L-sugars? How can you tell?

b. Name each form of allose:___-___allose and ___-___allose.

c. Do the Haworth projections show these monosaccharides to be pyranoses or furanoses? Explain.

d. Are these two forms of mannose enantiomers or diastereomers?

a. *The position of the group on C6 is_____ the ring.*
    These sugars are: _____.

b. *In the sugar on the left, the _____ on the anomeric carbon lies _____ the ring. The sugar on the left is ___-___allose. In the sugar on the right, the _____ on the anomeric carbon lies _____ the ring. The sugar on the right is ___-___allose.*

c. *The structures shown are _____-membered rings with an oxygen atom. They are: _____.*

d. *Are the structures mirror images?_____*
    They are: _____.

In a Nutshell: Modified Monosaccharides

Some molecules contain modified monosaccharides. These are monosaccharides with a modification to one of the hydroxyl groups on the ring.   Amino sugars have an amine functional group instead of an alcohol at one of the carbon atoms. Phosphosugars contain a phosphate ester at one of the hydroxyl groups. Deoxysugars are missing a hydroxyl group at one position. Glycosides contain an OR group instead of a hydroxyl group at the anomeric carbon. Glycosides are formed by the reaction of a monosaccharide with an alcohol in the presence of a catalyst.

Worked Example #4

Follicle stimulating hormone and luteinizing hormone both contain a modified monosaccharide, galactosamine. Galactosamine is similar to D-galactose except that the C2 hydroxyl group is replaced by an amine group (NH_2). Write the structure of galactosamine.

Solution

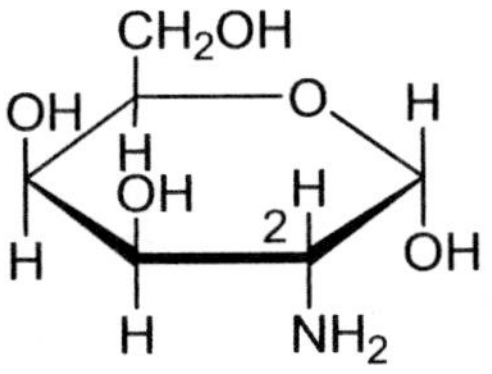

galactosamine

Try It Yourself #4

Glucose-6-phosphate, a phosphosugar, plays an important role in glycolysis. Glucose-6-phosphate is similar to D-glucose except that the C6 hydroxyl group has been replaced by a phosphate group (OPO_3^{2-}). Write the structure of glucose-6-phosphate.

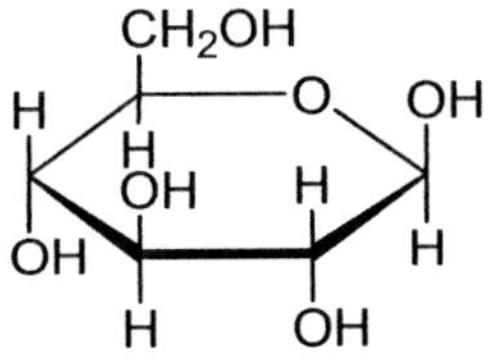

glucose

Structure of glucose-6-phosphate:

Practice Problems for Monosaccharides

1. The Fischer projections for three monosaccharides are shown below.

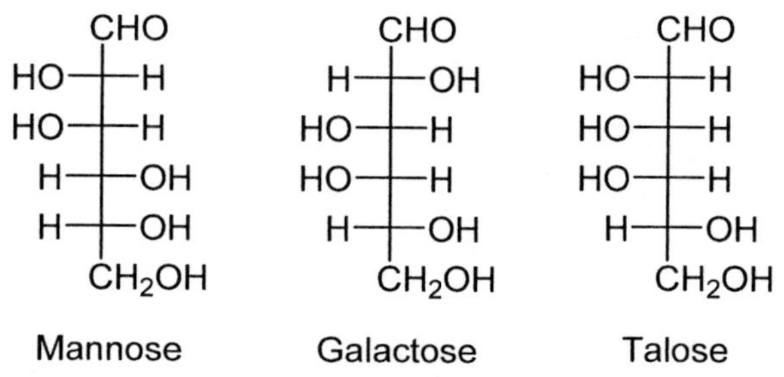

$$\begin{array}{ccc}
\text{CHO} & \text{CHO} & \text{CHO} \\
\text{HO}-\text{H} & \text{H}-\text{OH} & \text{HO}-\text{H} \\
\text{HO}-\text{H} & \text{HO}-\text{H} & \text{HO}-\text{H} \\
\text{H}-\text{OH} & \text{HO}-\text{H} & \text{HO}-\text{H} \\
\text{H}-\text{OH} & \text{H}-\text{OH} & \text{H}-\text{OH} \\
\text{CH}_2\text{OH} & \text{CH}_2\text{OH} & \text{CH}_2\text{OH} \\
\text{Mannose} & \text{Galactose} & \text{Talose}
\end{array}$$

a.   Are these sugars D-sugars or L-sugars? How can you tell?

b.   Write the structure for L-Mannose.

c.   Are L-mannose and D-mannose enantiomers or diastereomers?

d.   Are D-galatose and D-talose enantiomers or diastereomers?

2.   The Haworth projection of ribose is shown below.

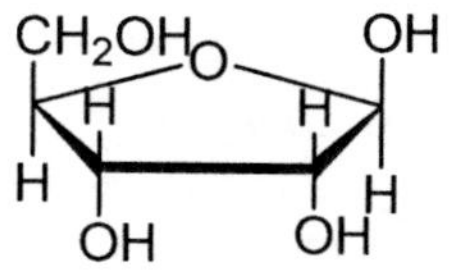

ribose

a.   Does the Haworth projection show this sugar to be a pyranose or a furanose?

b. Is the structure shown a D- or L-sugar? Explain.

c. Which anomer of ribose is shown?

3. Identify the aminosugar, the phosphosugar, and the glycoside shown below.

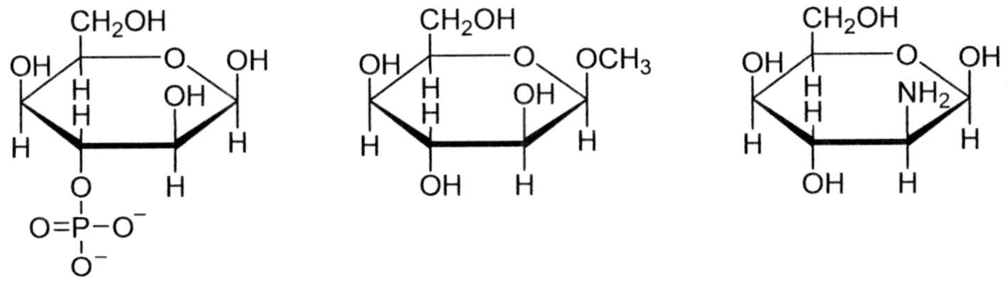

Section 12.3 Complex Carbohydrates

In a Nutshell: Disaccharides

Complex carbohydrates are composed of more than one monosaccharide linked together by a covalent bond. Disaccharides are carbohydrates that when hydrolyzed yield two monosaccharides. The two monosaccharides that make up a disaccharide may be identical or different. Three structural features identify the disaccharide: 1) the identity of the two monosaccharide components; 2) the carbon atoms that contain the hydroxyl groups that join the two monosaccharides in a glycosidic bond, which are identified by carbon number; and 3) the stereochemistry of the anomeric carbon atom(s), α- or β-, involved in the glycosidic bond.

The covalent bond that joins two monosaccharides is known as a glycosidic bond or glycosidic linkage. A glycosidic bond always forms between the anomeric carbon atom of one monosaccharide and the hydroxyl group on a carbon atom of another monosaccharide. The carbon number of the linked carbon atoms is one characteristic that is used to identify a disaccharide. A monosaccharide that is part of a glycosidic linkage does not undergo mutarotation. A catalyst or an enzyme is required to break a glycosidic linkage. The enzymes that facilitate the hydrolysis of a glycosidic linkage are known as glycosidases. Disaccharides are readily hydrolyzed into their monosaccharide components in the presence of water and a catalyst.

Worked Example #5

Mannobiose is a disaccharide composed of only mannose in an $\alpha 1 \rightarrow 4$ linkage.

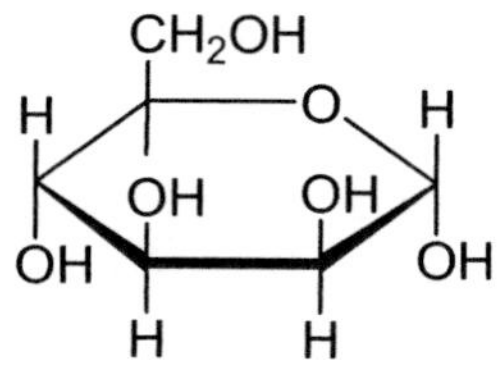

mannose

    a. Write the structure of mannobiose.

    b. Label the anomeric carbons in your structure.

    c. What monosaccharide(s) are produced upon hydrolysis of mannobiose?

a-b.

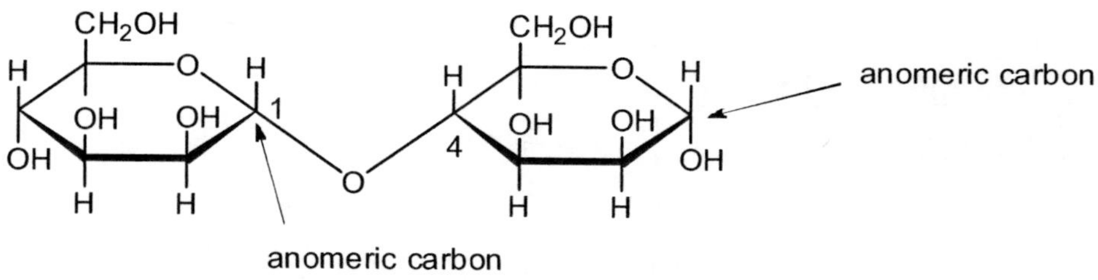

anomeric carbon

anomeric carbon

    c. The monosaccharide mannose is produced upon hydrolysis of mannobiose.

Try It Yourself #5

The structure of isomaltose is shown below.

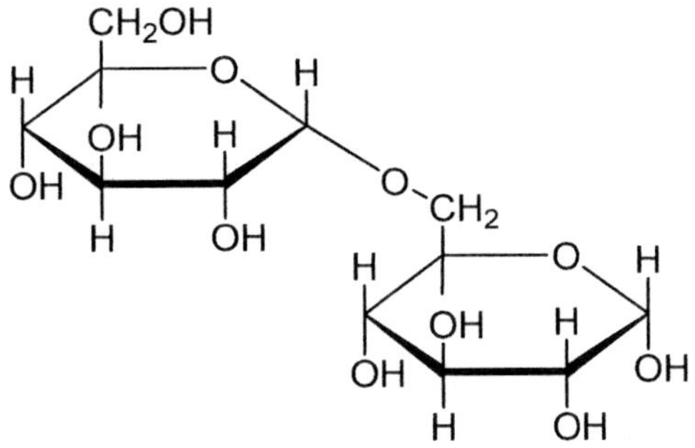

a.   What are the monosaccharide components of isomaltose?

b.   What type of linkage exists between the monosaccharides in isomaltose?

c.   Why is isomaltose classified as a disaccharide?

a.   *The monosaccharide components are:* _____

b.   *The linkage is:* _____

c.   *Isomaltose is a dissacharide because:* _____

In a Nutshell: Polysaccharides

Carbohydrates containing from 3 to 100 monosaccharides are known as oligosaccharides, and those containing more than a hundred are known as polysaccharides. When the same monosaccharide is used to build a polysaccharide, the polysaccharide is considered a polymer. A polymer is any large molecule with the same repeating structural component. An individual unit of the repeating component is known as a monomer. The important polysaccharides found in plants or animals are starch, cellulose, and glycogen.

Plants store energy in the form of starch, a mixture of 20% amylose and 80% amylopectin. Humans and animals store energy in the form of glycogen. Glycogen, amylose, and amylopectin are all polysaccharides composed of glucose monomers connected by $\alpha(1\rightarrow4)$ glycosidic bonds.

Amylose, the minor constituent of starch, is an unbranched polysaccharide; each glucose monomer is connected like a chain to the next monomer in the same type of linkage: $\alpha(1\rightarrow4)$. The overall molecule adopts helical shape, due to extensive hydrogen bonding between the hydroxyl groups.

Amylopectin, the major component of starch, is similar to amylase except that it is a branched polymer. The branches are created by $\alpha(1\rightarrow6)$ glycosidic bonds every 25–30 glucose units.

In humans and animals, glucose is stored as glycogen in liver and muscle cells. Glycogen is similar to amylopectin, except that it is even more highly branched; $\alpha(1\rightarrow6)$ glycosidic linkages are present every 8–12 glucose units.

Cellulose, the main component of wood, paper, and cotton, is another polysaccharide composed of glucose. The glycosidic bonds in cellulose are $\beta(1\rightarrow4)$ linkages. These distinctive linkages give cellulose a different overall shape. Cellulose has a flat sheet-like appearance, which provides structural rigidity for the plant.

In the laboratory, cellulose, starch, and glycogen can all be hydrolyzed into D-glucose in the presence of an acid. In the body, starch taken in through the diet is hydrolyzed in the mouth and small intestine by enzymes that specifically hydrolyze $\alpha(1\rightarrow4)$ linkages. These enzymes are known as *α-glycosidases*. Since humans do not have enzymes that hydrolyze $\beta(1\rightarrow4)$ linkages, cellulose cannot be digested. Some animals, however, can digest cellulose because bacteria that live in their digestive tracts produce *β-glycosidases*. These animals include horses, cows, giraffes, and other ruminants.

Worked Example #6

For each of the statements below, indicate the polysaccharides for which the statement is true. (More than one selection may be chosen.)

amylose   amylopectin   glycogen   cellulose

   a. _____ Provides structure in plants

b. _____ Is hydrolyzed by the enzyme *β-glycosidase*

c. _____ is helical in shape

d. _____ Is branched in appearance

e. _____ Has $\alpha(1\rightarrow4)$ linkages

a. *cellulose*

b. *cellulose*

c. *amylose*

d. *amylopectin and glycogen*

e. *amylose, amylopectin, and glycogen.*

Try It Yourself #6

For each of the statements below, indicate the polysaccharides for which the statement is true. (More than one selection may be chosen.)

amylose   amylopectin   glycogen   cellulose

a. _____ Serves as a glucose storage molecule in animals and humans

b. _____ Contains $\alpha(1\rightarrow6)$ linkages

c. _____ Has a layered sheet like appearance

d. _____ Can be digested by ruminants (horses, cows, and giraffes)

Solutions

a. _____

b. _____

c. _____

d. _____

Practice Problems for Complex Carbohydrates

1. Nigerose is a disaccharide that is a product of the hydrolysis of the polysaccharides in black mold. It contains two glucose monosaccharides joined in an $\alpha(1\rightarrow3)$ linkage. Write the structure of nigerose.

2. The structure of turanose is shown below.

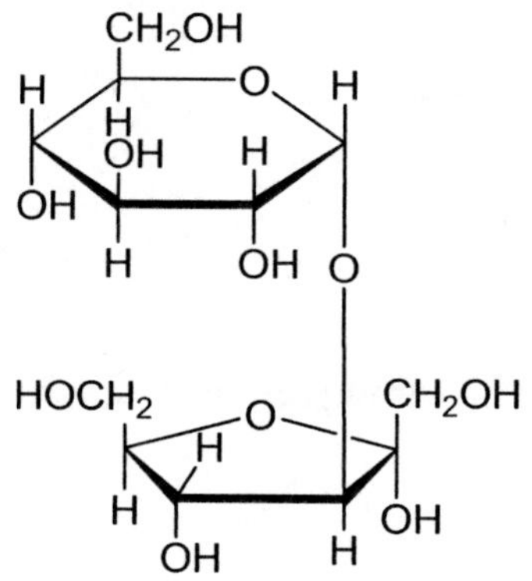

   a. Is turanose a disaccharide, oligosaccharide, or polysaccharide?

   b. What are the names of the two products when turanose is hydrolyzed?

   c. What type of linkage connects the monosaccharides in turanose?

   d. Label the anomeric carbons.

   e. Does turanose undergo mutarotation?

3. Why does cellulose have a layered, sheet-like appearance? Why does amylose have a helical shape?

Section 12.4 Carbohydrate Catabolism

The catabolism of carbohydrates—the extraction of energy—can be divided into three stages. In the first stage, polysaccharides taken in through diet are hydrolyzed into monosaccharides, primarily glucose. In the second stage, glucose is converted into pyruvate through a 10-step process called glycolysis, which produces ATP. If oxygen is available, then pyruvate is converted into acetyl CoA. In the third stage, acetyl CoA is converted into carbon dioxide in the citric acid cycle. During the citric acid cycle, electrons are extracted from acetyl CoA and carried into the electron-transport chain by NADH and $FADH_2$. In the electron-transport chain, these electrons are used to drive oxidative phosphorylation of ADP to ATP.

In a Nutshell: Stage 1: Polysaccharides Are Hydrolyzed into Monosaccharides

Digestion of carbohydrates begins in the mouth. The enzyme *amylase*, an α-*glycosidase* present in the saliva, catalyzes the hydrolysis of amylose and amylopectin into glucose, maltose, and dextrins (oligosaccharides composed of 3–13 glucose monomers in α(1→4) linkages. In the small intestine, maltose and dextrins are further hydrolyzed into glucose.

Hydrolysis of complex sugars occurs at the glycosidic bonds to produce monosaccharides. In the body, this reaction requires an α-*glycosidase,* an enzyme specific for α linkages. *Maltase* hydrolyzes maltose into two D-glucose monosaccharides; *lactase* hydrolyzes lactose into D-glucose and D-galactose; and *sucrase* hydrolyzes sucrose into D-glucose and D-fructose. When water breaks the glycosidic bond, the OH from water becomes the OH on the anomeric carbon on one monomer, and the H becomes the hydrogen atom for the alcohol functional group on the other monomer.

Monosaccharides diffuse through the small intestine and make their way into the blood stream. Glucose is distributed via the blood as fuel to cells. Fructose and galactose are carried to the liver, where they are converted into glucose and then reenter the blood. Once glucose enters a cell, the second stage of catabolism begins.

Worked Example #7

Isomaltulose, shown below has been studied as a possible sugar substitute.

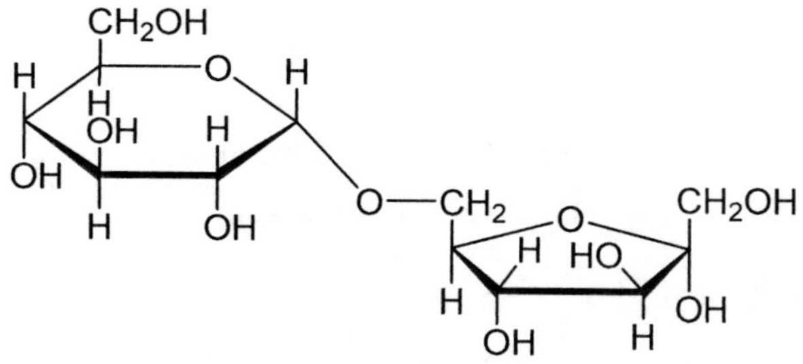

a. Is isomaltulose a mono-, di-, oligo-, or polysaccharide? Explain.
b. Label all the anomeric carbons in isomaltulose.
c. Label and identify the glycosidic bonds in isomaltulose.
d. What monosaccharides are produced from the complete hydrolysis of isomaltulose into its individual monosaccharide components?
e. What type of enzyme catalyzes the hydrolysis of isomaltulose?

a. *Isomaltulose is a disaccharide. There are two monomer components: glucose and fructose.*

b-c.

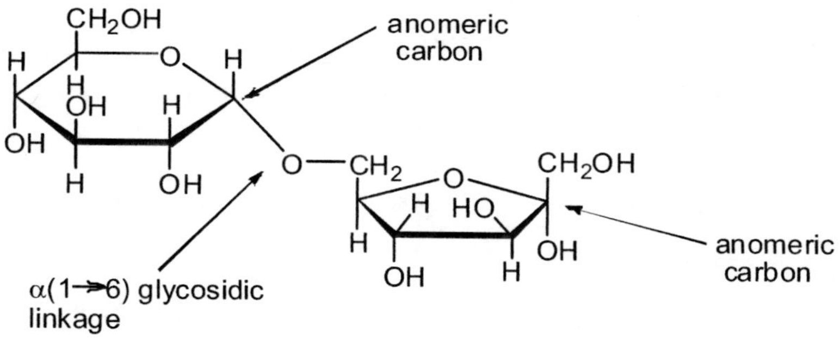

d. *D-glucose and D-fructose are produced from the hydrolysis of isomaltulose.*
e. *An α-glycosidase catalyzes the hydrolysis of isomaltulose.*

Try It Yourself #7

Maltotriose, a product of the hydrolysis of amylose and amylopectin is shown below.

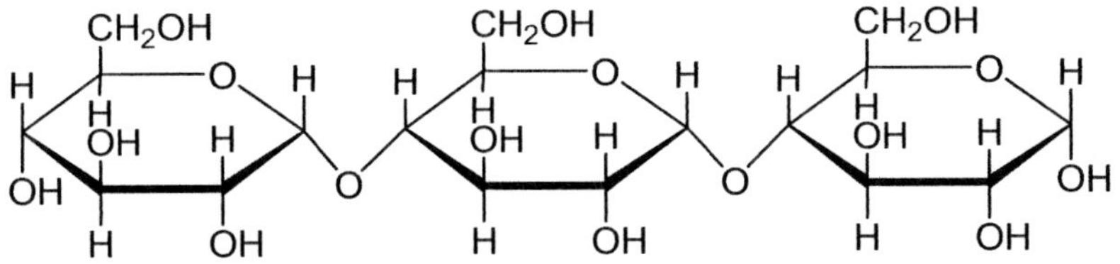

a. Is maltotriose a mono-, di-, oligo-, or poly saccharide? Explain.

b. Label all the anomeric carbons in isomaltulose.

c. Label and identify the glycosidic bonds in isomaltulose.

d. What monosaccharides are produced from the complete hydrolysis of isomaltulose into its individual monosaccharide components?

e. What type of enzyme catalyzes the hydrolysis of isomaltulose?

a. *Maltotriose is a:* _____.

b-c.   *The anomeric carbons and glycosidic bonds:*

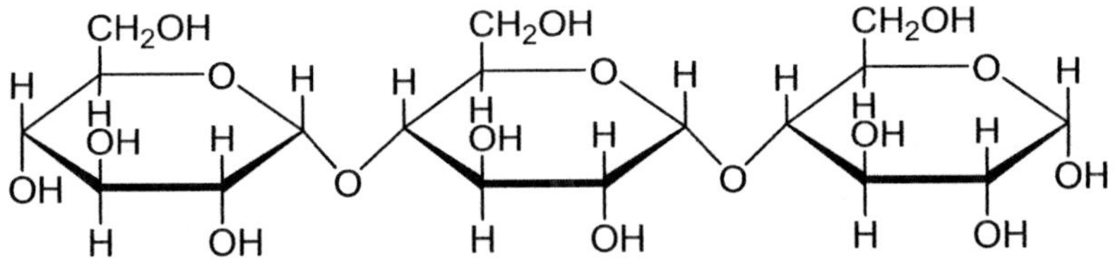

d. *The monosaccharides are:* _____

e. _____ *catalyzes the hydrolysis of isomaltulose.*

In a Nutshell: Stage 2: Glycolysis

After passing through the cell membrane of a cell, glucose undergoes further degradation in the cytoplasm. Glycolysis converts glucose into two molecules of pyruvate. The purpose of glycolysis is to harvest energy from glucose. Glycolysis is an anaerobic pathway; it does not require oxygen.

Glycolysis is a 10-step biochemical sequence. Each step requires an enzyme. The organic compounds in the sequence of reactions that defines a particular biochemical pathway are called intermediates. Every intermediate in glycolysis contains one or more phosphate groups. The first half of glycolysis produces two molecules of glyceraldehyde-3-phosphate. The second half of glycolysis converts these two glyceraldehyde-3-phosphate molecules into two pyruvate molecules.

The first half of glycolysis requires an input of energy: two ATP molecules are required for every one glucose molecule that enters the pathway. The last half of glycolysis (steps 6–10) has an energy output of four ATP molecules: two ATP for each glyceraldehyde-3-phosphate that is converted into pyruvate. The second half of glycolysis also produces 2 NADH molecules. The net result of glycolysis is an output of energy in the form of two ATP molecules and 2 NADH molecules.

Worked Example #8
Which steps in glycolysis are isomerization reactions?

Solution
Steps 2 and 5 are isomerization reactions. In step 2, a pyranose ring is isomerized to a furanose ring. In step 5, dihydroxyacetone phosphate is isomerized into glyceraldehyde-3-phosphate.

Try It Yourself #8
Which steps of glycolysis require energy? Which steps of glycolysis produce energy?

Solution:

In a Nutshell: The Fate of Pyruvate
In human cells, pyruvate can follow one of two pathways, depending on whether oxygen is present or not. If oxygen is present (aerobic conditions), pyruvate is oxidized into acetyl

CoA, the entry point into catabolism. If oxygen is not present (anaerobic conditions), pyruvate is reduced to lactic acid.

In a Nutshell: Oxidation of Pyruvate to Acetyl CoA

The reaction that converts pyruvate into acetyl CoA requires coenzyme A, and NAD^+ is the electron acceptor. Pyruvate is oxidized with the simultaneous loss of CO_2, producing a thioester with coenzyme A. Oxygen is necessary for the regeneration of NAD^+ from NADH in the electron-transport chain.

In a Nutshell: Reduction of Pyruvate to Lactic Acid

Under anaerobic conditions—when oxygen is not present, glucose is converted to pyruvate and then reduced to L-lactate in a process known as lactic acid fermentation. Reduction of pyruvate to lactate ensures that NAD^+ is regenerated so that glycolysis can continue.

Worked Example #9

Does the formation of acetyl CoA from pyruvate or the formation of lactic acid from pyruvate produce NAD^+? Is pyruvate oxidized or reduced in the reaction that produces NAD^+?

Solution

The formation of lactic acid from pyruvate produces NAD^+. Pyruvate is reduced in that reaction.

Try It Yourself #9

What structural changes occur when pyruvate is converted to acetyl CoA?

Solution:

In a Nutshell: Other Metabolic Roles of Glucose

Other biochemical pathways that begin or end with glucose include: glycogenesis, the formation of glycogen from glucose; glycogenolysis, the degradation of glycogen to form glucose; gluconeogenesis, the formation of glucose from pyruvate, and the pentose phosphate pathway, the formation of pentose phosphate from glucose.

Glucose plays a central role in metabolism. In the pentose phosphate pathway, glucose serves as the starting material for the synthesis of ribose, an important sugar that provides the backbone for RNA. The synthesis of the polysaccharide, glycogen, begins with glucose in a process known as glycogenesis, which requires glucose and insulin. The reverse process, the degradation of glycogen into glucose, follows a pathway known as glycogenolysis. When all glycogen stores are depleted and no glucose is available, the body can synthesize glucose in a biochemical pathway known as gluconeogenesis. The body uses gluconeogenesis to synthesize glucose from proteins, a process that requires breakdown of muscle tissue.

Worked Example #10

Which of the following processes are anabolic: gluconeogenesis, glycogenesis, glycogenolysis, or glycolysis?

Solution

An anabolic process synthesizes larger molecules from smaller ones. Therefore, gluconeogenesis and glycogenesis are anabolic pathways.

Try It Yourself #10

Which two processes produce glucose as a product: gluconeogenesis, glycogenesis, glycogenolysis, or glycolysis?

Solution:

Practice Problems for Carbohydrate Catabolism

1. Gentiobiose, shown below, is incorporated into the chemical compound that gives the spice saffron its color.

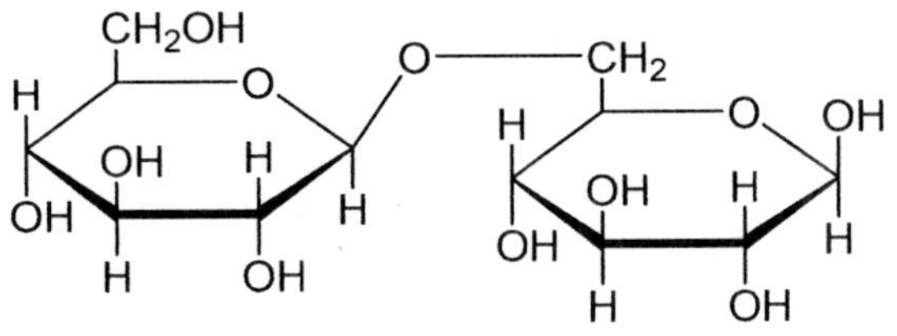

   a. Is gentiobiose a mono-, di-, oligo-, or polysaccharide? Explain.

   b. Label all the anomeric carbons in gentiobiose.

   c. Label and identify the glycosidic bonds in gentiobiose.

   d. What monosaccharides are produced from the complete hydrolysis of gentiobiose into its individual monosaccharide components?

   e. What type of enzyme catalyzes the hydrolysis of gentiobiose?

2. What molecules are the end products of glycolysis?

3. What happens to pyruvate under aerobic conditions?

4. Which of the following processes use glucose to produce other compounds: gluconeogenesis, glycogenesis, glycogenolysis, or glycolysis?

Section 12.5 Oligosaccharides as Cell Markers

Carbohydrates serve an important function as cell markers, which are chemical tags that identify a cell. Cell markers allow a cell to distinguish host from foreigner. Cell markers are typically oligosaccharides covalently bonded to the cell membrane or protein within the cell membrane.

Red blood cells contain oligosaccharide markers that define the blood groups A, B, AB, or O. The markers of all the blood groups contain the same core trisaccharide portion, but differ in whether they have a fourth monosaccharide and what that monosaccharide is. The monosaccharide components of the blood type oligosaccharides include: D-galactose, *N*-acetyl-D-glucosamine, *N*-acetyl-D-galactosamine, and L-fucose.

A person's blood type is determined by the oligosaccharide markers on their red blood cells. The core trisaccharide for all blood types consists of three monosaccharides: *N*-acetyl-D-glucosamine, D-galactose, and L-fucose. Type O blood has no additional monosaccharides attached to the core trisaccharide. Type A has the additional monosaccharide, *N*-acetyl-D-galactosamine covalently bonded to the D-galactose unit of the core trisaccharide. Type B blood has an additional D-galactose covalently bonded to the D-galactose unit of the core trisaccharide. Type AB blood has both type A and type B markers.

Type O blood is the universal donor. People with type O blood can donate blood to any recipient, but they can only receive blood from type O donors. People with type O blood have the trisaccharide component of all four blood types, so their blood is recognized as part of the recipient's blood. However, when people with type O blood receive type A, B, or AB blood, their body rejects the donated blood because the fourth monosaccharide is unfamiliar. Type AB blood is called the universal recipient.

Worked Example #11

Can people with type A blood accept a blood donation from someone with type AB blood? Can people with type AB blood receive a blood donation from someone with type A blood? Explain.

Solution

People with type A blood cannot accept blood from someone with type AB blood. The bodies of people with type A blood would not recognize the type B oligosaccharide that is present in type AB blood. People with AB blood may receive a blood donation from someone with type A blood because the type A oligosaccharide is one of the oligosaccharides that makes up type AB blood. The body of the person with type AB blood would recognize the type A oligosaccharide.

Try It Yourself #11

How is type O blood different than type A blood? Can people with type O blood accept a blood donation from someone with type A blood? Explain.

Solution:

Practice Problems for Oligosaccharides as Cell Markers

1. What are the three monosaccharides that are common to all four blood types?

2. Can someone with type B blood donate blood to someone with type O blood? Explain.

3. Can someone with type O blood donate blood to someone with type AB blood? Explain.

Chapter 12 Quiz

1. The Fischer projection for allose is shown below.

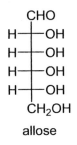

   CHO
   H——OH
   H——OH
   H——OH
   H——OH
   CH₂OH

   allose

   a. Is this sugar a D- or L-sugar? How can you tell?

   b. Draw the enantiomer of this sugar.

   c. What is the name of the enantiomer of this sugar?

2. The Fischer projections for three sugars are shown below:

CHO	CHO	CHO
HO——H	HO——H	H——OH
HO——H	HO——H	H——OH
H——OH	H——OH	HO——H
HO——H	H——OH	H——OH
CH₂OH	CH₂OH	CH₂OH
L-gulose	D-mannose	D-gulose

   a. Which sugars are enantiomers?

   b. Which sugars are diastereomers?

3. The Haworth projection for a sugar is shown below:

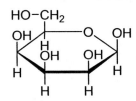

   a. Is the sugar a D- or L- sugar? How can you tell?

   b. Label the anomeric carbon.

   c. Is the α-anomer or β-anomer shown?

   d. Is the sugar shown a furanose or pyranose?

4. What kind of modified monosaccharide is shown below?

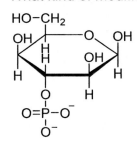

5. For each of the statements below, indicate the polysaccharides for which the statement
   is true. (More than one selection may be chosen.)

   amylose   amylopectin   glycogen   cellulose

   a. _____ Is hydrolyzed by the enzyme β-glycosidase

b. _____ Is hydrolyzed by an *α-glycosidase* enzyme

6. Gentianose, shown below, is found in the plant gentian.

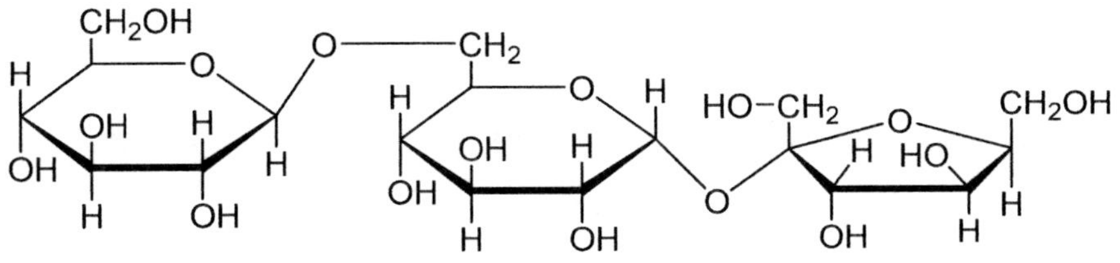

a. Is gentianose a mono-, di-, oligo-, or polysaccharide? Explain.

b. Label all the anomeric carbons in gentianose.

c. Label and identify the glycosidic bonds in gentianose.

d. What monosaccharides are produced from the complete hydrolysis of gentianose into its individual monosaccharide components?

e. What type of enzyme catalyzes the hydrolysis of gentianose?

f. Does gentianose undergo mutarotation?

7. a. Why do all the intermediates in glycolysis contain phosphate groups?

    b. How many ATP molecules and NADH molecules are produced in glycolysis?

8. a. Does the reaction that forms lactic acid from pyruvate use or produce NAD^+?

    b. Does the reaction that forms acetyl CoA from pyruvate use or produce NAD^+?

9. When the glycogen stores are depleted and no glucose is available, how does the body produce glucose?

10. Can someone with type AB blood donate blood to someone with type A blood? Explain.

Chapter 12

Answers to Additional Exercises

12.35 Starch provides energy for the plant; cellulose provides structure for the plant.

12.37 a. The Fischer projections shown are D-sugars. The OH group that is farthest away from the CHO group is pointing to the right.

   b. L-arabinose

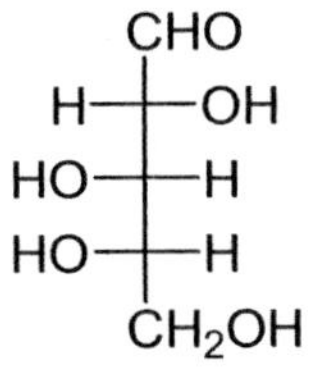

   c. D-arabinose and D-xylose are diastereomers of each other. The OH group and H groups are reversed on the second and third crosshairs (C-2 and C-3).

12.39 In the α anomer, the OH group on the anomeric carbon is pointing down. In the β anomer, the OH group on the anomeric carbon is pointing up.

12.41 a. furanose, five-membered ring, β anomer, OH group is pointing up

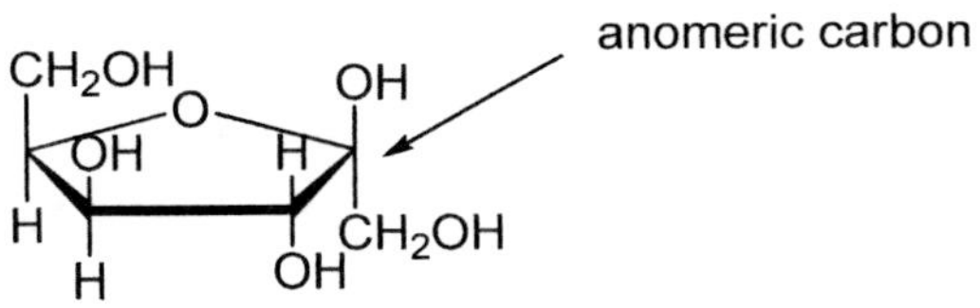

   b. pyranose, six-membered ring, β anomer, OH group is pointing up

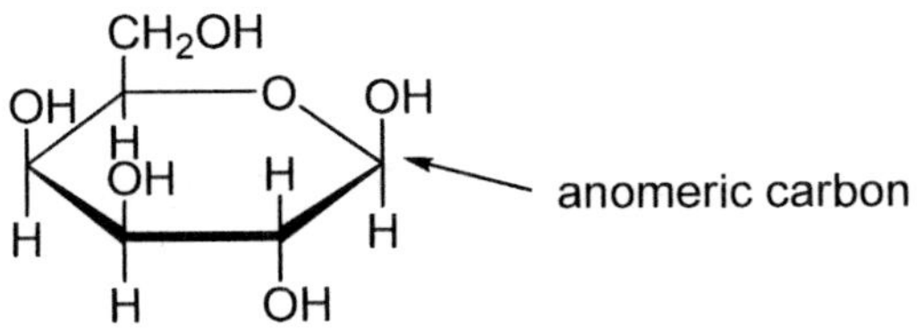

12-43 a.

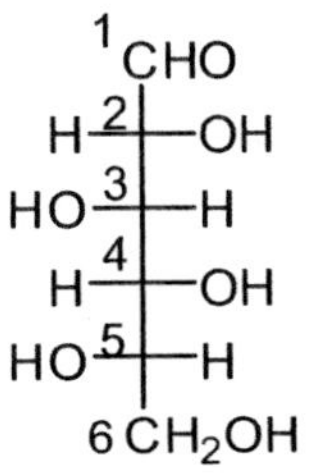

   L-idose

   b. The OH group on carbon 5 determines that it is an L-sugar.

c. D-idose and L-idose are enantiomers, non-superposable mirror images.

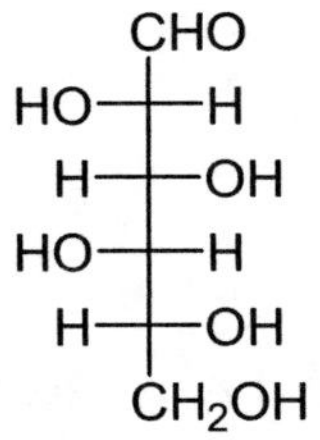

$$
\begin{array}{c}
CHO \\
HO-\!\!\!-\!\!\!-H \\
H-\!\!\!-\!\!\!-OH \\
HO-\!\!\!-\!\!\!-H \\
H-\!\!\!-\!\!\!-OH \\
CH_2OH
\end{array}
$$

D-idose

d. D-idose and D-glucose are diastereomers.

12.45 a-b.

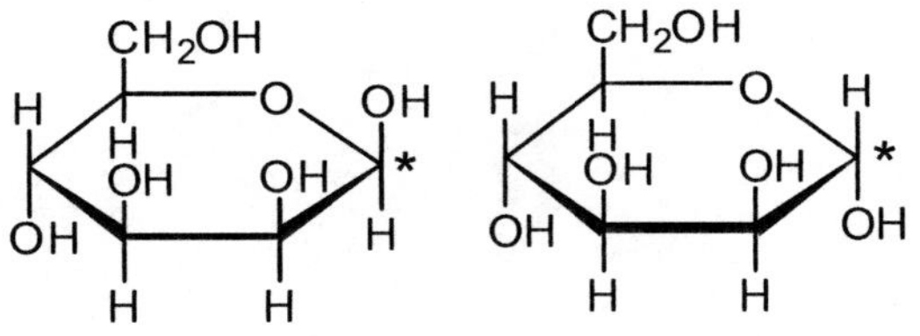

β anomer   α anomer

c. These two structures are diastereomers.

d. These two sugars interconvert through a process called mutarotation. The ring of one anomer opens to give the open chain form and then recloses to give the other anomer.

12.47

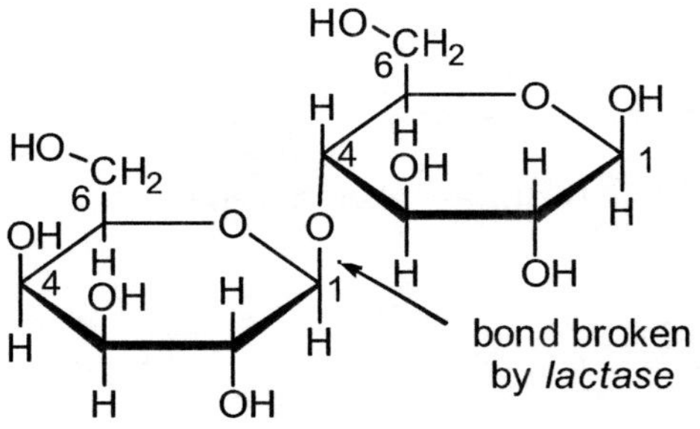

bond broken
by *lactase*

12.49

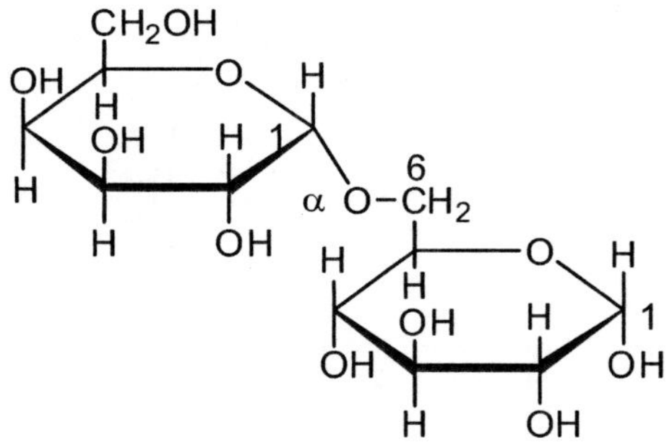

isomaltose

12.51 Giraffes have bacteria in their digestive tracts that produce β-glycosidases. These β-glycosidases can hydrolyze the β(1→4) glycosidic bonds in cellulose and produce glucose.

12.53 In D-glucosamine, the OH group on C2 has been replaced by an amine group.

12.55 a. Maltulose is a disaccharide, since two monosaccharides make up maltulose.

b-c.

glycosidic bond

d.

The hydrolysis of maltulose produces a glucose monomer and a fructose monomer.

12.57 Glycolysis is a process that converts *glucose* into *pyruvate*.

12.59 Every intermediate in glycolysis contains a *phosphate* functional group. The 2–charge on the phosphate group prevents the intermediates from diffusing out of the cell.

12.61 The primary role of carbohydrates in the body is to produce energy.

12.63 Step 4 converts fructose-1,6-bisphosphate, a molecule containing six carbon atoms into glyceraldehyde-3-phosphate and dihydroxyacetone, two molecules that contain three carbon atoms each..

12.65 In step 2, a pyranose ring, glucose-6-phosphate, is isomerized to a furanose ring, fructose-6-phosphate.

12.67 People with AB blood are called universal recipients because they can receive any type of blood: A, B, AB, or O. Type A and type B blood components are found in type AB blood, so it is recognized as part of the type AB recipient's blood. Type O blood has the trisaccharide component of type AB blood, so it is also recognized by the recipient.

12.69 a. He or she is hypoglycemic.

b. Glycogenolysis, the conversion of glycogen into glucose, is activated when glucose levels fall.

c. Glucose levels typically fall in between meals and during exercise.

d. Glucagon is secreted when glucose levels fall.

12.71 Normal fasting glucose levels are 70–110 mg/dL.

12.73 Someone with Type I diabetes must manage his or her diet carefully, controlling the amount of carbohydrates consumed, so that glucose levels do not rise significantly.

12.75 The hormone glucagon has a complementary role to insulin.

Chapter 13

Lipids: Structure and Function

Chapter Summary

In this chapter you learned about lipids as a class of biological molecules that are insoluble in water but soluble in nonpolar organic solvents. You learned that the function of a lipid depends on its structure. Lipids make up the cell membrane and serve as important hormones. The most abundant lipids are the triglycerides that serve as a supply of long-term energy for the cells. You examined the biochemical pathways involved in the first two stages of triglyceride catabolism that produce energy for the cells.

Section 13.1 Energy Storage Lipids: Triglycerides

Lipids include all the biological molecules that are insoluble in water; they are soluble in nonpolar organic solvents. Lipids serve a variety of functions, which depend on their structure. The different types of lipids include triglycerides; phospholipids and glycolipid; steroids; and eicosanoids.

The most abundant lipids in the body are triacylglycerols, also known as triglycerides, or as "fats." Triglycerides are obtained through diet. They are burned in muscle cells or stored in fat cells, known as adipocytes. Triglycerides serve as a reservoir for the body's long-term energy needs.

Triglycerides are derived from fatty acids. A fatty acid is a long, unbranched hydrocarbon chain with a carboxylic acid at one end. Fatty acids contain an even number of anywhere from 12 to 24 carbon atoms.

In a Nutshell: Saturated and Unsaturated Fatty Acids

A saturated fatty acid contains no double bonds, whereas an unsaturated fatty acid contains one or more double bonds. Fatty acids with one double bond are referred to as monounsaturated fatty acids. Polyunsaturated fatty acids contain more than one double bond. Naturally occurring unsaturated fatty acids have a cis double bond. The shape of a cis double bond gives the fatty acid a bend or a kink in its otherwise linear shape. Fatty acids with trans double bonds have a shape similar to saturated fatty acids.

The unsaturated fatty acids linoleic and linolenic acid are essential fatty acids, which must be obtained from the diet. The body cannot synthesize these two fatty acids from other metabolic intermediates.

Two different numbering systems are used to indicate the location of the double bond in a fatty acid: the omega system and the delta system. In the delta system, numbering begins with the carbonyl carbon (C1), and the location of each double bond is indicated by a number written as a superscript following the Greek letter delta. In the omega system, numbering begins at the end of the hydrocarbon chain, farthest away from the carbonyl group, and the location of only the first double bond is given. That location is indicated by a number following the Greek letter omega.

In a Nutshell: Melting Points of Fatty Acids

The melting point of a fatty acid increases as the number of carbon atoms in its formula increases. Since dispersion forces increase with surface area, this trend is expected. More energy is needed to separate the longer fatty acids chains when changing from the solid to the liquid state. The melting point drops significantly with the addition of a cis double bond. The bend created by the cis double bond(s) reduces the number of points of contact between the fatty acid molecules, leading to fewer dispersion forces and, therefore, a lower melting point. The unsaturated (cis) fatty acids are less tightly packed than saturated fatty acids.

Worked Example #1

The structure of myristoleic acid is shown below.

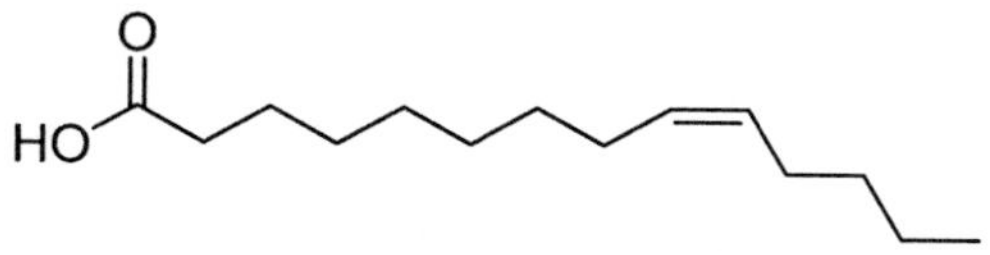

a. What type of fatty acid is myristoleic acid? Answer using the delta naming system and the omega naming system.

b. Myristic acid is a related fatty acid that also has 14 carbon atoms, but myristic acid is a saturated fatty acid. Which fatty acid, myristoleic acid or myristic acid, has the higher melting point? Explain.

c. Is myristoleic acid soluble in water?

a. *Myristoleic acid is a Δ^9 fatty acid. It is also an ω^5 fatty acid.*

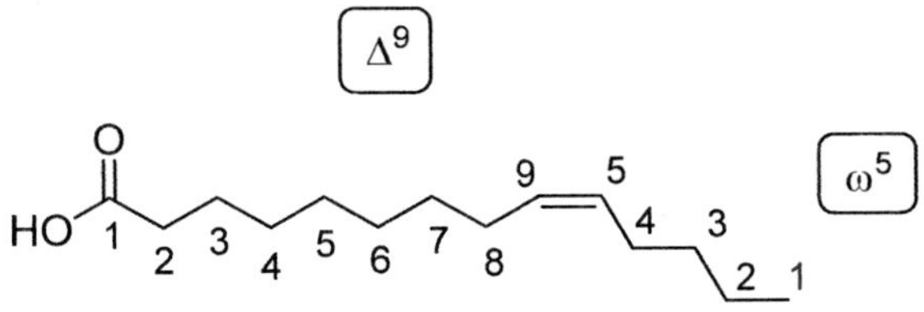

b. *Myristic acid has the higher boiling point. It is a saturated fatty acid; therefore, there are more dispersion forces present.*

c. *Myristoleic acid is not soluble in water. It is a nonpolar molecule.*

Try It Yourself #1

Docosahexaenoic acid, commonly known as DHA, is found in fish oil. Its structure is shown below.

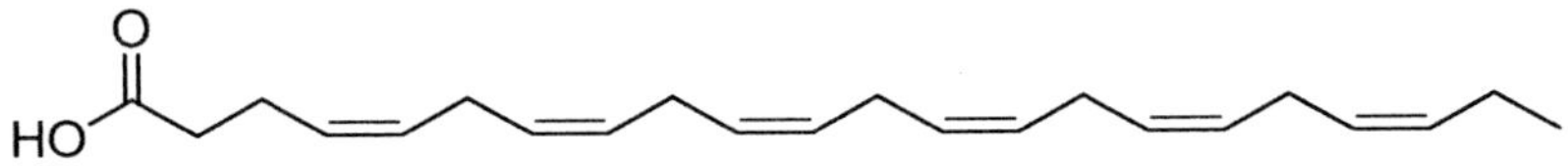

a. What type of fatty acid is DHA? Answer using the delta naming system and the omega naming system.

b. Is DHA a trans fatty acid?

c. Do you expect DHA to have a higher or lower melting point than a saturated fatty acid with 22 carbon atoms?

Solutions

a. *The delta naming system starts with the carbonyl carbon. The omega naming system starts with the end of the hydrocarbon chain farthest away from the carbonyl group.*

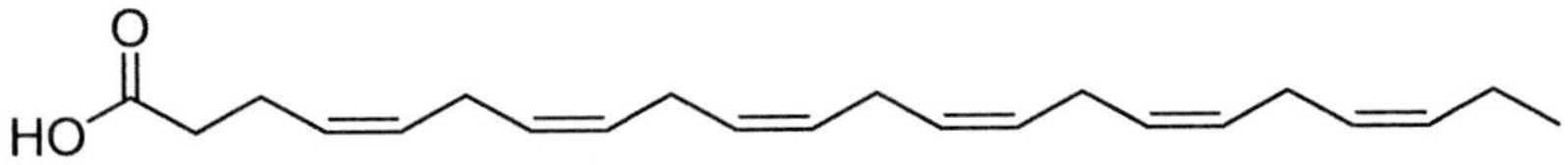

DHA is a _____ fatty acid. It is also an _____ fatty acid.

    b. *The double bonds in DHA are _____ double bonds; therefore, DHA is a _____ fatty acid.*

    c. *DHA has more or fewer dispersion forces than a saturated fatty acid with 22 carbon atoms. The melting point of DHA should be _____ than the saturated fatty acid.*

In a Nutshell: Triglycerides

Triacylglycerols, also known as triglycerides, are assembled from one glycerol molecule and three fatty acid molecules. Glycerol is a three-carbon-atom chain with a hydroxyl group on each carbon atom. The fatty acids are joined to the glycerol backbone at the hydroxyl group by three ester linkages. Triglycerides are nonpolar molecules. Simple triglycerides contain only one kind of fatty acid. Mixed triglycerides contain two or three different types of fatty acids.

In a Nutshell: Physical Properties of Triglycerides

Triglycerides derived from saturated fatty acids tend to be solids at room temperature and are commonly referred to as fats. These triglycerides tend to come from animal sources, though there are a few plant sources that contain saturated fatty acids. Triglycerides derived from unsaturated fatty acids, on the other hand, tend to be liquids at room temperature and are commonly referred to as oils. They are usually found in plants.

Saturated fats are more tightly packed than unsaturated fats because the bend created by the cis double bond(s) limits the number of points of contact in an unsaturated fat. Thus, unsaturated triglycerides have lower melting points, which explains why they are oils are room temperature.

Trans fats are not generally found naturally; they are produced artificially by the catalytic hydrogenation of unsaturated cis triglycerides. While some double bonds are reduced to single bonds in the reaction, others are isomerized from the cis to the trans geometric isomer. Diets high in saturated fats and trans fats have been linked to cardiovascular disease.

Worked Example #2

Write the structure of the triglyceride composed of two molecules of oleic acid and one molecule of linoleic acid.

    a. Would this triglyceride be considered a saturated or unsaturated fat?

    b. Would this triglyceride be considered a simple or a mixed triglyceride?

Solutions

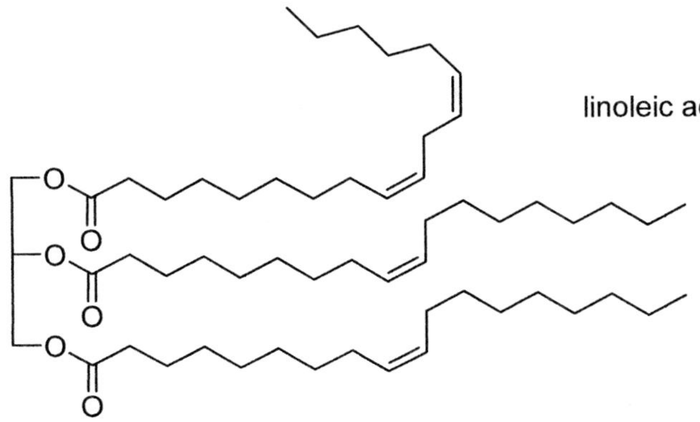

linoleic acid

oleic acid

oleic acid

    a. *This triglyceride would be considered an unsaturated fat. The fatty acids are unsaturated fatty acids.*

    b. *This triglyceride would be considered a mixed triglyceride because it is derived from two different fatty acids.*

Try It Yourself #2

Write the structure of a triglyceride composed of myristic acid, stearic acid, and palmitic acid.

    a. Would this triglyceride be considered a saturated or an unsaturated fat?

    b. Would this triglyceride be considered a simple or a mixed triglyceride?

    c. Would this triglyceride be considered a fat or an oil?

Structure of triglyceride:

a. Myristic acid, stearic acid, and palmitic acid are _____ fatty acids; therefore, this triglyceride is a _____ fat.

b. It is a _____ triglyceride.

c. It would be considered a _____ .

Practice Problems for Energy Storage Lipids: Triglycerides

1. The structure of linolenic acid is shown below.

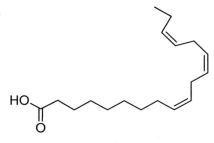

a. What type of fatty acid is linolenic acid? Answer using the delta naming system and the omega naming system.

b. Is linolenic acid soluble in nonpolar solvents?

c. Do you expect stearic acid (a saturated fatty acid with 18 carbon atoms) to have a higher or lower melting point than linolenic acid?

d. Is linolenic acid an essential fatty acid?

2. Which is more likely to be a liquid at room temperature: oleic acid or palmitic acid?

3. Write the structure of a triglyceride composed of stearic acid, oleic acid, and palmitic acid.

a. Circle the ester functional groups in the structure.

b. is this a simple or mixed triglyceride?

c. Are the double bonds cis or trans?

Section 13.2 Membrane Lipids: Phospholipids and Glycolipids

The cell membrane is part of all living cells. This semipermeable barrier separates two environments: the inside and the outside of the cell. The cell membrane controls the flow of ions and molecules in and out of the cell. The underlying components of cell membranes are lipids.

In a Nutshell: Phospholipids and Glycolipids

The two types of lipids found in human cell membranes are phospholipids and glycolipids. Phospholipids are the most common type of membrane lipids. They contain a phosphate group, whereas glycolipids contain a carbohydrate.

The carbon backbone of membrane lipids is derived from either glycerol or sphingosine. The sphingosine structure resembles glycerol except that it contains a long hydrocarbon chain as part of the backbone and an amine functional group in place of the middle hydroxyl group in glycerol. A phospholipid with a sphingosine backbone is called a sphingomyelin, while a phospholipid with a glycerol backbone is called a glycerophospholipid. Glycolipids found in human cells all contain a sphingosine backbone; therefore, they are called sphingolipids.

When the backbone of a membrane lipid is glycerol, two long-chain fatty acids are linked to the first two alcohol functional groups by ester linkages. When the backbone is sphingosine, one fatty acid is attached by an amide linkage. Therefore, regardless of the backbone, membrane lipids have two long hydrocarbon chains as a key part of their chemical composition. Since the hydrocarbon chains are nonpolar, this part of a membrane lipid is often referred to as the nonpolar "tails."

The remaining alcohol group on each backbone is attached to either a carbohydrate, in the case of a glycolipid, or a phosphate group, in the case of a phospholipid.

In a Nutshell: Phospholipids

Phospholipids contain a phosphate group at the terminal alcohol of either sphingosine or glycerol. The phosphate group is also attached to an amino alcohol. The three amino alcohols most commonly attached to phospholipids are ethanolamine, choline, and in glycerophospholipids, serine. The amines are positively charged. The phosphate group is linked to the terminal hydroxyl group on glycerol or sphingosine and the hydroxyl group of the amino alcohol, thereby creating a diphosphate ester.

In a Nutshell: Glycolipids

In human cells, the backbone of a glycolipid is always sphingosine. The hydroxyl group farthest from the hydrocarbon chain is linked to a carbohydrate. The carbohydrate component of a glycolipid is a mono-, di-, or oligosaccharide. Glycolipids containing a monosaccharide are known as cerebrosides. Glycosides containing an oligosaccharide are known as gangliosides.

In a Nutshell: Membrane Lipids Are Amphipathic Molecules

Membrane lipids differ from triglycerides in that they are amphipathic molecules: polar on one end and nonpolar on the other. Membrane lipids are amphipathic because they have two nonpolar hydrocarbon chains and a polar group. In a glycolipid, the polar group is the carbohydrate. In a phospholipid, the polar group contains the negatively charged phosphate group together with the positively charged amine. The polar end of the membrane lipid is commonly referred to as the "polar head," and the two nonpolar hydrocarbon chains are referred to as the "nonpolar tails." Since membrane lipids contain a polar head and two nonpolar tails, they are often depicted as a sphere, the polar head, attached to two wavy lines, the nonpolar tails.

Worked Example #3

For the membrane lipid below, answer the following questions:

a. Is it a phospholipid or a glycolipid? How can you tell?

b. Is the backbone derived from sphingosine or glycerol? Circle the backbone.

c. What is the fatty acid from which the lipid is derived? Is it saturated or unsaturated? What functional group connects it to the backbone?

d. Is it a sphingomyelin, cerebroside, or ganglioside?

e. If it is a phospholipid, what is the amino alcohol: serine, choline, or ethanolamine?

f. Circle the part of the molecule that defines the "polar head."

g. Circle the part of the molecule that defines the two tails.

Solutions

a. *It is a phospholipid because it contains a phosphate group and not a carbohydrate.*

b. *It is a sphingosine.*

Sphingosine backbone

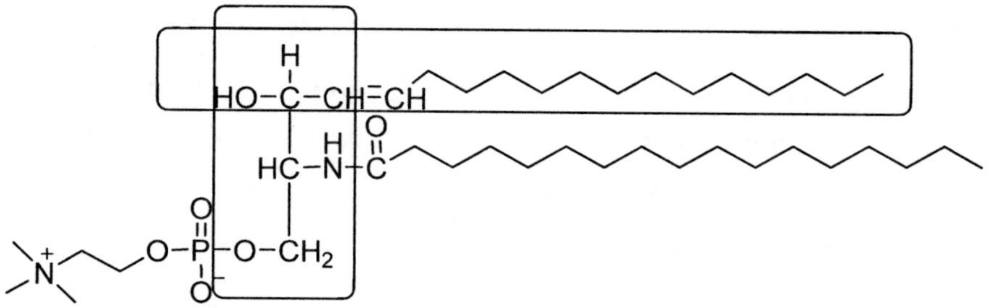

c. The fatty acid is stearic acid—an 18-carbon saturated fatty acid. It is joined to the amine in sphingosine by an amide linkage.

d. It is a sphingomyelin. It is a phospholipid with a sphingosine backbone.

e. The amino alcohol is choline.

f-g.

Two tails

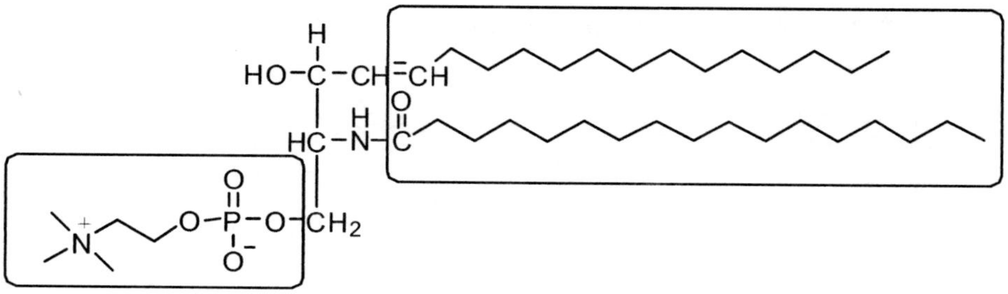

Polar head

Try It Yourself #3

For the membrane lipid below, answer the following questions:

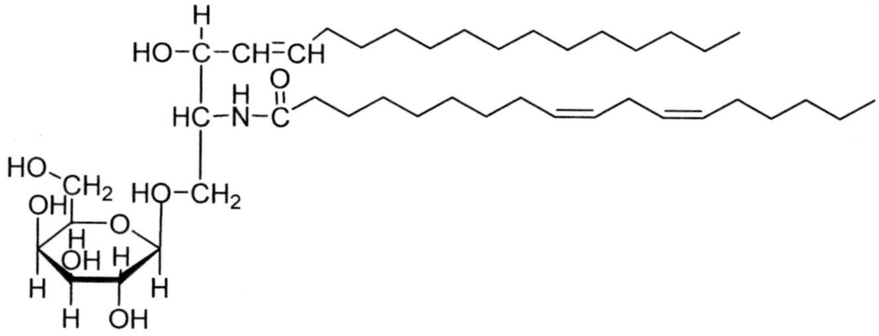

a. Is it a phospholipid or a glycolipid? How can you tell?

b. Is the backbone derived from sphingosine or glycerol? Circle the backbone.

c. What is the fatty acid from which the lipid is derived? Is it saturated or unsaturated? What functional group connects it to the backbone?

d. Is it a sphingomyelin, cerebroside, or ganglioside? How can you tell?

e. If it is a glycolipid, is it a glucose or galactose derivative? Is the glycosidic linkage α or β?

a. *Is there a phosphate group or a carbohydrate attached to the backbone?*

_____ *It is a _____ lipid.*

b. *The backbone is derived from: _____*

c. *The name of the fatty acid is: _____. Does the fatty acid contain any double bonds? _____ The fatty acid is (unsaturated or saturated) _____. The fatty acid is connected to the back bone by a _____ functional group.*

d. *If there is a carbohydrate attached to the backbone, is it a monosaccharide or an oligosaccharide? _____ Therefore, the membrane lipid is a (sphingomyelin, cerebroside, or ganglioside) _____.*

e. *The carbohydrate is a_____ derivative. The OH group on the anomeric carbon is (above or below) _____ the ring. The glycosidic linkage is _____.*

In a Nutshell: Membrane Lipids and the Structure of the Cell Membrane

The cell membrane is assembled from two layers of phospholipids and glycolipids called a bilayer. The two layers in a bilayer are aligned in a way that brings polar heads in contact with water and places the nonpolar tails in a nonpolar environment. The lipids are organized tail to tail, creating a hydrophobic environment in the interior of the membrane. The polar heads are on the outer edge of the membrane.

The lipid bilayer is often described as a "fluid mosaic" because the individual phospholipids and glycolipids are not covalently bonded to one another, but interact through weaker noncovalent intermolecular forces of attraction. Embedded within the cell membrane are cholesterol molecules to provide added rigidity. Various proteins are also found throughout the cell membrane.

In a Nutshell: Membrane Transport

Small nonpolar molecules, such as oxygen and carbon dioxide, pass freely through the cell membrane by simple diffusion. Water passes through the cell membrane by the process of osmosis. Ions and polar organic molecules cannot readily pass through the nonpolar interior of the cell membrane. Special transport proteins are necessary to move ions from one side of the membrane to the other. The energy needed to operate these "ion pumps" comes from ATP.

Worked Example #4

Draw an illustration of 14 membrane lipids arranged to form a bilayer such as the cell membrane. What molecule provides rigidity to the cell membrane?

Solution

Cholesterol provides rigidity to the cell membrane.

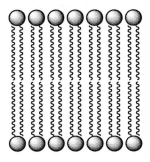

Try It Yourself #4

Draw an illustration of eight membrane lipids arranged to form a bilayer such as the cell membrane. Label the nonpolar part of the cell membrane. What kind of biomolecule moves ions across the cell membrane?

Solution:

Practice Problems for Membrane Lipids: Phospholipids and Glycolipids

1. For the membrane lipid below, answer the following questions:

a. Is it a phospholipid or a glycolipid? How can you tell?

b. Is the backbone derived from sphingosine or glycerol? Circle the backbone.

c. What are the fatty acids from which the lipid is derived? Are they saturated or unsaturated? What functional group connects them to the backbone?

d. Is it a sphingomyelin, glycerophospholipid, cerebroside, or ganglioside?

e. If it is a phospholipid, what is the amino alcohol: serine, choline, or ethanolamine?

f.   Circle the part of the molecule that defines the "polar head."

g.   Circle the part of the molecule that defines the two tails.

2. For the membrane lipid below, answer the following questions:

a.   Is it a phospholipid or a glycolipid? How can you tell?

b.   Is the backbone derived from sphingosine or glycerol? Circle the backbone.

c.   What is the fatty acid from which the lipid is derived? Is it saturated or unsaturated? What functional group connects them to the backbone?

d.   Is it a sphingomyelin, glycerophospholipid, cerebroside, or ganglioside?

e.   If it is a glycolipid, is it a glucose or galactose derivative? Is the glycosidic linkage α or β?

    f.   Circle the part of the molecule that defines the "polar head."

    g.   Circle the part of the molecule that defines the two tails.

3.   Draw an illustration of 10 membrane lipids arranged to form a bilayer such as the cell membrane. Label the polar part of the cell membrane.

Section 13.3 Fatty Acid Catabolism

Triglycerides are stored in fat cells and serve as the body's fuel during periods when other sources of energy are not available. The catabolism of triglycerides to provide energy in the form of ATP proceeds in three stages: 1) a triglyceride is hydrolyzed into glycerol and three fatty acids, 2) fatty acids are oxidized to acetyl CoA, 3) acetyl CoA is converted into carbon dioxide and energy via the citric acid cycle.

Fatty acid oxidation is the central energy-producing pathway for many cells. Fatty acids are obtained from dietary fat (triglycerides from diet) and body fat, stored in adipocytes.

In a Nutshell: Stage 1: Triglyceride Transport and Hydrolysis

Dietary fat must be first degraded into fatty acids and glycerol in order to deliver these molecules to the cells that store them or use them for energy. However, because they are not soluble in the aqueous environment of the blood, they require specialized aggregates to transport them. Lipoproteins transport triglycerides to their target cells.

The degradation of dietary fat begins in the small intestine. Triglycerides from the diet are initially present in large globules of insoluble fat.   Before lipid degradation can begin, these globules of insoluble fats must be turned into a colloid composed of finely dispersed microscopic fat droplets suspended in water, a process called emulsification. This process is aided by bile acids, which are released into the small intestine from the gall bladder. Emulsification is necessary to bring the water-insoluble triglycerides in contact with the water soluble enzymes called *lipases*. *Lipases* hydrolyze triglycerides into smaller components, such as fatty acids, glycerol, and partially hydrolyzed triacylglycerols, that can cross cell membranes.

Once these molecules have crossed the cell membrane of the cells lining the intestine, they undergo esterification reactions to reform triglycerides. These triglycerides are packaged along with cholesterol into chylomicrons for transport through the lymph system and bloodstream to cells that need energy or store triglycerides. Chylomicrons are spheres formed from a single layer of phospholipids. Since proteins are embedded throughout the lipid layer, chylomicrons are classified as a type of lipoprotein. The polar head of the lipid is positioned on the exterior of the chylomicron, where it interacts with the aqueous external environment. The nonpolar tails form a hydrophobic interior, which encapsulates the triglycerides.

Upon arrival at the capillaries of their target tissues, enzymes located outside the cell hydrolyze the triglycerides back into fatty acids and glycerol and substances that are small enough to pass through the cell membrane. Inside an adipocyte, fatty acids and glycerol are converted into triglycerides and stored. In muscle cells, fatty acids undergo the second stage of triglyceride catabolism.

In a Nutshell: Lipoproteins

There are five categories of lipoproteins that transport lipids through the blood: 1) chylomicrons, 2) very-low-density lipoproteins (VLDL), 3) intermediate-density lipoproteins (IDL), 4) low-density lipoproteins (LDL), and 5) high-density lipoproteins (HDL). These categories are based on the density of the lipoprotein. Chylomicrons are the largest of the lipoproteins. The lipoproteins differ in their specific transport function and their lipid-to-protein ratio. The higher the protein/lipid ratio, the more dense the lipoprotein. Chylomicrons have the lowest density because they have the highest lipid content. Low - density lipoproteins (LDLs) are also high in triglycerides and cholesterol compared to proteins. High-density lipoproteins (HDLs), on the other hand, have the highest protein content.

After the chylomicrons deliver most of their contents to the cells, they then return to the liver as chlyomicron remnants (leftovers) where they are degraded or, when there is an excess of fatty acids, turned into very low density lipoproteins (VLDLs). VLDLs also travel through the bloodstream to deliver triglycerides to adipose and muscle cells. Eventually, as they unload cholesterol and triglycerides into cells, they become intermediate density lipoproteins (IDLs) and low density lipoproteins (LDLs). The role of HDLs is to scavenge cholesterol in the blood and return it to the liver.

Worked Example #5

Why are triglycerides hydrolyzed into smaller molecules, such as fatty acids and glycerols, before crossing a cell membrane? What enzyme hydrolyzes the triglycerides?

Triglycerides are too big to pass through the membrane. Lipases hydrolyze the triglycerides..

Try It Yourself #5

Why do fat molecules undergo emulsification? What compound helps the emulsification process?

Solution:

In a Nutshell: Stage 2: Fatty Acid Oxidation

Oxidation of a single fatty acid produces several acetyl CoA molecules that feed into the third stage of catabolism, as well as several molecules of both NADH and $FADH_2$. Both the acetyl CoA molecules and the electrons carried by NADH and $FADH_2$ supply the energy for the formation of ATP from ADP.

In a Nutshell: Initial Activation Step

Fatty acid catabolism begins in the cytoplasm of muscle cells when a fatty acid reacts with a molecule of coenzyme A to produce fatty acyl CoA. This reaction activates the fatty acid molecule, making it ready for degradation by the biochemical process of β-oxidation. Activation of a fatty acid requires ATP, which is converted into AMP and 2 P_i. The activated fatty acid, fatty acyl CoA, is then transported into the mitochondria.

In a Nutshell: β-oxidation

The β-oxidation of a fatty acid is a biochemical sequence of four reactions that are repeated over and over again. With each pass through β-oxidation, a two-carbon fragment is removed from the fatty acid. By the repeated removal of two-carbon fragments, a long-chain fatty acid is degraded into several two-carbon units of acetyl CoA.

Two of the steps of β-oxidation are oxidation/reduction reactions and another step is a hydration reaction. Since half of these reactions involve oxidation of the third carbon atom in the chain, the β-carbon, the pathway is termed β-oxidation.

In step 1, the carbon-carbon single bond between the α- and the β- carbon atoms is oxidized into a carbon-carbon double bond. This reaction uses FAD as a coenzyme. In step 2, a hydration reaction, an OH group and an H atom from water are added to the double bond in the normal fashion: the OH group is added to the β-carbon and the H atom to the α-carbon. In step 3, the newly installed hydroxyl group at the β-carbon is oxidized to a ketone. This reaction uses NAD^+ as a coenzyme. In step 4, the single bond between the α- and β-carbon atoms breaks and the acyl group is transferred to coenzyme A. Two products are formed: acetyl CoA and a new fatty acyl CoA, which is two carbon atoms shorter than the original fatty acyl CoA molecule. The cycle then repeats itself until in the final step a four-carbon fatty acyl CoA molecule is converted into two acetyl CoA molecules.

In a Nutshell: Energy from β-oxidation

All of the acetyl CoA molecules produced through β-oxidation can enter the citric acid cycle, where each acetyl CoA molecule can produce two carbon dioxide molecules and 12 ATP. Later, in the electron-transport chain, $FADH_2$ formed in step 1 will be converted back into

FAD, converting 2 ADP into two ATP in the process. Also, in the electron-transport chain, NADH formed in step 3 will be converted back into NAD^+, producing 3 ATP in the process.

To calculate how many ATP molecules will be produced from the complete catabolism of a particular fatty acid, first determine the number of β-oxidation cycles that the fatty acid will undergo and the number of acetyl CoA molecules produced. For every β-oxidation cycle, 5 ATP molecules will be produced. Every cycle produces one NADH and one $FADH_2$, which generate 3 ATP molecules and 2 ATP molecules, respectively, in the electron-transport chain. Add 12 ATP for each acetyl CoA molecule produced. Subtract the 2 ATP molecules used to activate the initial fatty acid. Note: the number of β-oxidation steps is half the total number of carbon atoms minus one.

Worked Example #6

Calculate the number of ATP molecules produced from complete catabolism of myristoleic acid, a fatty acid that contains 14 carbon atoms.

Myristoleic acid: C-14

6 cycles	× 5	=	*30 ATP*
7 acetyl CoA molecules	× 12	=	*84 ATP*
One time activation of myristoleic acid		=	*−2 ATP*
Total			*112 ATP*

Try It Yourself #6

Calculate the number of ATP molecules produced from complete catabolism of arachidonic acid, a fatty acid that contains 20 carbon atoms.

Arachidonic acid: C-20

_____ *cycles*	× 5	=	_____ *ATP*
_____ *acetyl CoA molecules*	× 12	=	_____*ATP*
One time activation of arachidonic acid		=	_____ *ATP*
Total			_____*ATP*

Practice Problems for Fatty Acid Catabolism

1. What role do chylomicrons play in fatty acid catabolism?

2. Calculate the number of ATP molecules produced from complete catabolism of linolenic acid, a fatty acid that contains 18 carbon atoms.

3. Which step of β-oxidation produces $FADH_2$? Which step produces NADH?

Section 13.4 Cholesterol and Other Steroid Hormones

Cholesterol and the steroid hormones derived from cholesterol are characterized by a common core ring system: three six-membered rings and one five-membered ring fused together. Steroids are classified as lipids because they are insoluble in water and soluble in nonpolar organic solvents.

Cholesterol is an amphipathic compound with a polar head (OH) and a nonpolar tail. Its presence in the cell membrane provides rigidity to the membrane structure. Cholesterol serves as the biochemical precursor for bile acids, vitamin D, and many important hormones.

Bile acids are a group of amphipathic molecules that behave as emulsifying agents. Vitamin D_3 is one of four fat-soluble vitamins. It is produced in the skin by the action of the energy of sunlight on a cholesterol derivative. The five classes of steroid hormones produced from cholesterol include: glucocorticoids, mineralocorticoids, progestins, androgens, and estrogens.

The glucocorticoids regulate carbohydrate, protein, and lipid metabolism. They also have powerful anti-inflammatory and immunosuppressant activity. The mineralocorticoids are produced in the adrenal gland and regulate ion (Na^+, K^+, Cl^-) balance in tissues. Progestins, androgens, and estrogens collectively represent the major sex hormones.

Worked Example #7

What are the five classes of steroid hormones derived from cholesterol? What is the function of the mineralocorticoids?

Solution

The five classes of steroid hormones produced from cholesterol include: glucocorticoids, mineralocorticoids, progestins, androgens, and estrogens. The mineralocorticoids regulate ion balance in tissues.

Try It Yourself #7

What are the functions of the glucocorticoids and estrogens?

Solution:

Practice Problems for Cholesterol and Other Steroid Hormones

1. From what molecule are bile acids, vitamin D, and the steroid hormones derived?

2. What structural feature characterizes cholesterol and the steroid hormones?

3. Medrol, shown below, is commonly used to treat rheumatoid arthritis and severe allergic reactions, among other things. Is medrol a steroid?

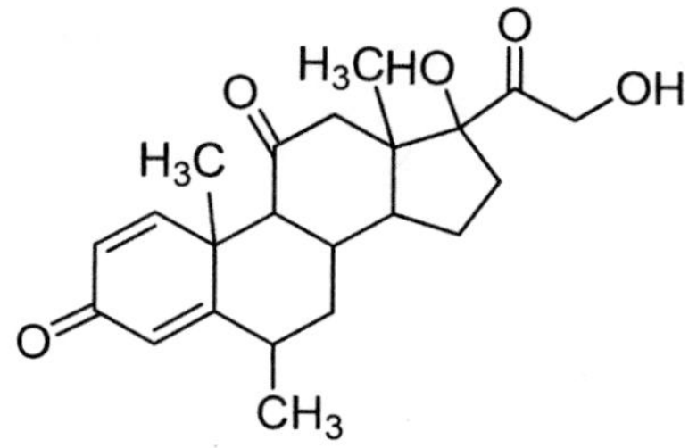

Chapter 13 Quiz

1. The structure of erucic acid, found in canola oil and mustard seed, is shown below.

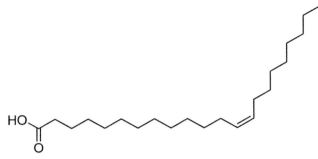

    a. What type of fatty acid is erucic acid? Answer using the delta naming system and the omega naming system.

    b. Is erucic acid a saturated or unsaturated fatty acid?

    c. What type of double bond is found in erucic acid?

    c. Is erucic acid soluble in water?

2. Explain why oleic acid has a lower melting point than stearic acid, even though both fatty acids contain 18 carbon atoms.

3. Write the structure of the triglyceride composed of one molecule of linolenic acid, one molecule of oleic acid, and one molecule of linoleic acid.

    a. Is this triglyceride a mixed or a simple triglyceride?

    b. Do you expect this triglyceride to be a fat or an oil?   Explain.

    c. Is this triglyceride unsaturated or saturated?

4. For the membrane lipid below, answer the following questions:

    a. Is it a phospholipid or a glycolipid? How can you tell?

    b. Is the backbone derived from sphingosine or glycerol? Circle the backbone.

    c. What is the fatty acid from which the lipid is derived? Is it a saturated or unsaturated fatty acid? What functional group connects it to the backbone?

    d. Is it a sphingomyelin, glycerophospholipid, cerebroside, or ganglioside?

    e. If it is a phospholipid, what is the amino alcohol: serine, choline, or ethanolamine?

    f. Circle the part of the molecule that defines the "polar head."

    g. Circle the part of the molecule that defines the two tails.

5. For the membrane lipid below, answer the following questions:

a. Is it a phospholipid or a glycolipid? How can you tell?

b. Is the backbone derived from sphingosine or glycerol? Circle the backbone.

c. What is the fatty acid from which the lipid is derived? Is it a saturated or unsaturated fatty acid? What functional group connects it to the backbone?

d. Is it a sphingomyelin, glycerophospholipid, cerebroside, or ganglioside?

e. If it is a glycolipid, is it a glucose or galactose derivative? Is the glycosidic linkage α or β?

f. Circle the part of the molecule that defines the "polar head."

g. Circle the part of the molecule that defines the two tails.

6. Draw an illustration of 12 membrane lipids arranged to form a bilayer, such as the cell membrane. Label the polar and nonpolar parts of the bilayer.

7. Is the interior or exterior of a chylomicron polar or nonpolar? Explain.

8. Does a high-density lipoprotein contain more proteins than lipids or more lipids than proteins?

9. Calculate the number of ATP molecules produced from complete catabolism of docosahexaenoic acid, DHA, a fatty acid that contains 22 carbon atoms.

10. What do the structures of cholesterol and the steroid hormones have in common?

Chapter 13
Answers to Additional Exercises

13.33   Lipids are insoluble in water and soluble in nonpolar organic solvents.

13.35   Lipids have an extensive hydrocarbon structure; therefore, they are nonpolar and insoluble in water.

13.37   Oleic acid is a Δ^9 fatty acid and ω-9 fatty acid.

13.39   Linolenic acid is an ω-3 fatty acid.

13.41   The presence of a cis double bond lowers the melting point of a fatty acid. The bend created by the double bond reduces the number of points of contact between the fatty acid molecules, leading to fewer dispersion forces.

13.43   Fats are derived primarily from saturated fatty acids. Saturated fatty acids pack together tightly, causing greater dispersion forces and therefore the melting point is higher.

13.45   Coconut oil should be a solid at room temperature because it is mostly saturated fats, which pack together tightly.

13.47

Olive oil is considered a healthy fat because it consists mostly of unsaturated fats.

13.49   A triglyceride would not be soluble in water. It contains mostly nonpolar hydrogen and carbon atoms.

13.51   Safflower oil is lower in saturated fats.

13.53   Olive oil is a healthier food because it contains fewer saturated fats.

13.55   Unsaturated trans fats should pack more closely. The trans double bond does not cause a bend in the chain, whereas the cis double bond causes a bend. The unsaturated trans fats would have a higher melting point because they can pack together tightly and have more dispersion forces, leading to a higher melting point.

13.57   Lipids are the main component of the cell membrane.

13.59   Glycerophospholipids and sphingomyelins contain a phosphate functional group.

13.61   The amine and the phosphate form the polar head of the molecule.

13.63   a.   It is a glycolipid. It has a carbohydrate attached to the OH group farthest from the hydrocarbon chain.

   b.   It is a sphingolipid. It has a sphingosine backbone.

sphingosine backbone

HO–CH

HC—N—C
      H

HO–CH₂   O—CH₂
  H

HO
OHH

OH   H

H   OH

   c.   It was derived from the fatty acid stearic acid. The fatty acid component is saturated. An amide functional group connects the fatty acid to the backbone.

   d.   This lipid is a cerebroside. It has a monosaccharide as the carbohydrate component.

   e.   It has a glucose derivative. The glycosidic linkage is β; the oxygen atom at the anomeric carbon is above the ring.

   f.

double bond

alcohol   HO—CH

HC—N—C
      H

alcohol HO—CH₂   O—CH₂ amide

H

H OH H

OH   H   glycosidic
                     bond

H   OH

alcohols

13.65   Dispersion forces and hydrogen bonding are responsible for the way membrane lipids organize to form a cell membrane.

13.66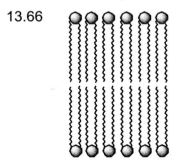

13.67   Cholesterol adds rigidity to the cell membrane.

13.69   Small nonpolar molecules such as oxygen and carbon dioxide pass freely through the cell membrane.

13.71   The body obtains fatty acids from dietary fat and body fat.

13.73   Triglycerides are packaged into chylomicrons to be transported through the lymph system and the bloodstream.

13.75   The triglycerides are found on the interior of the lipoprotein because triglycerides are nonpolar and the interior of the lipoprotein is hydrophobic.

13.77   Oleic acid: C-18

8 cycles	× 5	=	40 ATP
9 acetyl CoA molecules	× 12	=	108 ATP
One time activation of oleic acid		=	−2 ATP
Total			146 ATP

13.79   The first and third steps of β-oxidation are oxidation steps.

13.81   Cholesterol is similar to a fatty acid in that both molecules are amphipathic. They both have a polar head and a nonpolar tail.

13.83   Glucocorticoids regulate carbohydrate, protein, and lipid metabolism. They also have powerful anti-inflammatory and immunosuppressant activity.

13.85   The six-membered ring attached to an OH group has been converted into an aromatic ring. The double bond in the adjacent six-membered ring has been reduced to a single bond. The hydrocarbon chain attached to the five-membered ring has been converted into an OH group.

13.87   Eicosanoids are a class of lipids derived from arachidonic acid (eicosanoic acid) that are signaling molecules that stimulate signs of inflammation.

13.89   The symptoms of inflammation are redness, swelling, pain, and warmth.

13.91   a.   Leukotriene A_4 will not be produced when a *lipoxygenase* inhibitor is present.

       b.   Prostaglandin H_2, thromboxane A_2, and prostaglandin E_2 will not be produced when a *cycloxygenase* inhibitor is present.

13.93   Aspirin inhibits *cyclooxygenase*, the key enzyme involved in the formation of prostaglandin H_2.

13.95   An NSAID is a non-steroidal anti-inflammatory drug. NSAIDs include aspirin, ibuprofen (Motrin[®] and Advil[®]), and naproxen (Aleve[®]).

Chapter 14

Metabolism and Bioenergetics

Chapter Summary

In this chapter, you learned about the central role of acetyl coenzyme A in metabolism and the citric acid cycle. You learned how the citric acid cycle produces NADH and $FADH_2$. You observed how the electrons carried by these coenzymes drive the formation of ATP from ADP through a series of oxidation-reduction reactions known as oxidative phosphorylation. You studied the role of energy in metabolism (bioenergetics) and were introduced to the energy concept called entropy.

Section 14.1 Acetyl Coenzyme A and the Citric Acid Cycle

In a Nutshell: Acetyl CoA

Acetyl coenzyme A, abbreviated acetyl CoA, plays a central role in anabolic and catabolic pathways. Three biomolecules obtained from the diet can be converted into acetyl CoA: glucose, fatty acids, and several amino acids. Acetyl CoA has a chemical structure derived from adenine, panothenic acid, the thiol amine $HSCH_2CH_2NH_2$, and acetic acid. The hydrolysis of acetyl CoA produces acetic acid and the thiol coenzyme A. Coenzyme A can be attached to any acyl group via a thioester linkage and carries out the biochemical function of transferring acyl groups.

In a Nutshell: The Citric Acid Cycle

The third stage of catabolism begins with the citric acid cycle. The citric acid cycle is an eight-step biochemical pathway that begins with the reaction of acetyl CoA with oxaloacetate. Since the last step of the citric acid cycle regenerates oxaloacetate, the citric acid cycle has a unique and characteristic circular appearance when written.

By the end of the cycle, three molecules of NAD^+ are reduced to NADH, one molecule of FAD is reduced to $FADH_2$, and one molecule of GDP is phosphorylated to GTP. Reduction of these coenzymes is the most important outcome of the citric acid cycle because most of the energy harvested in the citric acid cycle comes from the electrons delivered to the electron-transport chain by these reduced coenzymes.

In the electron-transport chain, the electrons carried by one NADH molecule supply the energy to phosphorylate three ADP molecules to form three ATP molecules, and those from one $FADH_2$ molecule supply the energy to phosphorylate two ADP molecules to form two ATP molecules. In addition, the molecule of GTP that is produced in the citric acid cycle is converted into one molecule of ATP. Therefore, one round of the citric acid cycle can phosphorylate 12 ADP molecules to 12 ATP molecules during the biochemical process of oxidative phosphorylation.

In the first step of the citric acid cycle, the two carbon atoms of the acetyl group carried by coenzyme A form a carbon-carbon bond to the four-carbon oxaloacetate molecule to produce the six-carbon citrate molecule. The first half of the citric acid cycle generates citrate (six carbons) and then breaks it down into succinyl CoA (four carbons) and two molecules of carbon dioxide, CO_2. The second half of the citric acid cycle regenerates oxaloacetate from succinyl CoA. The two carbon atoms that are released as CO_2 come from two of the four carbon atoms of oxaloacetate and not the original acetyl CoA molecule.

Worked Example #1

The unionized form of oxaloacetate is shown below:

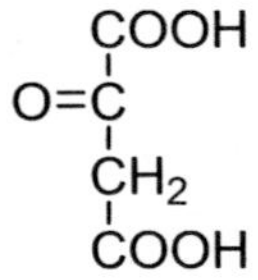

```
      COOH
      |
O=C
      |
      CH2
      |
      COOH
```

a. What structure would this compound have at physiological pH?

b. During the citric acid cycle what happens to two of the carbon atoms of oxaloacetate?

a. *At physiological pH, the carboxylic acid functional groups would be in their ionized form.*

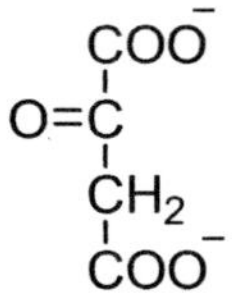

```
      COO⁻
      |
O=C
      |
      CH2
      |
      COO⁻
```

b. *During the citric acid cycle, two of the four carbon atoms of oxaloacetate are released as two molecules of carbon dioxide.*

Try It Yourself #1

The structure of α-ketoglutarate, produced in the third step of the citric acid cycle, in its unionized form is shown below.

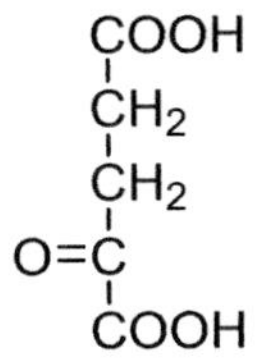

```
    COOH
     |
    CH₂
     |
    CH₂
     |
O=C
     |
    COOH
```

a. What structure would this compound have at physiological pH?

b. How many carbon atoms are in α-ketoglutarate?

a. *The structure of α-ketoglutarate at physiological pH:*

b. *There are _____ carbon atoms in α-ketoglutarate.*

Practice Problems for Acetyl Coenzyme A and the Citric Acid Cycle

1. How many molecules of ADP are converted into ATP as the result of one pass through the citric acid cycle, assuming that NADH and $FADH_2$ are used to phosphorylate ADP to ATP?

2. The structure of citrate is shown below.

$$
\begin{array}{c}
COO^- \\
| \\
CH_2 \\
| \\
HO-C-COO^- \\
| \\
CH_2 \\
| \\
COO^-
\end{array}
$$

a. What structure would this compound have at low pH? Write the carboxylic acid functional groups in their unionized form.

b. What other functional group is present in this molecule?

3. What acyl group is carried by acetyl CoA?

Extension Topic 14-1: The Chemical Reactions of the Citric Acid Cycle

In step 1, acetyl CoA reacts with oxaloacetate to form a new carbon-carbon bond and produce citrate. The thioester functional group is hydrolyzed releasing coenzyme A in an acyl transfer reaction.

In step 2, an isomerization reaction converts citrate into its structural isomer isocitrate. The isomerization reaction is accomplished by a dehydration reaction followed by a hydration reaction. Dehydration of citrate removes an OH group and a hydrogen atom from adjacent carbon atoms, expelling water. Then hydration reintroduces the OH group and hydrogen atom but on different carbon atoms.

Step 3 produces the first molecule of carbon dioxide and reduces the first NAD^+ molecule to NADH. The first reaction in this step is an oxidation-reduction reaction that transfers electrons to NAD^+ as the hydroxyl group is oxidized to a ketone. The second reaction in this step is a decarboxylation reaction in which carbon dioxide is lost from the substrate and replaced with a proton from solution. The overall reaction in step 3 converts isocitrate into α-ketoglutarate and carbon dioxide.

Step 4 produces the second molecule of CO_2 and reduces a second molecule of NAD^+ to NADH. This step involves three reactions catalyzed by one enzyme. The first reaction is a decarboxylation reaction that expels one molecule of CO_2. The intermediate aldehyde is then oxidized to a carboxylic acid by NAD^+. In the final reaction, the acyl group is transferred to the thiol group on coenzyme A in a thioesterification reaction to produce the thioester succinyl CoA.

In step 5, succinyl CoA is hydrolyzed into succinate and coenzyme A. This reaction releases enough energy to simultaneously phosphorylate GDP to GTP.

Step 6 generates the one molecule of $FADH_2$ from FAD by oxidizing the carbon-carbon single bond in succinate to a carbon-carbon double bond, forming fumarate.

In step 7, a hydration reaction produces the chiral alcohol L-malate from fumarate. In this step water is added to the double bond of fumarate to form L-malate.

The final step of the citric acid cycle is an oxidation-reduction reaction that produces the third molecule of NADH. In this step, the alcohol in L-malate is oxidized to the achiral ketone oxaloacetate.

Worked Example #2
Step 5 of the citric acid cycle is shown below.

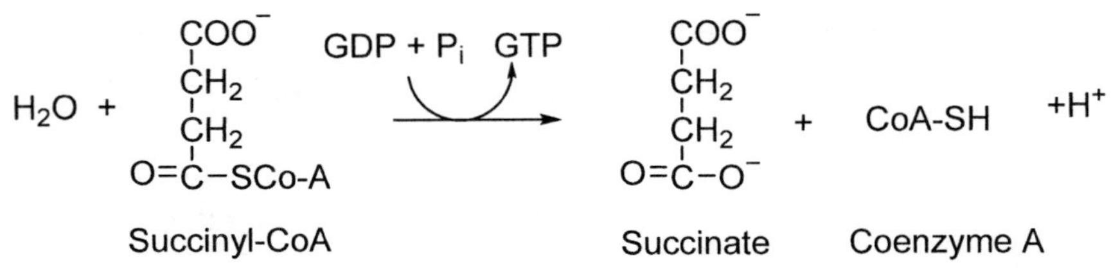

$$H_2O + \begin{matrix} COO^- \\ | \\ CH_2 \\ | \\ CH_2 \\ | \\ O=C-SCo\text{-}A \end{matrix} \xrightarrow[\text{GDP + P}_i \quad \text{GTP}]{} \begin{matrix} COO^- \\ | \\ CH_2 \\ | \\ CH_2 \\ | \\ O=C-O^- \end{matrix} + CoA\text{-}SH \quad +H^+$$

Succinyl-CoA   Succinate   Coenzyme A

a. What type of reaction does succinyl CoA undergo?

b. What functional groups are unchanged in this reaction?

c. How many carbon atoms are in succinate?

a. *Succinyl CoA undergoes a hydrolysis reaction to produce succinate and coenzyme A.*

b. *The carboxylate group at the top of succinyl CoA remains unchanged in this reaction.*

c. *There are four carbon atoms in succinate.*

Try It Yourself #2

Step 3 of the citric acid cycle is shown below.

$$\begin{matrix} COO^- \\ | \\ CH_2 \\ | \\ H-C-COO^- \\ | \\ HO-C-H \\ | \\ COO^- \end{matrix} \xrightarrow[\text{NAD}^+ \quad \text{NADH + H}^+]{} \begin{matrix} COO^- \\ | \\ CH_2 \quad O \\ | \quad \diagup\!\!\!\diagdown \\ H-C-C \\ | \quad \diagdown O^- \\ O=C \\ | \\ COO^- \end{matrix} \xrightarrow[\text{O=C=O}]{} \begin{matrix} COO^- \\ | \\ CH_2 \\ | \\ H-C-H \\ | \\ O=C \\ | \\ COO^- \end{matrix}$$

Isocitrate   α-ketoglutarate

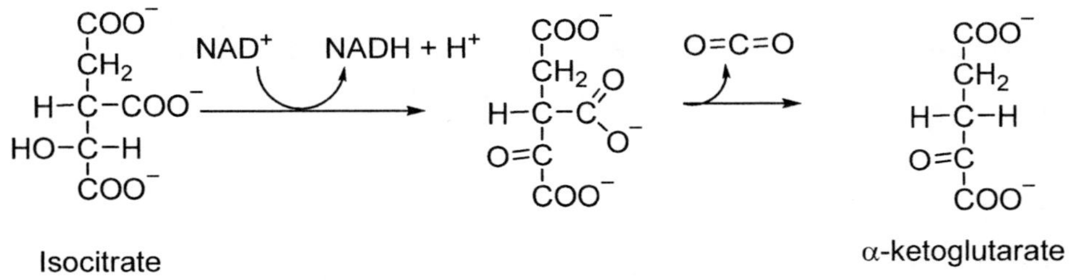

a. What type of reaction happens in the first reaction of this step?

b. Is isocitrate oxidized or reduced?

c. Is NAD$^+$ oxidized or reduced?

a. *The reaction is a:* _____.

b. *Isocitrate is:* _____

c. *NAD$^+$ is:* _____

Practice Problems for The Chemical Reactions of the Citric Acid Cycle

1. Step 6 of the citric acid cycle is shown below.

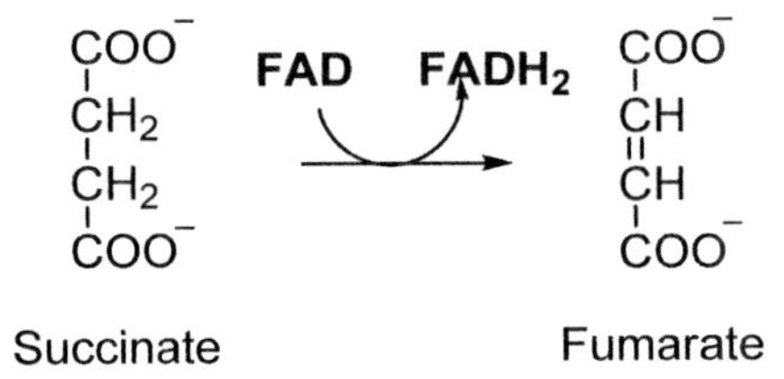

Succinate   Fumarate

a. What type of reaction occurs in this step?

b. What functional group change occurred to the succinate molecule?

c. Is succinate oxidized or reduced?

d. Is FAD oxidized or reduced?

2. Step 4 of the citric acid cycle is shown below.

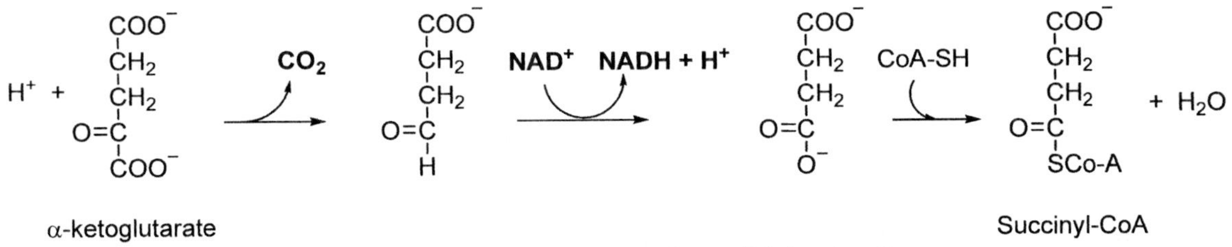

α-ketoglutarate   Succinyl-CoA

a. What type of reaction occurs in the first reaction of this step?

b. How many carbon atoms does α-ketoglutarate contain? How many carbon atoms does succinyl CoA contain?

c. What is the significance of the NADH produced in the second reaction of this step?

Section 14.2 Oxidative Phosphorylation

In a Nutshell: Where It All Happens: The Mitochondria

The electron-transport chain and phosphorylation of ADP to ATP, known collectively as oxidative phosphorylation, occur in the final stage of cellular respiration. Glycolysis—the conversion of glucose to pyruvate—occurs in the cytoplasm of the cell. Most of the other energy-producing biochemical pathways occur in the mitochondria. The key parts of a mitochondrion are the outer membrane, the intermembrane space, the inner mitochondrial membrane, and the matrix.

The outer membrane encloses and defines the mitochondrion and is permeable to many ions and small molecules. Inside the mitochondrion is a second selectively permeable membrane, called the inner mitochondrial membrane. This highly folded inner membrane separates the interior of the mitochondrion into two distinct aqueous regions: 1) the intermembrane space and 2) the matrix.

Ions, including protons, are unable to diffuse through the inner membrane by simple diffusion. Since the inner membrane is not permeable to protons, proton concentrations on either side of the membrane are not able to equalize. Therefore, the concentration of protons—the pH—is different on the two sides of this membrane. The matrix has a lower concentration of protons (higher pH) than the intermembrane space. A difference in proton concentration between adjacent regions is known as a proton gradient.

Most of the enzymes involved in the citric acid cycle are found in the matrix. The proteins involved in oxidative phosphorylation are located or embedded within the inner mitochondrial membrane. The important and large enzyme, *ATP synthase*, is found in the inner membrane with a significant portion extending into the matrix.

Worked Example #3

Which has a lower pH, the intermembrane space or the matrix? Which has a higher concentration of protons?

Solution

The intermembrane space has a higher concentration of protons and a lower pH than the matrix.

Try It Yourself #3

Arrange the following parts of a mitochondrion from the inside to the outside: inner mitochondrial membrane, outer membrane, and the matrix.

The inside of a mitochondrion is: _____

The middle of a mitochondrion is: _____

The outside of a mitochondrion is: _____

In a Nutshell: The Electron-Transport Chain

Electrons carried by the coenzymes NADH and $FADH_2$ are produced in the citric acid cycle and other biochemical pathways are relayed through a series of protein complexes to molecular oxygen, O_2, which is reduced to water. The protein complexes are known as the electron-transport chain. The electron-transport chain consists of four large multienzyme protein complexes identified as Complexes I, II, III, IV, along with two mobile proteins— coenzyme Q and cytochrome *c*.

Each protein complex is a membrane protein that spans the inner mitochondrial membrane, extending from the matrix to the intermembrane space. Sites within each protein complex accept electrons then transmit them. The mobile carriers transfer electrons between protein complexes.

In a Nutshell: The Proton-Motive Force

The flow of electrons through protein complexes I, II, and IV supplies the energy necessary to pump protons from the matrix across the inner membrane and into the intermembrane space, generating a proton gradient. Since the movement of protons is against the proton gradient, the process is referred to as a proton pump.

A proton gradient represents potential energy, which ultimately provides the energy for oxidative phosphorylation. The accumulation of protons in the intermembrane space represents both chemical and electrical potential energy. Since the protons are charged, there is a difference in charge between the two spaces. The potential energy store in the unequal distribution of protons is called the proton-motive force.

In a Nutshell: Oxidation-Reduction in the Electron-Transport Chain

Electrons are transferred through a series of oxidation-reduction reactions that occur at various metal atom centers (copper and iron) in the protein complexes and at cofactors, such as coenzyme Q. A metal or molecule's intrinsic affinity for electrons determines the direction of electron flow in an oxidation-reduction reaction. In the electron-transport chain, electrons are passed to centers with increasingly greater affinity for electrons. Thus, electron-transport chain begins with NADH or $FADH_2$, which have the least electron affinity and ends with the acceptance of electrons by oxygen, which has the greatest electron affinity.

Centers with greater electron affinity represent species with lower potential energy. The energy needed to create a proton gradient comes from the energy released by the stepwise oxidation-reduction reactions that occur as electrons are transferred to centers of even lower potential. That energy is used to pump protons against the proton gradient from the matrix through the inner membrane to the intermembrane space.

When two electrons are received from NADH in the matrix by Complex I, they pass through a series of oxidation-reduction sequences to coenzyme Q, a mobile electron carrier within the inner membrane. In the process, two electrons and two protons are received by coenzyme Q, and four protons are pumped from the matrix to the intermembrane space. $FADH_2$ enters the electron-transport chain at Complex II, which is lower in energy than NADH and does not pump protons into the intermembrane space. Therefore $FADH_2$ is not capable of producing as much ATP as NADH. From Complex II, electrons are passed to the mobile carrier coenzyme Q.

Coenzyme Q transfers electrons to Complex III, whereupon protons are also pumped from the matrix to the intermembrane space. Electrons from Complex III are transferred to the water-soluble mobile electron carrier protein: cytochrome *c*.

Cytochrome *c* delivers electrons to Complex IV, which transfers electrons from cytochrome *c* to molecular oxygen. Reduction of oxygen turns it into water. Protein Complex IV also pumps protons across the inner membrane in the process.

In a Nutshell: Cyanide Poisoning

The last oxidation-reduction step that occurs in Complex IV is the transfer of electrons from the iron atom in the enzyme, *cytochrome c oxidase,* to oxygen. Cyanide, CN^-, is a fatal poison because it is an irreversible inhibitor of *cytochrome c oxidase.* Consequently, cyanide poisoning brings the electron-transport chain to a halt, thereby preventing phosphorylation of ADP and depriving the cell of energy it needs to function. One treatment for cyanide poisoning is hyperbaric oxygen treatment. This treatment floods the mitochondria with oxygen in an attempt to displace cyanide from the enzyme and allow the electron-transport chain to resume function.

Worked Example #4

Fill in the blanks for the following chain of events, which describes the path electrons take through the electron-transport chain when introduced from $FADH_2$:

$FADH_2$ → Complex ____ → Coenzyme ____ → Complex ____ → Cytochrome ____ → Complex ____ → O_2.

$FADH_2$ → Complex II → Coenzyme Q → Complex III → Cytochrome c → Complex IV → O_2.

Try It Yourself #4

Fill in the blanks for the following chain of events, which describes the path electrons take through the electron-transport chain when introduced from NADH:

NADH → Complex ____ → Coenzyme ____ → Complex ____ → Cytochrome ____ → Complex ____ → O_2.

Solution

NADH → Complex ____ → Coenzyme ____ → Complex ____ → Cytochrome ____ →
Complex ____ → O_2.

In a Nutshell: ATP Synthesis

ATP synthase uses the proton-motive force to drive the phosphorylation of ADP to ATP.
ATP synthase is a multienzyme protein complex that looks like a cylindrical motor on a stick.
The stick portion of the enzyme is embedded in the inner membrane and the cylinder
extends into the matrix.

When the concentration of protons in the intermembrane space reaches a certain level, and
ADP and P_i are both bound to *ATP synthase*, an ion channel within the stick portion of *ATP
synthase* opens, allowing protons in the intermembrane space to flow back into the matrix
where the proton concentration is lower. Since protons flow through *ATP synthase* from a
region of higher proton concentration to a region of lower concentration, energy is released.
As protons flow with the proton gradient through the proton channel created by *ATP
synthase*, the energy released causes part of the enzyme to rotate like a turnstile. Rotation
brings the reactants ADP and P_i together so that phosphorylation can occur. Another turn of
the turnstile releases ATP.

In a Nutshell: Summary of Metabolism

Energy is extracted from the biomolecules—proteins, carbohydrates, and triglycerides—in
our diet. The combustion of these molecules in cellular respiration produces more than 90%
of the ATP required by humans. Catabolism of these molecules to produce energy is
divided into three stages. In the first stage, proteins are hydrolyzed into amino acids,
carbohydrates are hydrolyzed into monosaccharides, and triglycerides are hydrolyzed into
glycerol and fatty acids. In the second stage, each of these smaller biomolecular building
blocks is degraded by well-known biochemical pathways to produce acetyl CoA. Amino
acids undergo transamination, where they are converted to acetyl CoA or other
intermediates of the citric acid cycle. Glucose undergoes glycolysis to produce pyruvate,
which is converted to acetyl CoA. Fatty acids undergo β-oxidation to produce acetyl CoA.

The acetyl CoA produced from all three biomolecules enters the third stage of catabolism: the citric acid cycle and oxidative phosphorylation. In this stage the electrons from acetyl CoA are carried by the coenzymes NADH and $FADH_2$ to the electron-transport chain. Electrons pass through the electron-transport chain through a series of oxidation-reduction reactions that simultaneously cause protons to be pumped from the matrix into the intermembrane space, creating a proton gradient. *ATP synthase* periodically opens a channel within the proton, which allows protons to return to the matrix, releasing energy that drives the phosphorylation of ADP to ATP. Ultimately, one acetyl CoA molecule phosphorylates 12 ADP to ATP.

Worked Example #5
Where does the energy come from that drives the phosphorylation of ADP to ATP catalyzed by *ATP synthase*?

Solution
The energy comes from the flow of protons in the intermembrane space to the matrix. This flow of protons is in the same direction as the proton gradient (from a region of high proton concentration to a region of lower concentration), so energy is released.

Try It Yourself #5
Why do protons in the intermembrane space not diffuse into the matrix where the concentration is lower?

Solution:

Practice Problems for Oxidative Phosphorylation
1. When protons move from the matrix across the inner membrane into the intermembrane space, are they moving with or against the proton gradient?

2. When *ATP synthase* allows protons to move from the intermembrane space back into the matrix, are they moving with or against the proton gradient?

3. How many ADP molecules are phosphorylated by the energy supplied by the electrons from one molecule of NADH?

Section 14.3 Entropy and Bioenergetics

Bioenergetics is the study of energy transfer in reactions occurring in living cells. Heat energy is transferred in chemical reactions and is known as the change in enthalpy, ΔH. To understand energy transfer in living cells you must consider two additional energy concepts: entropy and free energy.

In a Nutshell: Spontaneous and Nonspontaneous Reactions

A reaction is said to be spontaneous if it continues on its own once started. Conversely, a reaction is said to be nonspontaneous if energy must be continuously supplied to the reaction for it to proceed. A spontaneous chemical reaction is not necessarily a fast one, but one that can proceed on its own in the direction written. By definition, if a reaction is spontaneous, the reverse reaction is nonspontaneous.

Biochemical reactions in a living cell occur in an aqueous environment and under conditions of constant temperature and pressure. Under these conditions, whether a reaction is spontaneous or nonspontaneous can be determined from the change in free energy (ΔG) for the reaction. The change in free energy depends on several factors: the change in enthalpy, ΔH; the change in entropy, ΔS; and the temperature, T.

Worked Example #6

Which of the following processes are spontaneous?

    a. sledding downhill

    b. carrying a sled uphill

    a. Sledding downhill is spontaneous. After you get on the sled and give it an initial push to go downhill, it keeps going downhill.

    b. Carrying a sled uphill is a nonspontaneous process. You must continually supply energy to get the sled uphill.

Try It Yourself #6

Which of the following processes are nonspontaneous?

    a. a plant growing in sunlight

    b. plants and leaves decaying in a forest

    a. A plant growing in sunlight is: _____

    b. Plants and leaves decaying in a forest is: _____

In a Nutshell: Entropy

The change in entropy, ΔS, is a measure of the change in the degree of randomness or disorder in a reaction or physical change. Entropy increases as disorder increases. Entropy may increase or decrease in a chemical reaction. If the product molecules exhibit greater disorder than the reactant molecules, the reaction results in an increase in entropy, $\Delta S_{rxn} > 0$. The reverse is also true: more ordered product molecules mean the results in a decrease of entropy, $\Delta S_{rxn} < 0$.

The change in entropy in a chemical reaction is defined as the difference in the entropy of the products and the reactants. Thus, when the products of a reaction are more disordered than the reactants, the reaction results in an increase in entropy and ΔS is positive. When the products of a reaction are less disordered than the reactants, the reaction results in a decrease in entropy and ΔS is negative.

The change in entropy in a chemical reaction depends on three factors: 1) the changes in state of reactants and products, 2) the difference in complexity of reactants and products, and 3) the difference in number of reactant and product molecules. In a physical change, entropy changes with a change of state. Entropy increases going from a solid to a liquid to a gas. In a chemical reaction, entropy increases when the products are less complex than the reactants because there are more degrees of translational and rotational freedom. In a chemical reaction, an increase in the number of product molecules compared to reactant molecules increases the entropy.

The second law of thermodynamics states that the entropy of the universe is always increasing, $\Delta S_{universe} > 0$. The entropy of the universe is the sum of the change in entropy of a reaction (or physical change) and the change in entropy of the surroundings.

Worked Example #7
Which of the following represent an increase in entropy?
   a. CO_2 (s) $\rightarrow$ CO_2 (g)
   b.   alanine + glycine + leucine $\rightarrow$ Ala-Gly-Leu (tripeptide)
   c.   A reaction that has more product molecules than reactant molecules.

   a.   *This reaction has an increase in entropy. The carbon dioxide molecules in the gas phase have more freedom of motion than the molecules in the solid state.*
   b.   *This reaction has a decrease in entropy. The tripeptide product is a more complex molecule than the individual amino acid reactants. The tripeptide has less freedom of motion than the individual amino acids.*
   c.   *This reaction has an increase of entropy. Because there are more product molecules, there is more freedom of motion.*

Try It Yourself #7
Which of the following represent an increase in entropy?

a. CH_3CH_2OH, ethanol (l) $\rightarrow$ CH_3CH_2OH, ethanol (s)

b. Enzyme (E) + Substrate (S) $\rightarrow$ E-S complex

c. 4 Fe (s) + 3 O_2 (g) $\rightarrow$ 2 Fe_2O_3 (s)

a. *The (products or reactants) _____ have more freedom of motion.*

This reaction represents _____ in entropy.

b. *The (products or reactants) _____ have more freedom of motion.*

This reaction represents _____ in entropy.

c. *The (products or reactants _____ have more freedom of motion.*

This reaction represents _____ in entropy.

In a Nutshell: Free Energy, *ΔG*

The change in free energy, ΔG, is determined by both enthalpy and entropy changes and can be used to predict whether a reaction is spontaneous or nonspontaneous. The change in free energy, ΔG, is defined as the difference between the free energy of the products and the reactants. The change in free energy, ΔG, can be determined from the change in enthalpy, ΔH; the change in entropy, ΔS; and the absolute temperature, T, of a reaction or a physical change. The mathematical relationship between these energy terms is shown below:

$$\Delta G = \Delta H - T \times \Delta S$$

The change in free energy of a reaction allows one to predict whether a reaction will occur. If the change in free energy, ΔG, has a negative value, the reaction is spontaneous and is classified as an exergonic reaction. If the change in free energy, ΔG, has a positive value, the reaction is nonspontaneous and is classified as an endergonic reaction.

The sign of ΔH and ΔS together determine the sign of ΔG. Table 14-4 summarizes the four possible combinations for the signs of ΔH and ΔS and shows how they affect the overall sign of ΔG.

Table 14-4: The effect of ΔH and ΔS on the sign of ΔG.

Case	Sign of ΔH	Sign of ΔS	Sign of ΔG	Comment
1	−(exothermic)	+ (increase in entropy)	−exergonic	Spontaneous
2	+ (endothermic)	−(decrease in entropy)	+ endergonic,	Nonspontaneous
3	+ (endothermic	+ (increase in entropy)	More information is needed*	More information is needed*
4	−(exothermic)	−(decrease in entropy)	More information is needed*	More information is needed*

* The result depends on the relative values of ΔH and $T \times \Delta S$

Worked Example #8

Using Table 14-4, determine whether the following reactions are spontaneous or nonspontaneous.

 a. $CH_3CH_2OH + 3 O_2 \rightarrow 2 CO_2 + 3 H_2O$ + heat; $\Delta S > 0$

 b. Glucose-6-phosphate $\rightarrow$ fructose-6-phosphate; $\Delta G = + 0.40$ kcal/mol

Solutions

 a. *Heat is released,: therefore, ΔH is negative and ΔS is positive; therefore $\Delta G < 0$ and the reaction is spontaneous.*

 b. *The change in free energy, $\Delta G < 0$, therefore the reaction is nonspontaneous.*

Try It Yourself #8

Using Table 14-4, determine whether the following reactions are spontaneous or nonspontaneous.

 a. an endothermic reaction where $\Delta S < 0$

 b. phosphoenolpyruvate $\rightarrow$ pyruvate + P_i; $\Delta G = - 14.8$ kcal/mol

Solutions

 a. *ΔG is: _____ , therefore the reaction is: _____*

 b. *ΔG is: _____ , therefore the reaction is: _____*

In a Nutshell: Energy Management in Metabolism: Coupled Reactions

Many of the individual chemical reactions involved in metabolism are endergonic, and therefore require energy to proceed in the forward direction. Endergonic reactions use the energy from exergonic reactions to go forward. The hydrolysis of ATP to ADP is the most common exergonic reaction available to cells. Energy is transferred from an exergonic reaction to an endergonic reaction by coupling the two reactions. The net free energy change for a coupled reaction is the sum of the free energies of each individual reaction.

Worked Example #9

When glycogen is broken down in muscle, the following reactions occur, resulting in the transformation of glucose-1-phosphate into fructose-6-phosphate.

Reaction	ΔG kcal/mol
Glucose-1-phosphate $\rightarrow$ Glucose-6-phosphate	-1.74
Glucose-6-phosphate $\rightarrow$ Fructose-6-phosphate	$+0.40$
Glucose-1-phosphate $\rightarrow$ Fructose-6-phosphate	

a. Is the reaction of glucose-6-phosphate forming fructose-6-phosphate exergonic or endergonic?

b. Is the coupled reaction exergonic or endergonic? Show your calculation.

c. Is the overall reaction spontaneous or nonspontaneous?

Solutions

a. *The reaction of glucose-6-phosphate to form fructose-6-phosphate is endergonic; ΔG is positive.*

b. *Reaction*   *ΔG kcal/mol*

Glucose-1-phosphate $\rightarrow$ Glucose-6-phosphate	*-1.74*
Glucose-6-phosphate $\rightarrow$ Fructose-6-phosphate	*$+0.40$*
Glucose-1-phosphate $\rightarrow$ Fructose-6-phosphate	*-1.34*

The overall coupled reaction is exergonic. The sum of the individual free energies for each reaction results in a negative value for ΔG.

c. *The overall coupled reaction is spontaneous because the change in free energy is negative.*

Try It Yourself #9

In glycolysis, 1,3-bisphosphoglycerate is converted to 3-phosphoglycerate. This reaction is coupled to the formation of ATP to form ADP. The reactions are shown below.

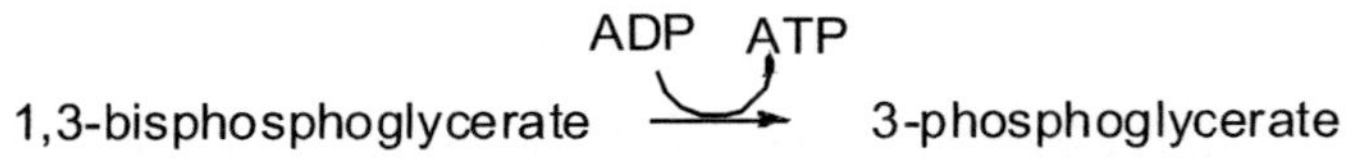

ADP   ATP
1,3-bisphosphoglycerate ⤳➔ 3-phosphoglycerate

a. If $\Delta G = -11.9$ kcal/mol for the formation of 3-phosphoglycerate, is this reaction endergonic or exergonic?

b. Is the overall coupled reaction endergonic or exergonic? Note: $\Delta G = +7.3$ kcal/mol for the formation of ATP.

c. What is $\Delta G_{overall}$?

d. Which reaction provided the energy to drive the coupled reaction?

a. *The sign of ΔG is:* _____

   The reaction is: _____

b. *Calculate $\Delta G_{overall}$.*

c. *The value of $\Delta G_{overall}$:* _____

d. *The reaction that provided the energy to drive the coupled reaction is:*

   _____

Practice Problems for Entropy and Bioenergetics

1. Which of the following represents an increase in entropy?

   a. H_2O (g) → H_2O (l)

   b. E-S complex → E (enzyme) + P (product)

c. PCl_3 (l) + Cl_2 (g) $\rightarrow$ PCl_5 (s)

2. Which of the following reactions are spontaneous?

a. glucose-1-phosphate $\rightarrow$ glucose-6-phosphate; $\Delta G = -1.74$ kcal/mol

b. An endothermic reaction with a decrease in entropy ($\Delta S < 0$)

c. Palmitic acid, $C_{16}H_{32}O_2$ + 23 O_2 $\rightarrow$ 16 CO_2 + 16 H_2O; $\Delta G = -2338$ kcal/mol

3. In the construction of cell membranes the conversion of glycerol to glycerol-3-phosphate is coupled to the hydrolysis of ATP to ADP.

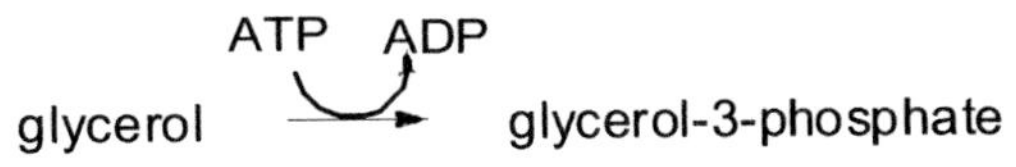

a. If $\Delta G = +2.2$ kcal/mol for the formation of glycerol-3-phosphate, is this reaction endergonic or exergonic reaction?

b. Is the overall coupled reaction endergonic or exergonic?

c. What is $\Delta G_{overall}$?

d. Which reaction provided the energy to drive the coupled reaction?

Chapter 14 Quiz

1. What molecules produced in the citric acid cycle provide the energy for the phosphorylation of ADP into ATP?

2. Do the carbon atoms in the carbon dioxide molecules produced in the citric acid cycle come from acetyl CoA or oxaloacetate?

3. The structure of isocitrate in its unionized form is shown below.

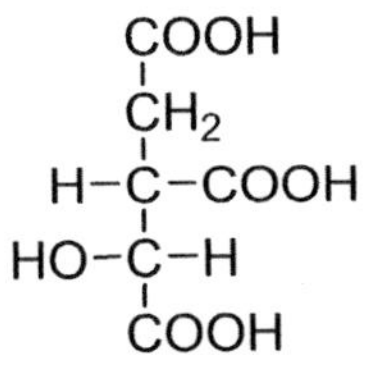

```
    COOH
    |
    CH₂
    |
  H-C-COOH
    |
HO-C-H
    |
    COOH
```

   a. What structure would this compound have at physiological pH?

   b. What other functional group is present in this molecule?

   c. How many carbon atoms are in isocitrate?

4. Which has a higher pH, the intermembrane space or the matrix? Which has a lower concentration of protons?

5. What is the proton-motive force?

6. How does *ATP synthase* generate energy to phosphorylate ADP into ATP?

7. Which of the following processes are nonspontaneous?
   a. a tomato plant growing in the sun

   b. water falling down a waterfall

   c. cleaning your messy room

8. Which of the following represents an increase in entropy?
   a. liquid water turning into steam

   b. ten amino acids reacting together to form a decapeptide

   c. N_2 (g) + $3H_2$ (g) →2 NH_3 (g)

9. Which of the following reactions are spontaneous or nonspontaneous?
   a. glucose + P_i → glucose-6-phosphate; ΔG = +3.0 kcal/mol

   b. ATP → ADP + P_i; ΔG = −7.3 kcal/mol

c. α-ketoglutarate → succinyl-CoA; ΔG = −8.0 kcal/mol

d. an exothermic reaction with an increase in the change of entropy

10. In step 5 of the citric acid cycle, succinyl-CoA is converted into succinate. This reaction is coupled to the formation of GTP as shown below.

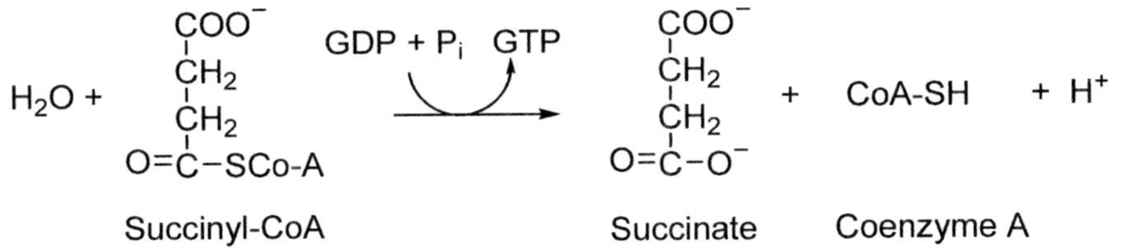

H_2O +

COO⁻
|
CH_2
|
CH_2
|
O=C−SCo-A

GDP + P_i GTP

COO⁻
|
CH_2
|
CH_2
|
O=C−O⁻

+   CoA-SH   + H⁺

Succinyl-CoA   Succinate   Coenzyme A

a. If ΔG = 8 kcal/mol for the formation of succinate, is this reaction endergonic or exergonic reaction?

b. What is $\Delta G_{overall}$? Note: For the formation of GTP from GDP, ΔG = −8.8 kcal/mol.

c. Is the overall coupled reaction endergonic or exergonic?

d. Which reaction provided the energy to drive the coupled reaction?

Chapter 14
Answers to Additional Exercises

14.23 Catabolic pathways break down larger molecules into smaller molecules and produce energy. Anabolic pathways build larger molecules from smaller molecules and consume energy.

14.25 Glycolysis, β-oxidation of fatty acids, and the degradation of several amino acids produce acetyl coA.

14.27 The citric acid cycle begins with the reaction of acetyl CoA with oxaloacetate. The last step of the biochemical pathway regenerates oxaloacetate.

14.29 a.

$$\begin{array}{c} COOH \\ | \\ CH_2 \\ | \\ H-C-COOH \\ | \\ HO-C-H \\ | \\ COOH \end{array}$$

b. Isocitrate and citrate have three carboxylic acid groups, COOH, and a hydroxyl group, OH, attached to a three-carbon atom chain. The difference between the two isomers is the location of the hydroxyl group.

c. They are structural isomers.

14.31 After half of the citric acid cycle, two carbon atoms have been removed from citrate. Two carbon dioxide have been expelled in order to remove two carbon atoms. Two molecules of NADH are produced in the first half of the citric acid cycle.

14.33 The citric acid cycle produces *three* NADH molecules. Every NADH molecule produces 3 ATP molecules.

14.35 Steps 3 and 4 produce the carbon dioxide that we exhale. The carbon atom in carbon dioxide comes from oxaloacetate.

14.37 A new carbon-carbon bond is formed. Oxaloacetate and acetyl CoA react to form this new bond.

14.39 a. An oxidation-reduction reaction occurs in this step.

b. Succinate is oxidized.

c. FAD is reduced.

d. A carbon-carbon single bond is oxidized to a carbon-carbon double bond.

14.41 a. An oxidation-reduction reaction occurs in step 8.

      b. A secondary alcohol is converted into a ketone and NAD$^+$ is converted into NADH.

      c. L-malate is oxidized.

      d. The oxaloacetate produced in this step can react with another molecule of acetyl CoA, starting another round of the citric acid cycle.

14.43 Glycolysis–the conversion of glucose to pyruvate–occurs in the cytoplasm of the cell.

14.45 Ions can pass easily through the outer membrane.

14.47 The matrix has a higher pH.

14.49 The potential energy stored in the unequal distribution of protons is called the proton-motive force.

14.51 Cyanide poisoning is lethal because cyanide is an irreversible inhibitor of *cytochrome c oxidase,* a key enzyme in the last step of the electron transport chain. Cyanide brings the electron-transport chain to a halt, by preventing phosphorylation of ADP and depriving the cell of the energy needed to function.

14.53 *ATP synthase* uses the proton-motive force to drive phosphorylation of ADP to ATP.

14.55 A spontaneous reaction continues on its own once it's started, whereas in a nonspontaneous reaction energy needs to be continuously supplied to the reaction for it to proceed.

14.57 b. and c. A plant growing in sunlight needs the constant energy from the sun to keep growing. Swinging on a swing needs constant energy to keep the swing in motion.

14.59 b., d., and e. represent an increase in entropy. When a reaction produces more product molecules than reactant molecules, the product molecules can move in more ways than the reactant molecules. Molecules in liquid water have more freedom of motion than those in snow. Molecules in steam have more freedom of motion than those in liquid water.

14.61 The binding of a substrate to an enzyme to form an enzyme substrate complex is a decrease in entropy. The enzyme–substrate complex has less freedom of motion than the individual enzyme and substrate. The entropy of the surroundings increases more than the enzyme-substrate entropy decreases; therefore, the entropy of the universe increases.

14.63 a. spontaneous, ΔG is negative

      b. nonspontaneous, ΔG is positive

      c. spontaneous, ΔG is negative

      d. spontaneous, ΔG is negative

14.65 The reaction is spontaneous, ΔG is negative.

14.67 a. Spontaneous. There is an increase in entropy because there are more product

molecules than reactant molecules. $\Delta S > 0$ and $\Delta H < 0$, therefore $\Delta G < 0$.

b. Nonspontaneous. $\Delta H > 0$, $\Delta S < 0$, therefore $\Delta G > 0$.

c. Spontaneous. $\Delta H < 0$, $\Delta S > 0$, therefore, $\Delta G < 0$.

14.69 Temperature plays a key role in determining ΔG in two cases: (1) when ΔH is

negative and ΔS is negative and (2) when ΔH is positive and ΔS is positive.

Temperature is multiplied by ΔS in the equation for free energy; therefore, entropy

has a greater influence at higher temperatures.

14.71 a. The phosphorylation of glucose is endergonic.

b.

Reaction	ΔG, Kcal/mol
fructose $\rightarrow$ fructose-6-phosphate	+3.8
ATP $\rightarrow$ ADP	−7.3

fructose $\xrightarrow{\text{ATP ADP}}$ fructose-6-phosphate	−3.5

The overall coupled reaction is exergonic.

c. $\Delta G_{overall} = -3.5$ kcal/mol

14.73 a. The formation of pyruvate is exergonic, ΔG is negative.

b.

Reaction	ΔG, Kcal/mol
phosphoenolpyruvate $\rightarrow$ pyruvate	−14.8
ADP $\rightarrow$ ATP	+7.3

phosphoenolpyruvate $\xrightarrow{\text{ADP ATP}}$ pyruvate	−7.5

The overall coupled reaction is exergonic.

c. $\Delta G_{overall} = -7.5$ kcal/mol

d. The reaction of phosphoenolpyruvate forming pyruvate provided the energy to

drive the phosphorylation of ADP.

14.75 If left untreated, PKU leads to mental retardation and regular seizures.

14.77 In a person who does not have PKU, phenylalanine is either converted into tyrosine

or incorporated into polypeptide chains.

14.79   If *phenylalanine hydroxylase, PAH,* were ingested in pill form, the enzymes and acid in the digestive tract would degrade the enzyme..

Chapter 15

Nucleic Acids: DNA and RNA

Chapter Summary

In this chapter, you studied DNA and RNA. You learned about the nucleotides that make-up the nucleic acids, DNA and RNA. You observed how DNA is held together. You learned how DNA replicates itself so that the genetic information it contains can be passed onto daughter cells. You gained insight into how information flows from DNA to RNA to construct a protein from its individual amino acids. Understanding the structure and function of DNA helps people in the health field fight diseases.

Section 15.1 Nucleotides and Nucleic Acids

Nucleic acids are constructed from nucleotides. There are only four nucleotides used to build a nucleic acid. A nucleotide contains three basic parts: 1) a nitrogen containing ring called a base, 2) a monosaccharide, and 3) a phosphate group. A nucleoside only contains the nitrogen base and the monosaccharide.

In a Nutshell: The Monosaccharide Component

The monosaccharide component of a nucleotide is *D*-ribose, when it is part of RNA and the related sugar 2-deoxy-*D*-ribose, when it is part of DNA. These two monosaccharides differ only at the second carbon atom, where 2-deoxy-*D*-ribose lacks an OH group. The carbons in the monosaccharide ring are numbered beginning at the anomeric carbon from 1' to 5' proceeding clockwise around the ring.

In a Nutshell: The Nitrogen Base

The nitrogen bases found in the nucleotides that make up DNA and RNA are derived from purines and pyrimidines. DNA contains the four bases, adenine, A, guanine, G, cytosine, C and thymine, T. RNA contains the same bases except that thymine is replaced by uracil, U. In a nucleoside, the anomeric carbon, the 1' carbon atom, on the monosaccharide is covalently bonded to one of these four possible nitrogen bases. The covalent bond between the monosaccharide and the base is called a β-*N*-glycosidic bond.

In a Nutshell: The Phosphate Group

To create a nucleotide from a nucleoside, a phosphate group is attached to the 5' alcohol of the monosaccharide.

Worked Example #1

For the nucleotide shown below, circle and label the following:

a.   The nitrogen base. What is the name of the nitrogen base?

b.   The monosaccharide. What is the name of this monosaccharide?

c.   The phosphate group

d.   The 5' carbon

e.   The 3' carbon

f.   Would this nucleotide be found in DNA or RNA? Explain.

Solutions

a–e.

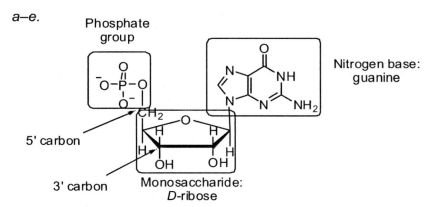

f.   *This nucleotide would be found in RNA because the monosaccharide is D-ribose.*

Try It Yourself #1

For the nucleotide that follows, circle and label the following:

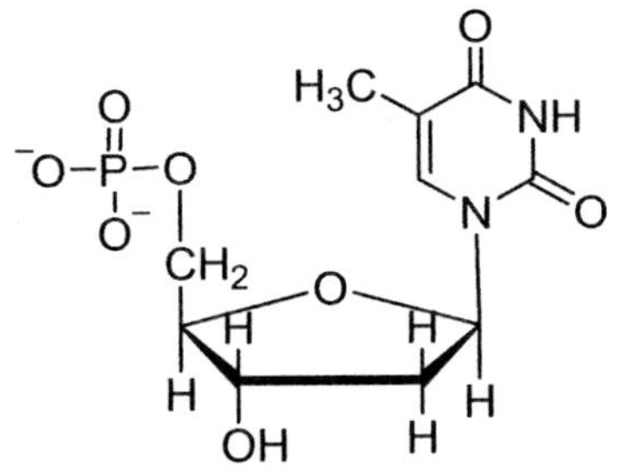

a. The nitrogen base. What is the name of the nitrogen base?

b. The monosaccharide. What is the name of this monosaccharide?

c. The phosphate group

d. The 2' carbon

e. The anomeric carbon

f. Would this nucleotide be found in DNA or RNA? Explain.

Solutions

a–e.

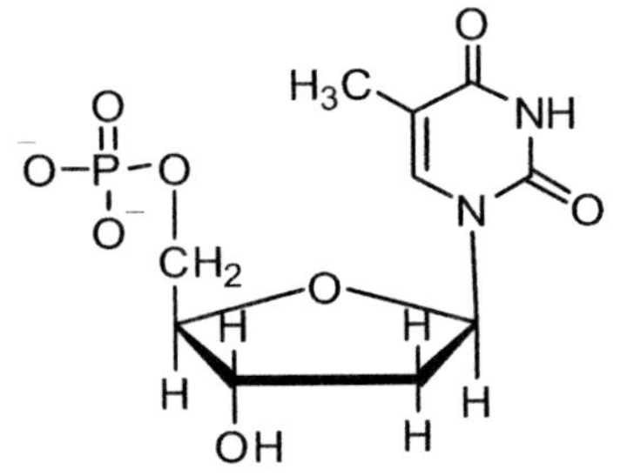

f. Does the 2' carbon atom have an OH group bonded to it? _____

The nucleotide is found in: _____

In a Nutshell: Nucleic Acids

The nucleic acids, 2-deoxyribonucleic acid (DNA) and ribonucleic acid (RNA) are formed when nucleotides are joined by covalent bonds to produce a linear sequence of nucleotides. The bond between two nucleotides is a phosphate ester bond formed between the 3' alcohol of one nucleotide and the phosphate group of the next nucleotide in the sequence. Except for the nucleotides at the end of the sequence, the 5' OH and the 3' OH groups of each nucleotide are joined to two different phosphate groups. At one end is a free phosphate group on the 5' carbon, known as the "five prime end." On the other end is a free OH group on the 3' carbon, known as the "three prime end." Thus, the backbone of the nucleic acids consists of repeating sugar-phosphate units, and the only difference is the identity of the nitrogen base. Therefore, a nucleic acid need only be identified by the sequence of its nitrogen base.

The convention is to list the sequence of nitrogen bases starting from the 5' end and working towards the 3' end (5' → 3'). If the monosaccharide is 2-deoxy-*D*-ribose, as in the case of DNA, a "d" is inserted as a prefix before the sequence. Statistically, the number of different nucleic acids that can be formed from a sequence of n nucleotides is 4^n.

Worked Example #2

Statistically, how many different heptanucleotides—a sequence of seven nucleotides—can be constructed? Show your calculation.

The number of possible nucleic acid structures that can be formed from four different nucleotides is $4^7 = 4 \times 4 \times 4 \times 4 \times 4 \times 4 \times 4 = 16,384$.

Try It Yourself #2

Statistically, how many different dinucleotides—a sequence of two nucleotides—can be constructed? Show your calculation.

Calculation:

The number of possible nucleic acids is: _____

Worked Example #3

The two nucleotides that follow are T and G. Write the structure of the dinucleotide dTG and circle the new phosphate ester bond. Label the 5' and the 3' ends of the dinucleotide. Is the dinucleotide dTG the same as or different from dGT?

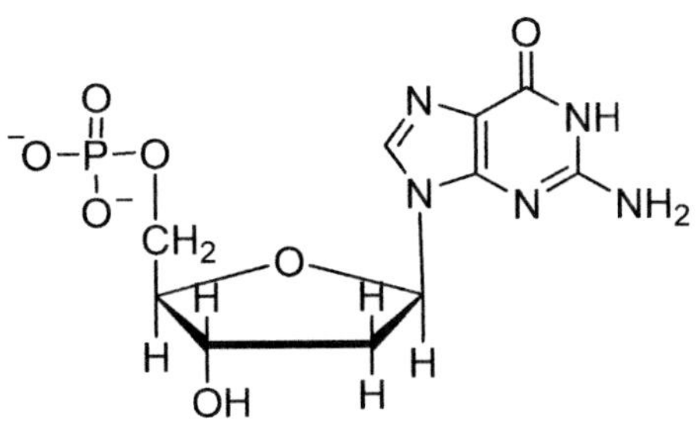

G

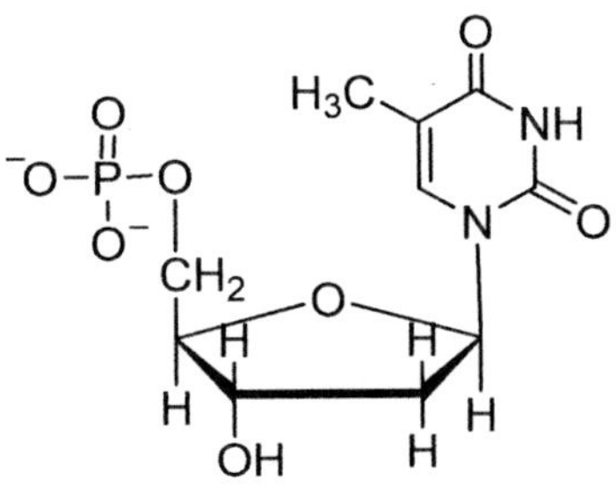

T

The structure of dTG is shown below:

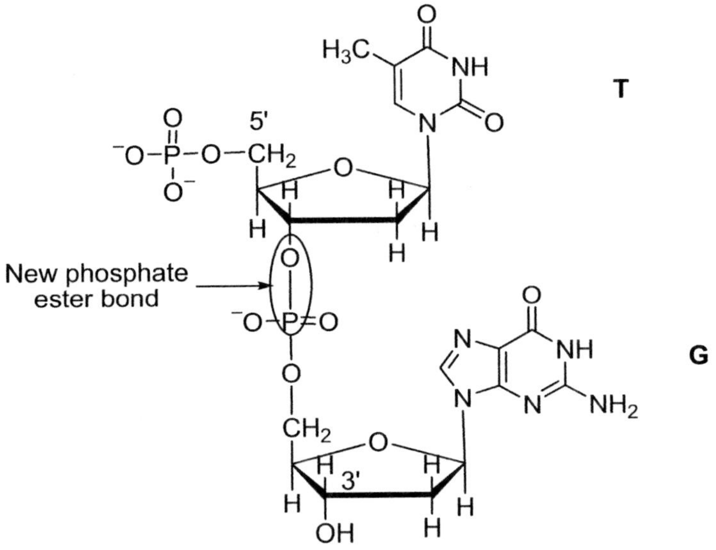

T

New phosphate ester bond

G

5'

3'

The molecule dTG is different from dGT.

Try It Yourself #3

The two nucleotides that follow are U and C. Write the structure of the dinucleotide CU and circle the new phosphate ester bond. Label the 5' and the 3' ends of the dinucleotide. Is the dinucleotide CU the same as or different from UC?

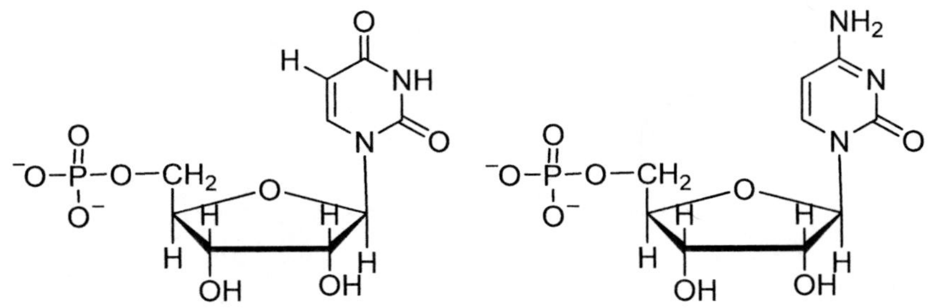

U

C

Structure of CU:

Is the nucleotide CU the same or different from UC? _____

Practice Problems for Nucleotides and Nucleic Acids

1. For the nucleotide shown below, circle and label the following:

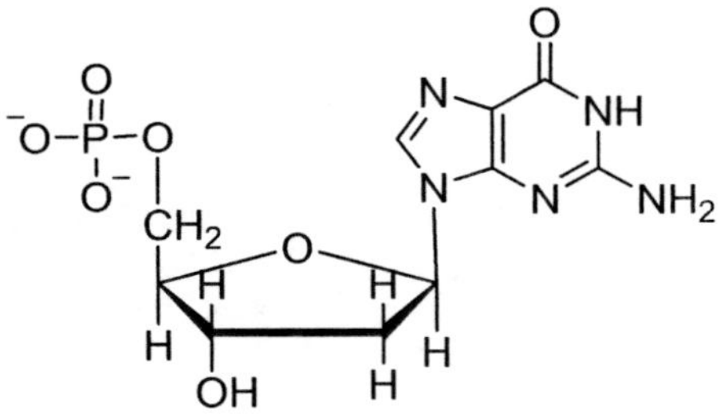

a. The nitrogen base. What is the name of the nitrogen base?

b. The monosaccharide. What is the name of this monosaccharide?

   c. The phosphate group

   d. The 5' carbon

   e. The anomeric carbon

   f. The 2' carbon

   g. The 3' carbon

   h. Would this nucleotide be found in DNA or RNA? Explain.

2. Statistically, how many different tetranucleotides—a sequence of four nucleotides—
    can be constructed? Show your calculation.

3. Two nucleotides are shown below: T and A. Write the structure of the dinucleotide
    dAT and circle the new phosphate ester bond. Label the 5' and the 3' ends of the
    dinucleotide. Is the dinucleotide dAT the same as or different from dTA?

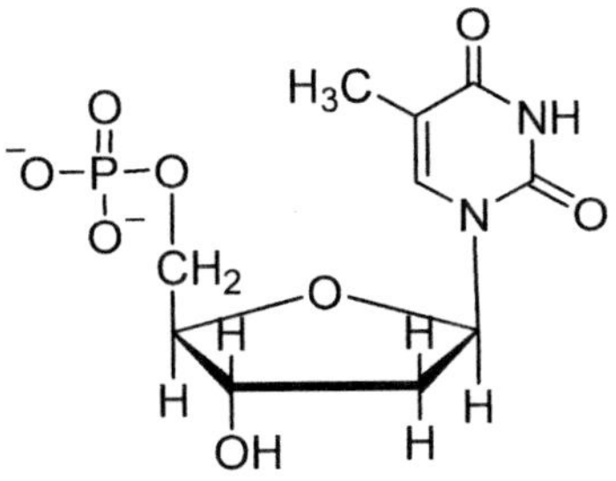

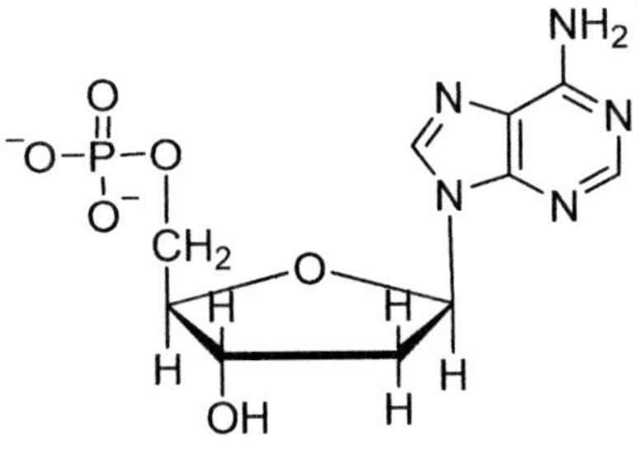

Section 15.2 DNA

In a Nutshell: Double Helix Structure of DNA

DNA consists of two nucleic acid molecules, each a linear sequence of millions of nucleotides. The two linear molecules are often referred to as two "strands" of DNA. These two strands are twisted around each other into a double helix. The sugar-phosphate backbone forms the outer portion of the DNA double helix. The monosaccharide and phosphate groups of the backbone are hydrophilic, so they are positioned on the outside of the three dimensional structure. The two strands of DNA are arranged antiparallel to each other, which means that their 5' and 3' ends are positioned at opposite ends. The relatively nonpolar nitrogen bases are projected toward the interior of the DNA double helix. Each nitrogen base forms hydrogen bonds to a complementary base on the adjacent strand, creating base pairs. The complementary base pairs in DNA are A-T and G-C. There are two hydrogen bonds that form between A and T and three hydrogen bonds that form between G and C.

Worked Example #4

The sequences of nucleotides located on a segment of one DNA strand are shown below. Indicate the complementary sequence of base pairs that would appear on the other DNA strand. How many hydrogen bonds hold this section of DNA together?

    dTAGC

The complementary base pairing would be:
    dTAGC
    dATCG

Each A-T pair has two hydrogen bonds and each G-C pair has three hydrogen bonds (2 A-T pairs × 2 hydrogen bonds) + (2 G-C pairs × 3 hydrogen bonds) = 10 hydrogen bonds.

Try It Yourself #4

The sequences of nucleotides located on a segment of one DNA strand are shown below. Indicate the complementary sequence of base pairs that would appear on the other DNA strand. How many hydrogen bonds hold this section of DNA together?

    dGGACT

DNA strand: dGGACT

Complementary strand: _____

Number of A-T base pairs: _____

Number of G-C base pairs: _____

Total number of hydrogen bonds: _____

Worked Example #5

The sequences of nucleotides located on a segment of one DNA strand are shown below. Indicate the complementary sequence of base pairs that would appear on the other DNA strand.

    a. dTAGGCCAT

    b. dGGCCAATT

The complementary base pairings would be:

    a. dTAGGCCAT

        dATCCGGTA

    b. dGGCCAATT

        dCCGGTTAA

Try It Yourself #5

The sequences of nucleotides located on a segment of one DNA strand are shown below. Indicate the complementary sequence of base pairs that would appear on the other DNA strand.

    a. dGACGTGGCA

    b. dAATCCGAGG

a. *DNA strand:*   dGACGTGGCA

    Complementary strand: _____

b. *DNA strand:*   dAATCCGAGG

    Complementary strand: _____

In a Nutshell: Higher-Order DNA Structure: Chromosomes

DNA is present in the chromosomes found in the nucleus of cells. Each chromosome contains one double stranded DNA double helix. The DNA within the chromosome is supercoiled to fit within the nucleus of the cell. The DNA is first coiled around a core of proteins, known as histones, creating a nucleosome. The electrostatic attraction between the positive charge on the histone proteins and the negative charge on the DNA phosphate groups facilitates the formation of nucleosomes. Nucleosomes coil upon themselves to create a chromatin fiber. The chromatin fibers coil even further forming the familiar X-shaped structure of the chromosome. The DNA double helix is about 2 nm wide, while a chromosome is about 1400 nm wide. Most human cells contain 46 chromosomes.

In a Nutshell: Genes and the Human Genome

The complete sequence of bases in your DNA, distributed over 46 chromosomes, is known as your genome. The human genome contains about 3 million base pairs. Only about 2% of the human genome contains sequences of proteins that code for proteins. A gene is a segment of DNA that contains the instructions for making a protein.

The entire nucleotide sequence of the human genome has been determined and the genes associated with many proteins were identified.

In a Nutshell: DNA Replication

DNA must be able to replicate itself before cell division in order for the information encoded within DNA to be passed on to the new cells. DNA replication begins with the unraveling of the supercoiled DNA to expose the double helix. Then, the enzyme *helicase*, along with ATP catalyzes the unwinding of the double helix. Each strand of DNA serves as a template for a new daughter strand that is complementary to the template.

Replication begins where a set of three nucleotides codes for the start of a gene, known as a start codon. The two strands of DNA come apart at a site called the replication fork. Nucleotides containing complementary bases then assemble along the template provided by the parent strand forming new hydrogen bonds with the complementary base on the parent strand. *DNA polymerase* catalyzes the formation of phosphate ester bonds between the assembled nucleotides. *DNA polymerase* also proofreads the daughter strands for mistakes. When errors are detected, it signals other enzymes to replace and repair incorrectly placed nucleotides.

Worked Example #6

Consider a portion of double stranded DNA with the following complementary sequence of base pairs:

dATATGCGGCCATA

dTATACGCCGGTAT

Write the sequence of nucleotides found in each daughter after replication and label the original parent strands and the two new daughter strands.

Parent strands:	*dATATGCGGCCATA*	*dTATACGCCGGTAT*
Daughter strands:	*dTATACGCCGGTAT*	*dATATGCGGCCATA*

Try It Yourself #6

Consider a portion of double stranded DNA with the following complementary sequence of base pairs:

dGAGCCTTCCAACG

dCTCGGAAGGTTGC

Write the sequence of nucleotides found in each daughter after replication and label the original parent strands and the two new daughter strands.

Parent strands: _____

Daughter strands: _____

Chapter 15

Practice Problems for DNA

1. The sequences of nucleotides located on a segment of one DNA strand are shown below. Indicate the complementary sequence of base pairs that would appear on the other DNA strand. How many hydrogen bonds hold this section of DNA together?
   dGCCAATGACT

2. Explain how DNA coils to fit inside the nucleus of a cell.

3. Consider a portion of double-stranded DNA with the following complementary sequence of base pairs:
   dGCCTCTAGAATGAG
   dCGGAGATCTTACTC

   Write the sequence of nucleotides found in each daughter after replication and label the original parent strands and the two new daughter strands.

Section 15.3 RNA and Protein Synthesis

RNA is needed to direct the synthesis of a protein—the assembly of amino acids in their correct order. RNA is found in structures outside the nucleus of the cell called ribosomes. There are three major forms of RNA: 1) ribosomal RNA (rRNA), 2) messenger RNA (mRNA), and 3) transfer RNA (tRNA). RNA is single strand and contains *D*-ribose instead of 2-deoxy-*D*-ribose.

Ribosomes, located in the cytoplasm, are the protein making factories of the cell, where amino acids are assembled into proteins. Ribosomes are composed of ribosomal RNA and about 50 different proteins. Messenger RNA (mRNA) is used to deliver the instructions encoded in DNA to the ribosomes because DNA never leaves the nucleus.

The process of protein synthesis involves two steps, transcription and translation. In transcription, the portion of DNA carrying the instructions for a particular protein—a gene—is copied to form a complementary strand of messenger RNA. Translation takes place at the ribosome where the amino acids are assembled into a protein, whose sequence of amino acids is determined by the instructions provided by mRNA. Amino acids are delivered to the ribosome by transfer RNA (tRNA).

In a Nutshell: Transcription: DNA to mRNA

When a particular protein is needed by the cell, the gene that codes for it is expressed. Gene expression begins with transcription where the nucleotide sequence on DNA—the gene—is copied as a complementary single-stranded mRNA molecule. mRNA begins at the three nucleotides on the DNA strand known as start codons. Only one of the two strands of DNA is copied. The strand that is copied is known as the template strand.

The double helix must be first unraveled to expose the nucleotide sequence of the gene that is to be copied. Then, *RNA polymerase* catalyzes the synthesis of a new mRNA molecule: It copies the sequence of nucleotides on the template strand of DNA from the 3' end to the 5' end. *RNA polymerase* catalyzes the formation of phosphate ester bonds. The base sequence of the newly created mRNA is complementary to the DNA sequence on the template strand. Once transcription is complete, DNA refolds into the double helix structure and the mRNA is exported out of the nucleus and into the cytoplasm.

In a Nutshell: Translation: mRNA to tRNA and Protein Synthesis

At the ribosome, the nucleotide sequence of mRNA is read and used to build a polypeptide. Every grouping of three nucleotides on an mRNA molecule is known as a codon and codes for one of the 20 natural amino acids. The set of three-nucleotide codons and the amino acids they specify is known as the genetic code. A few codons do not specify an amino acid, but instead mark the start and end of a gene.

Transfer RNA (tRNA) is responsible for matching a codon on mRNA with the amino acid that it encodes. tRNA contains two important regions: the anticodon loop and the 3' end covalently bound to one particular amino acid. The anticodon loop on a tRNA contains three nucleotides that recognize the complementary codon on mRNA.

At the ribosome, the first mRNA codon is read from the 5' end of the mRNA and a matching tRNA molecule is recruited, temporarily forming base pairs between the anticodon of tRNA and the codon of mRNA. Another matching tRNA arrives and an amide bond is formed between the two amino acids on adjacent tRNA molecules docked on the ribosome. The first tRNA is released in the process and diffuses away without its amino acid attached. The second tRNA still bound to the ribosome now contains a dipeptide. The ribosome then shifts in the 5' to 3' direction to read the next codon on the mRNA, a process known as translocation. The next codon is read and another tRNA molecule is recruited. These steps repeat in an overall process known as translation, which continues until a stop codon is reached. The polypeptide must still undergo additional modifications to become a functional protein, folding and joining any additional subunits to adopt the tertiary structure.

In a Nutshell: Genetic Mutations

A genetic mutation is any permanent chemical change that occurs at one or more nucleotides in the DNA sequence and affects the primary structure of a protein. A mutation may involve a substitution of one nucleotide or the deletion of a nucleotide. The effects of DNA mutations are minimized by the fact there is more than one codon for any given amino acid. If a mutation changes one or more amino acids in a metabolic enzyme, the enzyme may no longer be able to perform its function. Mutations can be caused by a number of factors including ultraviolet radiation from the sun; certain chemicals, such as those found in cigarettes; chemotherapy; and high-energy radiation such as x-rays and gamma rays.

Worked Example #7

Using Table 15-1, determine the amino acid specified by the following codons:

    a. UGG

    b. ACG

    c. GUG

    a. UGG on mRNA codes for tryptophan.

    b. ACG on mRNA codes for threonine.

    c. GUG on mRNA codes for valine.

Try It Yourself #7

Using Table 15-1, determine the amino acid specified by the following codons:

    a. AUC

    b. CAA

    c. CGC

    a. AUC on mRNA codes for: _____.

    b. CAA on mRNA codes for: _____.

    c. CGC on mRNA codes for: _____.

Worked Example #8

For the following nucleotide sequences on mRNA, indicate the anticodons on the three tRNA molecules recruited. What is the amino acid sequence of the tripeptide formed?

    a. UGUUACGAA

    b. UUCCAAGCG

    a. mRNA strand: *UGUUACGAA*

        Anticodons on tRNA molecules: *ACA AUG CUU*

        Amino acids indicated by codons on mRNA: *Cys Tyr Glu*

        The tripeptide will be Cys-Tyr-Glu.

    b. mRNA strand: *UUCCAAGCG*

        Anticodons on tRNA molecules: *AAG GUU CGC*

Amino acids indicated by codons on mRNA: Phe   Gln   Ala

The tripeptide will be Phe-Gln-Ala.

Try It Yourself #8

For the following nucleotide sequences on mRNA, indicate the anticodons on the three tRNA molecules recruited. What is the amino acid sequence of the tripeptide formed?

a. GAUAUUAUG

b. CAUCGACAA

a. mRNA strand: _____

Anticodons on tRNA molecules: _____   _____   _____

Amino acids indicated by codons on mRNA: _____   _____   _____

Tripeptide: _____

b. mRNA strand: _____

Anticodons on tRNA molecules: _____   _____   _____

Amino acids indicated by codons on mRNA: _____   _____   _____

Tripeptide: _____

Worked Example #9

For the following sequences on DNA, write the corresponding mRNA sequence. What tRNA molecules would be involved in building the proteins? What is the amino acid sequence?

a. dGAGTACCTACCC

b. dATGGCTCAACGG

a. DNA strand: dGAGTACCTACCC

mRNA strand: CUCAUGGAUGGG

Anticodons on tRNA: GAG UAC CUA CCC

Amino acids: Leu Met Asp Gly

Peptide chain: Leu-Met-Asp-Gly

b. DNA strand: dATGGCTCAACGG

mRNA strand: UACCGAGUUGCC

Anticodons on tRNA: AUG GCU CAA CGG

| Amino acids: | Tyr Arg Val Ala |
| Peptide: | Tyr-Arg-Val-Ala |

Try It Yourself #9

For the following sequences on DNA, write the corresponding mRNA sequence. What tRNA molecules would be involved in building the proteins? What is the amino acid sequence?

a. dTAAGGACTTTGG

b. dTTTTCATTGGTT

a. DNA strand: _____

   mRNA strand: _____

   Anticodons on tRNA: _____ _____ _____ _____

   Amino acids: _____ _____ _____ _____

   Peptide chain: _____

b. DNA strand: _____

   mRNA strand: _____

   Anticodons on tRNA: _____ _____ _____ _____

   Amino acids: _____ _____ _____ _____

   Peptide chain: _____

Worked Example #10

Indicate whether the following normal mRNA sequence would produce the same or a different dipeptide if the mutation shown occurred (assume all reading occurs from left to right):

| DNA | dAATGTA |
| mRNA | UUACAU |

a. The third nucleotide on DNA is substituted with C.

b. The fifth nucleotide on DNA is substituted with A

c. The fourth nucleotide is deleted.

Solutions

   DNA: *dAATGTA*

mRNA: UUACAU

Dipeptide: Leu-His

a. DNA mutation: dAA**C**GTA

   mRNA: UUGCAU

   Dipeptide: Leu-His

   There is no change in the dipeptide.

b. DNA mutation: dAAT**GA**A

   mRNA: UUACUU

   Dipeptide: Leu-Leu

   The dipeptide is different.

c. DNA mutation: dAATTA

   mRNA: UUGAU

   Dipeptide: Leu-?

   The dipeptide will be different because the first nucleotide for the codon for the second amino acid is not C. The codon for histidine begins with C.

Try It Yourself #10

Indicate whether the following normal mRNA sequence would produce the same or a different dipeptide if the mutation shown occurred (assume all reading occurs from left to right):

DNA   dTGATCA

mRNA   ACUAGU

a. The third nucleotide on DNA is substituted with C.

b. The sixth nucleotide on DNA is substituted with G.

c. The second nucleotide is deleted.

Solutions

DNA   dTGATCA

mRNA   ACUAGU

Dipeptide: _____

a. *DNA mutation:* _____

　　mRNA: _____

　　Dipeptide: _____

　　Effect on dipeptide: _____

b. *DNA mutation:* _____

　　mRNA: _____

　　Dipeptide: _____

　　Effect on dipeptide: _____

c. *DNA mutation:* _____

　　mRNA: _____

　　Dipeptide: _____

　　Effect on dipeptide: _____

Chapter 15 Quiz

1. The three basic parts of a nucleotide are given below.

    a. What are each of these molecules called?

    b. Construct a nucleotide from these three parts.

    c. Label the 3' and the 5' carbon atoms in your nucleotide.

    d. Would this nucleotide be found in DNA or RNA? How can you tell?

2. Statistically, how many different trinucleotides—a sequence of three nucleotides—can be constructed? Show your calculation.

3.   How many hydrogen bonds are found in the complementary base pair A-T in DNA?
      How many hydrogen bonds are found in the complementary base pair G-C in DNA?

4.   Below are the sequences of nucleotides located on a segment of one DNA strand.
      Indicate the complementary sequence of base pairs that would appear on the other
      DNA strand.

      a. dATCGGCAATT

      b. dGCATGCCATAG

5.   What is a gene?

6.   Consider a portion of double stranded DNA with the following complementary
      sequence of base pairs:

          dTGCCATCATG
          dACGGTAGTAC

      Write the sequence of nucleotides found in each daughter strand after replication and
      label the original parent strands and the two new daughter strands.

7.   What role does *DNA polymerase* play in DNA replication?

8. For the following nucleotide sequences on mRNA, indicate the anticodons on the three tRNA molecules recruited. What is the amino acid sequence of the tripeptide formed?

    a. CGCAUGACU

    b. GUCGAUAAG

9. For the following sequences on DNA, write the corresponding mRNA sequence. What tRNA molecules would be involved in building the proteins? What is the amino acid sequence?

    a. dGGGACAATGTAA

    b. dGTACGGCCCAAT

10. Indicate whether the following normal mRNA sequence would produce the same or a different dipeptide if the mutation shown occurred:

DNA   dGCCCGA

mRNA   CGGGCU

a. The first nucleotide on DNA is substituted with T.

b. The fifth nucleotide on DNA is substituted with A

c. The third nucleotide is deleted.

Chapter 15

Answers to Additional Exercises

15.29 A gene is a section of DNA that contains the instructions for making a protein. A mutation is a permanent alteration in one or more nucleotides of a gene, which in turn produces an altered protein.

15.31 A nucleotide contains a nitrogen containing ring called a base, a monosaccharide and a phosphate group. A nucleoside has a base and a monosaccharide, but no phosphate group.

15.33 a–f.

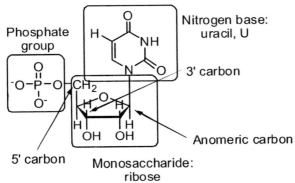

    g. The nucleotide would be found in RNA. The second carbon atom in the monosaccharide has an OH group.

15.35

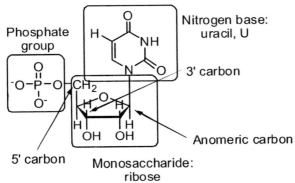

15.37 When two nucleotides are joined together to form a dinucleotide, the alcohol on 3' carbon of one nucleotide and the phosphate group on the 5' carbon of the other nucleotide. A phosphate ester bond is produced.

15.39 The number of possible hexanucleotides that can be formed from four different nucleotides is $4^6 = 4 \times 4 \times 4 \times 4 \times 4 \times 4 = 4096$.

15.41

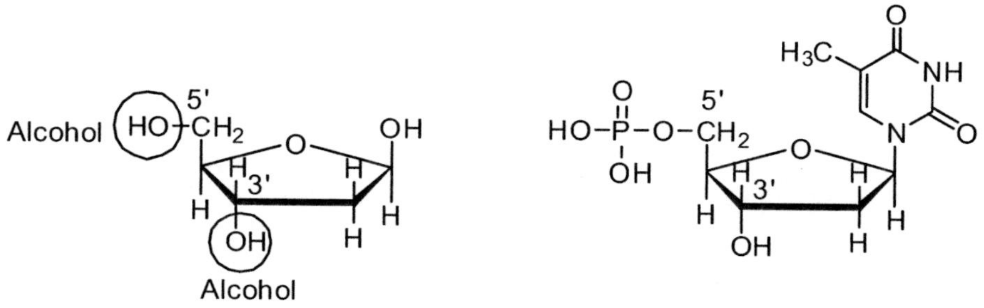

The nucleotide would be found in DNA. The second carbon atom in the monosaccharide is missing an OH group.

15.43

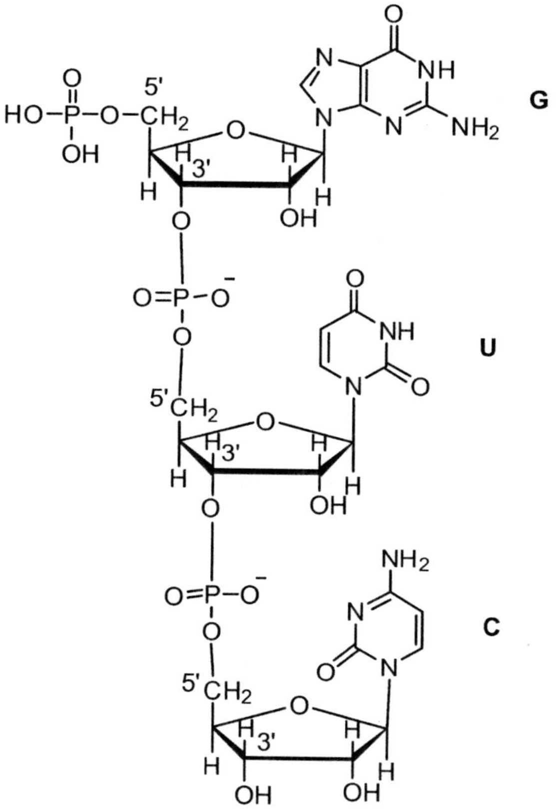

15.45 The monosaccharide and phosphate groups are located on outside of the three-dimensional DNA structure. These groups are hydrophilic and can interact with the polar aqueous environment.

15.47 Hydrogen bonds exist between base pairs.

15.49 A C–G base pair is linked by three hydrogen bonds.

15.51 The complementary sequence is dGCTATC. Each A–T pair has two hydrogen bonds and each C–G pair has three hydrogen bonds: (3 A–T pairs × 2 hydrogen bonds) + (3 C–G pairs × 3 hydrogen bonds) = 15 hydrogen bonds.

15.53 a. dATACGG

b. dTTGGAC

c. dGGATAA

d. dCATAGG

15.55 Most human cells contain 46 chromosomes.

15.57 See Figure 15-12. For example, diabetes mellitus and schizophrenia are associated with a protein whose gene is located on chromosome 6.

15.59 DNA replication begins with the unraveling of the supercoiled DNA to expose the double helix.

15.61

	DNA 1	DNA 2
Parent:	dGGTACGCTT	dCCATGCGAA
Daughter:	dCCATGCGAA	dGGTACGCTT

15.63 Hydrogen bonds are formed between the base pairs of the parent and daughter strands of DNA.

15.65 When DNA polymerase detects an error in the daughter strand, it signals other enzymes to replace and repair incorrectly placed nucleotides. The mistake rate is less than one in one billion molecules after proofreading.

15.67 a. dCGTACTGGA: Parent strand
dCCATGAACT: Daughter strand (Errors in daughter strand are bolded.)
dGCATGACCT: Corrected daughter strand

b. dGAGTATCT: Parent strand
dCTCCTCGA: Daughter strand (Errors in daughter strand are bolded.)
dCTCATAGA: Corrected daughter strand

15.69 RNA is single stranded and contains *D*-ribose. It also contains uracil, U, instead of thymine, T.

15.71 *RNA polymerase* catalyzes the synthesis of a new mRNA molecule.

15.73 a. dCCGGAATATA: DNA strand
GGCCUUAUAU: mRNA strand

b. dAAGGCCAATT: DNA strand
UUCCGGUUAA: mRNA strand

c. dGTACACGTCG: DNA strand
CAUGUGCAGC: mRNA strand

15.75 The two important regions of a tRNA molecule are the anticodon loop and the 3' end covalently bound to one particular amino acid.

15.77   The anticodon AAA will correspond to the codon UUU; therefore, the amino acid will be phenylalanine.

15.79   No more amino acids will be added to the growing peptide chain when a stop codon is reached. The stop codons are UAA, UAG, and UGA.

15.81   The codons for alanine are GCU, GCC, GCA, and GCG.

15.83   a.   mRNA: GUUGCUCGU

Anticodons on tRNA: CAA-CGA-GCA

Amino acids: Val Ala Arg

Anticodon CAA carrying valine, followed by anticodon CGA carrying alanine, followed by anticodon GCA carrying arginine. The tripeptide would be Val-Ala-Arg.

   b.   mRNA: CUACGCGGU:

Anticodons on tRNA: GAU   GCG   CCA

Amino acids: Leu Arg Gly

Anticodon GAU carrying leucine, followed by anticodon GCG carrying arginine, followed by anticodon CCA carrying glycine. The tripeptide would be Leu-Arg-Gly.

   c.   mRNA: AGUAACUCG

Anticodons on tRNA: UCA   UUG   AGC

Amino acids: Ser Asn Ser

Anticodon UCA carrying serine, followed by anticodon UUG carrying asparagine, followed by anticodon AGC carrying serine. The tripeptide would be Ser-Asn-Ser.

   d.   mRNA: UAUGAUACC

Anticodons on tRNA: AUA   CUA   UGG

Amino acids: Tyr Asp Thr

Anticodon AUA carrying tyrosine, followed by anticodon CUA carrying aspartic acid, followed by anticodon UGG carrying threonine. The tripeptide would be Tyr-Asp-Thr.

15.85   a.   DNA strand: dTTACCCGACGGC

mRNA strand: AAUGGGCUGCCG:

Anticodons on tRNA: UUA   CCC   GAC   GGC

Amino acid sequence: Asn-Gly-Leu-Pro

   b.   DNA strand: dGAAGCAACCATA

mRNA strand: CUUCGUUGGUAU

Anticodons on tRNA: GAA   GCA   ACC   AUA

Amino acid sequence: Leu-Arg-Trp-Tyr

c. DNA strand: dATGACCACAGAA

mRNA strand: UACUGGUGUCUU:

Anticodons on tRNA: AUG   ACC   ACA   GAA

Amino acid sequence: Tyr-Trp-Cys-Leu

d. DNA strand: dCGACATCCTCTA

mRNA strand: GCUGUAGGAGAU

Anticodons on tRNA:CGA   CAU   CCU   CUA

Amino acid sequence: Ala-Val-Gly-Asp

15.87 DNA: dAGTAAA

mRNA: UCAUUU

Dipeptide: Ser-Phe

a. DNA mutation: dAG**C**AAA

mRNA: UC**G**UUU

Dipeptide: Ser-Phe

The dipeptide would be the same.

b. DNA mutation: dAGTAA**T**

mRNA: UCAUU**A**

Dipeptide: Ser-Leu

The dipeptide would be different.

c. DNA mutation: dATAAA

mRNA: UAUUU

Peptide: Tyr-

The dipeptide would be different.

d. DNA mutation: dAGTAA

mRNA: UCAUU

Peptide: Ser-

If the next nucleotide in the mRNA sequence is U or C, then the dipeptide will be the same. If the next nucleotide in the mRNA sequence is A or G, then the dipeptide will be a different dipeptide, Ser-Leu.

15.89 The HIV virus contains two strands of RNA.

15.91   The HIV virus uses *reverse transcriptase*, *protease*, *integrase*, and *ribonuclease* to replicate.

15.93   HIV infects T-lymphocytes.

15.95   HIV uses its genomic RNA to code for DNA.

15.97   The HIV virus rapidly mutates and becomes resistant to any one type of drug therapy; therefore, the best treatment often is to use combination drug therapies.

15.99   *Protease* inhibitors prevent the release of essential enzymes, produced late in the life cycle of the virus, required for the virus to replicate.

Chapter 16

Nuclear Chemistry and Medicine

Chapter Summary

In this chapter, you studied nuclear reactions, changes to the nucleus of an atom. You learned about different radioisotopes and different types of radiation emitted in a nuclear reaction. You were introduced to the different types of nuclear decay and their effects on biological tissue. The radiation produced in a nuclear reaction can be used in the diagnosis and treatment of disease.

Section 16.1 Radioisotopes and Nuclear Reactions

Nuclear chemistry is the study of changes in the nucleus of an atom. Changes to the composition of the nucleus of an atom are known as nuclear reactions. Almost all elements have more than one isotope. All isotopes of a given element have the same number of protons, which is equal to the atomic number, Z, of the element. Each isotope will have a different number of neutrons, reflected in its mass number, A, the sum of the protons and neutrons in the nucleus.

In a Nutshell: Radioisotopes

For most isotopes, the optimal neutron/proton ratio is 1.0 to1.5, depending on the mass number. If this ratio is too high or too low, the nucleus will be unstable. Isotopes with an atomic number greater than 82 are unstable because they have too many protons and neutrons. Unstable isotopes are known as radioactive isotopes or radioisotopes.

The nucleus of a radioactive isotope undergoes a natural process known as radioactive decay to become a more stable nucleus. Radioactive decay typically produces an isotope with a different atomic number, and hence a different element. Radioactive decay is always accompanied by a release of a form of energy called radiation. Radiation from the decay of radioisotopes appears in either of two forms: (1) high-energy electromagnetic radiation, such as x rays or gamma rays; or (2) high-energy particles, such as α-particles, β-particles or positrons.

Worked Example #1

Fill in the empty cells in the table below for the given radioisotopes.

Radioisotope	Atomic number	Mass number	Number of neutrons	Number of protons
Lead-197				
	85	209		

Solutions

Tools: Periodic Table

Radioisotope	Atomic number	Mass number	Number of neutrons	Number of protons
Lead-197	*82*	*197*	*115*	*82*
Astatine-209	85	209	*124*	*85*

Remember the atomic number is equal to the number of protons, and the mass number is equal to the sum of the number of protons and the number of neutrons.

Try It Yourself #1

Fill in the empty cells in the table below for the given radioisotopes.

Radioisotope	Atomic number	Mass number	Number of neutrons	Number of protons
Americium-242				
	59	125		

In a Nutshell: Electromagnetic Radiation

Electromagnetic radiation is a form of energy that travels through space as a wave at the speed of light. The wavelength of electromagnetic radiation is defined as the distance between wave crests. Electromagnetic radiation is classified into regions according to wavelength. The electromagnetic spectrum is the range of electromagnetic radiation spanning all possible wavelengths. The different forms of electromagnetic radiation include gamma-rays, x-rays, ultraviolet, visible, infrared, microwaves, and radiowaves. Gamma rays and x-rays have short wavelengths, whereas radiowaves have long wavelengths.

The energy associated with a particular type of electromagnetic radiation depends on its wavelength. The wavelength of electromagnetic radiation is inversely proportional to its energy: the shorter the wavelength, the higher the energy. High-energy electromagnetic radiation can damage biological tissue. The most damaging forms of electromagnetic radiation are gamma-rays and x-rays.

Worked Example #2

Refer to Figure 16-3 and answer the questions below by choosing from among the following three forms of electromagnetic radiation.

   i. visible   ii. gamma-rays   iii. infrared

   a. Which form has the longest wavelength?
   b. Which form is the highest in energy?
   c. Which form causes the most biological damage?

   a. *Of those three, infrared has the longest wavelength.*
   b. *Gamma rays are the highest in energy.*
   c. *Gamma rays cause the most biological damage.*

Try It Yourself #2

Refer to Figure 16-3 and answer the questions below by choosing from among the following three forms of electromagnetic radiation.

   i. radiowaves   ii. ultraviolet   iii. x-rays

   a. Which form has the shortest wavelength?
   b. Which form is the lowest in energy?
   c. Which form causes the most biological damage?

   a. *The shortest wavelength:* _____
   b. *The lowest in energy:* _____
   c. *The biological damage:* _____

In a Nutshell: Types of Radioactive Decay

During radioactive decay, five types of radiation can be emitted as either high-energy particles and/or high-energy electromagnetic radiation. The high-energy particles are α-particle, β-particle, and positrons. The high-energy electromagnetic radiation includes x-rays and gamma-rays.

In a Nutshell: α Decay and the Balanced Nuclear Equation

Radioisotopes that undergo α-decay emit α-particles. An α-particle is a slow-moving, high-energy particle consisting of two protons and two neutrons. Its nuclear symbol is $_2^4\alpha$. The composition of an alpha particle is the same as a helium nucleus; hence its nuclear symbol is also written as $_2^4\text{He}$. α-Particles are extremely dense and carry a +2 charge. After emission of an α-particle, the mass number of the isotope decreases by four and the atomic number decreases by two.

Radioactive decay can be depicted by writing a nuclear equation. The convention is to write the radioisotope to the left of the arrow and the new isotope as well as the emitted radiation to the left of the arrow. The atomic number and mass number for all species in the equation are included. The radioisotope undergoing decay is often referred to as the parent nuclide. When a different element is formed, it is referred to as the daughter nuclide. In a balanced equation the sum of the atomic numbers (subscripts) of all species on the left side of the arrow must equal the sum of the atomic numbers of all species on the right side of the arrow; likewise for the mass numbers (superscripts).

Worked Example #3

Predict the daughter nuclide for α-decay of Ra-219.

Write the radioisotope with its atomic mass numbers to the left of the reaction arrow and write the α-particle and the daughter nuclide on the right of the reaction arrow. Determine the identity of the daughter nuclide, $_Z^A\text{X}$, by subtracting 4 from the mass number and subtracting 2 from the atomic number of the parent nuclide. In this case:

$$A = 219 - 4 = 215; \text{ and}$$
$$Z = 88 - 2 = 86.$$

Thus the daughter nuclide is $^{215}_{86}X$. *Use the periodic table to find the element that*

corresponds to atomic number 86; the element is radon. The balanced nuclear equation is:

$$^{219}_{88}\text{Ra} \rightarrow {}^{215}_{86}\text{Rn} + {}^{4}_{2}\alpha$$

Try It Yourself #3

Predict the daughter nuclide for α-decay of Po-209.

Write the radioisotope with its atomic mass numbers to the left of the reaction arrow and

write the α-particle and the daughter nuclide on the right of the reaction arrow.

Determine the identity of the daughter nuclide.

Mass number of daughter nuclide: _____

Atomic number of daughter nuclide: _____

Element that corresponds to atomic number of daughter nuclide: _____

Balanced nuclear equation: _____

In a Nutshell: β Decay

Beta decay occurs in an unstable nucleus that has too many neutrons compared to protons.

In the β-decay process, a neutron is converted into a proton and a high-energy electron.

The daughter nuclide will have the same mass number as the parent nuclide, but its atomic

number is increased by one; therefore, it is the next element on the periodic table. The

high-energy electron is known as a β-particle and written as: $^{0}_{-1}\beta$.

Worked Example #4

Predict the daughter nuclide for β-decay of Am-247 and write the balanced nuclear equation

for the reaction.

The daughter nuclide will be the next element on the periodic table. Alternatively, you can

write the daughter nuclide as X, then insert the values for Z and A after you balance the

nuclear equation, and finally look up the element symbol on the periodic table that

corresponds to the atomic number calculated.

$$^{247}_{95}\text{Am} \rightarrow {}^{247}_{96}\text{Cm} + {}^{0}_{-1}\beta$$

Try It Yourself #4

Predict the daughter nuclide for β-decay of U-237 and write the balanced nuclear equation for the reaction.

Atomic number of daughter nuclide: _____

Mass number of daughter nuclide: _____

Element that corresponds to atomic number of daughter nuclide: _____

Balanced nuclear equation: _____

In a Nutshell: Positron Decay

A radioisotope decays by positron emission when its nucleus contains too many protons compared to neutrons. A positron is similar to a β-particle in mass and energy, but it is positively charged. A positron has the symbol $^{0}_{+1}\beta$ and is formed when a proton is converted into a neutron and a positron. The daughter nuclide has the same mass number as the parent, but its atomic number is one less than the parent; thus is will precede the parent in the periodic table.

Worked Example #5

Write the balanced nuclear reaction for the radioactive decay of K-40, a positron emitter.

The daughter nuclide will be the prior element on the periodic table. Alternatively, you can write the daughter nuclide as X, then insert the values for Z and A after you balance the nuclear equation, and finally look up the element symbol on the periodic table that corresponds to the atomic number calculated.

$$^{40}_{19}\text{K} \rightarrow {}^{40}_{18}\text{Ar} + {}^{0}_{+1}\beta$$

Try It Yourself #5

Write the balanced nuclear reaction for the radioactive decay of N-13, a positron emitter.

Atomic number of daughter nuclide: _____

Mass number of daughter nuclide: _____

Element that corresponds to atomic number of daughter nuclide: _____

Balanced nuclear equation: _____

In a Nutshell: X-rays and Gamma Radiation

X-ray and gamma radiation are both short-wavelength, high-energy forms of electromagnetic radiation. X-rays accompany some forms of radioactive decay. Gamma-ray emission accompanies almost all forms of radioactive decay. However, the gamma emission is not usually shown in the nuclear equation because it does not affect the atomic number or the mass number.

After radioactive decay has occurred, the daughter nuclide is often in an excited state, a condition in which the nucleus still contains excess energy. An isotope in its excited state is referred to as a metastable isotope, and is notated by the abbreviation m following the mass number of an isotope. The daughter nuclide releases its excess energy in returning to the ground state by releasing a pulse of gamma-radiation.

Worked Example #6

Show the nuclear reaction that Fr-214m undergoes to become Fr-214.

Since no particle is emitted, the atomic number, mass number, and element symbol for the daughter nuclide is the same.

$$^{214m}_{87}\text{Fr} \rightarrow {}^{214}_{87}\text{Fr} + \gamma$$

Try It Yourself #6

Show the nuclear reaction that Cs-119m undergoes to become Cs-119.

Atomic number of daughter nuclide: _____

Mass number of daughter nuclide: _____

Element that corresponds to atomic number of daughter nuclide: _____

532   Chapter 16

Balanced nuclear equation: _____

In a Nutshell: Half-Life

The time it takes a macroscopic sample of the radioisotope to decay to one-half its original mass is known as its half-life. Half-lives range from a few seconds to billions of years.

Worked Example #7

For a 50.0 g sample of gallium-67:

    a. How much Ga-67 is left after four half-lives?

    b. How much time has elapsed after four half-lives?

    a. *Divide the original amount of material, 50.0 g, in half four consecutive times (four half-lives):*

$$50.0 \text{ g} \xrightarrow{1} 25.0 \text{ g} \xrightarrow{2} 12.5 \text{ g} \xrightarrow{3} 6.25 \text{ g} \xrightarrow{4} 3.1 \text{ g}$$

       Therefore, 3.1 g of the original sample of is left after four half-lives.

    b. *Table 16-4, shows that the half-life for Ga-67 is 78 hr. Determine how many hours four half-lives correspond to, given that one half-life is 78 hours.*

$$4 \ \cancel{\text{half-lives}} \times \frac{78 \text{ hr}}{1 \ \cancel{\text{half-life}}} = 312 \text{ hr} \text{ , } \textit{rounded to 310 hr}$$

Try It Yourself #7

For a 150. g sample of gold-198:

    a. How much Au-198 is left after three half-lives?

    b. How much time has elapsed after three half-lives?

    a. *Initial amount of Au-198:* _____

       Amount after one half-life: _____

       Amount after two half-lives: _____

       Amount after three half-lives: _____

b. *Length of one half-life for Au-198:* _____

Length of three half-lives: _____

In a Nutshell: Artificial Radioisotopes

None of the elements with atomic numbers greater than 92 are found naturally on earth. Elements 93-118 on the periodic table are all artificial radioisotopes. Artificial isotopes are created by bombarding nuclei with the high-energy particles generated in an accelerator.

Worked Example #8

When tellurium-130 is bombarded with a neutron, it produces a new radioisotope of tellurium which undergoes β-decay to produce a new element. What new radioisotopes are produced? Write the two nuclear equations for the process.

The new radioisotopes are $^{131}_{52}Te$ *and* $^{131}_{53}I$ *:*

$$^{130}_{52}Te + ^{1}_{0}n \rightarrow ^{131}_{52}Te$$

$$^{131}_{52}Te \rightarrow ^{131}_{53}I + ^{0}_{-1}\beta$$

Try It Yourself #8

When oxygen-18 is bombarded with a proton ($^{1}_{1}p$), a neutron and a radioisotope are formed. What is the radioisotope formed? Write a nuclear equation for the process.

New radioisotope: _____

Nuclear equation for the process:

Practice Problems for Radioisotopes and Nuclear Reactions

1. Refer to Figure 16-3 and answer the questions below by choosing from among the following three forms of electromagnetic radiation.

      i. visible   ii. ultraviolet   iii. radiowaves

a. Which form has the shortest wavelength?

b. Which form is the lowest in energy?

c. Which form causes the most biological damage?

2. Write the balanced nuclear equation for the following:
   a. α-Decay of Sg-261

   b. formation of Sm-134 through β-decay

   c. formation of Te-121 through positron emission

3. How much of a 20.0 g sample of fluorine-18 remains after 327 minutes? How many half-lives does 327 minutes represent? The half-life of fluorine-18 is 109 minutes.

Section 16.2 Biological Effects of Nuclear Radiation

The radiation emitted from a nuclear reaction is classified as ionizing radiation because it has the energy to dislodge an orbital electron from an atom creating an ion. When an atom in a molecule is ionized, it changes the molecule in a significant way. In living organisms this change can be quite destructive. The biological effects of nuclear radiation depend in large part on the type of radiation produced. The energy of the radiation and the penetrating power of the radiation determine the biological effects of the radiation.

Penetrating power varies with the type of radiation. An α-particle is relatively large, slow moving, and high in energy. Due to its size and slow speed, an α-particle has little penetrating power, though its high-energy makes it very destructive to human tissue. A piece of paper or light clothing is sufficient protection against α-particles. On the other hand inhalation or ingestion of an α-emitter can cause major damage to internal organs.

β-Particles and positrons have slightly less energy than α-particles and much more penetrating power because they are substantially lighter. Specialized heavy clothing or a thick piece of aluminum is required for protection against β-particles and positrons.

The energy of gamma-rays and x-rays is less than or equal to the energy of β-particles. However, gamma rays have the most penetrating power of all forms of radiation. Several inches of lead are required to protect against gamma-radiation. A thin sheet of lead is sufficient protection against x-rays.

Worked Example #9
Which form of radiation has greater penetrating power, an α-particle or an x-ray?

Solution
The x-ray has more penetrating power; it is lighter.

Try It Yourself #9
Which form of radiation has greater energy, an α-particle or an x-ray?

Solution:

In a Nutshell: Measurement of Radiation

There are different ways to measure radiation. The Geiger counter is an inexpensive instrument used in the field that can detect all forms of radiation. A radiation badge is used to monitor radiation exposure of personnel who work in areas or use instruments that produce radiation.

Several units of measurement are used to measure radiation. These units may indicate the number of radioactive emissions, the amount of energy absorbed (absorbed dose) or the biological effectiveness of the energy absorbed (effective dose).The amount of radioactive decay can be measured in becquerel (Bq) or curie (Ci), which indicate the number of emissions per second. These units do not distinguish between the different types of radiation.

An absorbed dose measurement indicates the energy of radiation absorbed per mass of tissue. The Gray (Gy) unit is the most common unit for absorbed dose used in medicine; another unit is the Rad. Absorbed dose measurements do not account for the difference in penetrating power of the different forms of radiation.

The effective dose encompasses both the penetrating power and the amount of energy to give the actual biological effect. The effective dose is calculated by multiplying the absorbed dose by a quality factor, Q, which varies for the different types of radiation. When the unit of absorbed dose is the Gray, the unit of effective dose is the sievert (Sv). Another common unit for the effective dose is rem.

Worked Example #10

An ionizing smoke detector contains a small amount of americium-241. As long as the americium stays in the detector, you receive about 0.01 mrem of radiation. The americium emits 1×10^{-6} Ci. What type of information does mrem and Ci convey?

Solution

A mrem is a unit for measuring the relative biological effectiveness of radiation exposure, while Ci is a unit for measuring the amount of radiation emitted by the americium.

Try It Yourself #10

A chest x-ray produces about 0.1 mSv per x-ray image. What type of information does the Sv convey?

Solution:

In a Nutshell: Radiation Sickness

Exposure to radiation can occur as a single dose (acute exposure) or as smaller doses over a longer period of time (chronic exposure). Radiation sickness results from acute exposure to radiation. The severity of the symptoms from radiation sickness is directly proportional to the effective dose received.   The effective dose is often measured in values of LD_x. LD_x refers to the lethal dose of the radiation in x% of the population after 30 days.

Worked Example #11

Identify the following situations as either acute or chronic exposure to radiation.
    a. A patient with hyperthyroidism (an overactive thyroid) is treated with a single dose of I-131.
    b. An airline pilot, who is exposed to cosmic radiation on each flight.

Solutions
    a. *Acute exposure. It is a single exposure to radiation.*
    b. *Chronic exposure. It is exposure to smaller doses of radiation over longer periods of time.*

Try It Yourself #11

Identify the following situations as either acute or chronic exposure to radiation.

    a. A hospital worker who carries out CT scans on patients regularly.

    b. A patient receiving a chest x-ray.

    a. Radiation exposure is: _____

    b. Radiation exposure is: _____

Worked Example #12

Using Table 16-6, what does it mean if the LD_{100} is 6–10 Sv?

An LD_{100} signifies that 100% of the population would die in 30 days after exposure to 6–10 Sv of radiation.

Try It Yourself #12

Using Table 16-6, what does it mean if the LD_{10} is 1.0–2.0 Sv? What are the symptoms of this dose of radiation?

An LD_{10} signifies: _____

The symptoms of exposure to 1.0–2.0 Sv of radiation are: _____

Practice Problems for Biological Effects of Nuclear Radiation

    1. Which form of radiation has greater penetrating power?

        a. α-particle or gamma ray

        b. positron or x-ray

        c. gamma ray or β-particle

2. Which form of radiation has greater energy?

    a. gamma ray or positron

    b. β-particle or α-particle

    c. x-ray or α-particle

3. A bone scan with 600 MBq (megabequerel) of technetium-99m produces 3 mSv. What do the units Bq and Sv convey?

4. Using Table 16-6, what does it mean if the LD_{60} is 4–6 Sv? What are the symptoms of this dose of radiation?

Chapter 16 Quiz

1. Fill in the empty cells in the table below for the given radioisotopes.

Radioisotope	Atomic number	Mass number	Number of neutrons	Number of protons
		269		108
	87		113	

2. Refer to Figure 16-3 and answer the questions below by choosing from among the following three forms of electromagnetic radiation.

   i. ultraviolet   ii. gamma-rays   iii. microwaves

   a. Which form has the shortest wavelength?

   b. Which form is the lowest in energy?

   c. Which form causes the most biological damage?

3. Rubidium-82, used to image the heart, undergoes positron emission. What is the daughter nuclide formed in this process? Write the nuclear equation for the positron emission of Ru-82.

4. Y-90 is used in treatment of cancer. It undergoes β-decay. Identify the daughter nuclide produced in this process. Write the nuclear equation for the β-decay of Y-90.

5.   Thallium-179 undergoes α decay. Write the nuclear equation for this process. What is the daughter nuclide produced in the process?

6.   Write the balanced nuclear equation for the following:
      a. formation of Rn-199 through α-decay

      b. formation of Eu-151 through β-decay

      c. formation of Nd-137 through positron emission

7.   How much of a 15.0 g sample of P-32 is left after three half-lives? How much time has elapsed in three half-lives? The half-life of P-32 is 14.3 days.

8.   How much of a 150 g sample of Ba-131 is left after 58 days? How many half-lives have elapsed in 58 days? The half-life of Ba-131 is 11.6 days.

9.   What type of radiation has a large penetrating power but a small amount of energy?

10.   Which units are used to measure effective doses of radiation? Which units are used to measure absorbed doses of radiation?

Chapter 16
Answers to Additional Exercises

16.37   The atomic number represents the number of protons in an isotope. The mass
          number represents the sum of the protons and neutrons in the nucleus.

16.39

Radioisotope	Atomic Number	Mass Number	Number of Neutrons	Number of Protons
Thallium-201	81	201	120	81
Selenium-75	34	75	41	34
Cobalt-60	27	60	33	27
Ba-131	56	131	75	56

16.41   Electromagnetic radiation is a form of energy that travels through space as a wave
          at the speed of light.

16.43   Radio waves have insufficient energy to damage biological tissues, while x-rays
          can cause significant damage to biological tissue.

16.45   a.   Radio-wave radiation has the longer wavelength.

          b.   X-ray radiation has the longer wavelength.

          c.   Visible radiation has the longer wavelength.

16.47   a.   Gamma-rays are more damaging to biological tissue because they are higher
                 in energy than x-rays.

          b.   Ultraviolet radiation is more damaging to biological tissue because it is higher
                 in energy than visible radiation.

16.49 $^{241}_{95}\text{Am} \rightarrow ^{237}_{93}\text{Np} + ^{4}_{2}\alpha$

16.51   Both positrons and β-particles are high-speed, high-energy particles. The
          difference between the two is the charge on the particles. The positron has a
          charge of +1, while a β-particle has a charge of −1.

16.53   The nuclear equation is $^{12}_{5}\text{B} \rightarrow ^{12}_{6}\text{C} + ^{0}_{-1}\beta$. The daughter nuclide is carbon-12.

16.55   a. $^{26}_{11}\text{Na} \rightarrow ^{26}_{12}\text{Mg} + ^{0}_{-1}\beta$

          b. $^{210}_{86}\text{Rn} \rightarrow ^{206}_{84}\text{Po} + ^{4}_{2}\alpha$

          c. $^{52}_{26}\text{Fe} \rightarrow ^{52}_{25}\text{Mn} + ^{0}_{+1}\beta$

16.57   Thirty-two days represents four half-lives.

$$32 \; \text{-days-} \times \frac{1 \; \text{half-life}}{8 \; \text{-days-}} = 4 \; \text{half-lives}$$

18.0 g (initial amount), 9.0 g (one half-life), 4.5 g (two half-lives), 2.25 g (three half-lives), 1.1 g (four half-lives)

16.59   The half-life of Tc-99m is only 6 hours, so the patient's exposure to radiation can be kept to a minimum because Tc-99m in a patient's body decays quickly.

16.61   100. g (initial amount), 50. g (one half-life). 75. g is half way between 100. and 50. g; therefore half of a half-life has elapsed, or 33 hours.

16.63   Artificial radioisotopes are created by bombarding nuclei with the high-energy particles generated in an accelerator.

16.65 $^{59}_{27}\text{Co} + ^{1}_{0}\text{n} \rightarrow ^{60}_{27}\text{Co}$ Cobalt-59 is used to produce cobalt-60.

16.67 $^{238}_{92}\text{U} + ^{4}_{2}\alpha \rightarrow ^{239}_{94}\text{Pu} + 3 \; ^{1}_{0}\text{n}$ U-238 is bombarded with an α-particle to produce Pu-239 and three neutrons.

16.69   particle accelerator; period 7; the actinoids

16.71 Ca^{2+} The atomic number is 20.

16.73   A mutation occurs when radiation breaks phosphodiester bonds sufficiently close to each other on opposite strands of DNA. These mutations are passed on when the cell reproduces, possibly causing cancer.

16.75   X-rays do not have enough penetrating power to pass through the lead apron.

16.77   a. β-particle

          b. gamma-ray

          c. gamma-ray

16.79 β-Particles and positrons have the same penetrating power and the same energy.

16.81   The bequerel and curie measure the same property of radiation. They measure the rate of radioactive emissions from a sample. The abbreviations are Bq for bequerel and Ci for curie. A millicurie would be abbreviated with mCi.

16.83   An absorbed dose measures the energy of radiation absorbed per mass of tissue, but does not take into account the penetrating power of the radiation. The effective dose takes into account both the penetrating power of radiation and the amount of energy to give a biological effect. The units for absorbed dose are the Gray and the Rad. The effective dose is measured in sieverts and rem.

16.85   An LD_{50} indicates a level of exposure that would result in death in 50% of the population in 30 days.

16.87   The CT scan exposed the man to a higher dose of radiation.

16.89   The quarter is higher in density than the tissue in the esophagus. The quarter absorbed more x-rays and thus appears lighter in color than the tissue of the esophagus.

16.91   The MRI is best for imaging soft tissue areas of the body. The technique uses low-energy radio waves that do not pose a risk of biological damage. An MRI is not ideal for imaging denser tissues in the body such as bones and joints.

16.93   Radio waves travel at the speed of light, 3×10^8 m/s.

$$394{,}403 \ \cancel{km} \times \frac{1000 \ \cancel{m}}{1 \ \cancel{km}} \times \frac{1 \ s}{3 \times 10^8 \ \cancel{m}} = 1.3 \ s$$

16.95   a. $^{192}_{77}Ir \rightarrow \ ^{192}_{78}Pt + \ ^{0}_{-1}\beta + \gamma$ The daughter nuclide is Pt-192.

        b.   A thick sheet of lead should be used to protect the healthy tissues from the β and γ radiation.